Molecular Biology Biochemistry and Biophysics 36

Edited by Marc Solioz

H.G. Callan

Lampbrush Chromosomes

With 67 Figures

Springer-Verlag
Berlin Heidelberg New York Tokyo

Professor Dr. HAROLD GARNET CALLAN
Emeritus Professor of Natural History
Gatty Marine Laboratory
The University
St. Andrews, KY16 8LB
Scotland

ISBN-13:978-3-642-82794-5   e-ISBN-13:978-3-642-82792-1
DOI: 10.1007/978-3-642-82792-1

Library of Congress Cataloging-in-Publication Data. Callan, Harold Garnet, 1917–
Lampbrush chromosomes. (Molecular biology, biochemistry, and biophysics; 36)
Includes bibliographical references and index. 1. Lampbrush chromosomes. I. Title.
II. Series: Molecular biology, biochemistry, and biophysics; v. 36. QH600.6.C35
1986   574.87′322   86-6533

© by Springer-Verlag Berlin Heidelberg 1986
Softcover reprint of the hardcover 1st edition 1986

2131/3130-543210

Preface

In 1965 I was asked by Dr. Konrad Springer whether I would consider writing a monograph on "Lampbrush chromosomes and their physiological meaning", and although I accepted in principle I refused to write there and then, or to meet a deadline. I wanted to continue with my own research, and I had other responsibilities that left me with little free time, but a much more important consideration was that in the 1960s the subject was beset by a host of unresolved questions. I felt that to write a review then would be premature, full of speculations many of which would be refuted, and indeed were refuted, within the next decade or two. Had I written at that time the only real advantage over the present would have been that few biologists were studying lampbrush chromosomes, and the published literature was therefore scanty. I am glad that I insisted on delay, and am grateful for Springer Verlag's patient acceptance of my decision.

The first chapter of this monograph describes the history of research on lampbrush chromosomes from their discovery towards the end of the 19th century until the early 1960s. By then several facts concerning their structure and chemistry had been firmly established, including the evidence that a lampbrush chromatid is unineme; it contains a single uninterrupted DNA duplex. This exposed a major problem, the C-value paradox; grades of complexity of organization in eukaryotes are unrelated to their genome sizes. Largely from suggestive features present in lampbrush chromosomes I was led to propose in 1967 that a Master/Slave organization of repetitive DNA sequences, conceivably universal in eukaryotes and applying to all their genes, might explain the paradox. This speculation was soon proved to be invalid.

In the early 1960s, and in common with most biologists who had studied them, I thought that transcriptionally active lateral loops, with their bases inserted in transcriptionally inactive chromomeres, were the diagnostic features of lampbrush chromosome organization, that this organization might well be universal in animal oocytes and perhaps equally universal in spermatocytes. As will be discussed in Chaps. 6 and 7, when not demonstrably false this proposition is at best a half-truth, and the organization of DNA/histone in chromatids as a series of loop domains, whether or not the loops be active in transcription, seems likely to prove characteristic of eukaryotic chromatids generally, in somatic as well as germ-line nuclei. It has therefore become well-nigh impossible to produce an exclusive definition of lampbrush chromosomes as present in oocytes, no bad thing in itself if one is seeking for universal principles, but making difficulties for someone attempting to discuss a topic whose boundaries have become so unclear. This is a penalty that I have had to pay for prevarication!

In the early 1960s it was generally supposed that the lateral loops of amphibian lampbrush chromosomes are single units of transcription, with constant polarities vis-à-vis their immediate neighbours, and this idea was largely responsible for my proposing the generalized Master/Slave theory. By the mid 1970s however it had been demonstrated that whereas some of the most conspicuous lateral loops are genuinely units of transcription, other loops include multiple units of similar or diverse polarities, some of these transcribing related but others transcribing unrelated DNA sequences. Had I been aware of this 10 years earlier the Master/Slave theory would have been quietly aborted. I mention this to explain why, insofar as possible, I have attempted while writing each chapter of this monograph to take account of the historical sequence of discoveries, and how they have influenced one another.

I would like to thank many friends for their kindness in sending me preprints and offprints of publications on lampbrush chromosomes, thus enabling me to keep reasonably abreast of developments in a now rapidly expanding field. These include in particular Dr. J. G. Gall, with whom I have regularly corresponded since 1951, my former student Prof. H. C. Macgregor, Prof. G. Mancino, Prof. J.-C. Lacroix, Dr. U. Scheer and Dr. J. Kezer. I am also indebted to Prof. O. Hess and Prof. W. Hennig for keeping me up-to-date with information about *Drosophila* spermatocytes; Prof. Hess read an early draft of Chap. 7, and the present version owes a great deal to his helpful criticism. Prof. J. G. Gall was kind enough to read the entire manuscript when it was nearing completion, enabling me to eliminate several errors and, fully as important, encouraging me to finish the work. Residual errors are my own responsibility. All the above-mentioned biologists, and several more, kindly supplied me with photographs and other illustrative material for the plates; these are acknowledged individually in the figure legends.

I am also grateful to my former research assistant, Mrs. L. Lloyd, for her devoted help over many years, in particular for her drawings recording information from temporary preparations, for help with photography and autoradiography, for help in raising families of newts and during the battle to save precious hybrids when our newt colony was devastated by parasites. Miss M. M. Moncrieff, former secretary of the Zoology Department, typed and re-typed various drafts and the final version of the manuscript; I applaud and thank her for patient labour on my behalf.

During the last few years my field of study has been illuminated by new and sophisticated approaches of molecular biologists, resolving some problems yet exposing further unanticipated complications. From a background of classical cytology I have found difficulty in adjusting to, and assessing, the implications of some of these most recent developments. I want to express thanks to my wife Amarillis for giving me continuous support while I have attempted to adjust to the climate of research on lampbrush chromosomes that prevails today. It strikes me as remarkably appropriate that my wife's father, Dr. Reinhard Dohrn, who was for many years Director of the Stazione Zoologica at Naples, was personally acquainted with Dr. J. Rückert, the investigator who first put lampbrush chromosomes "on the map" and gave them their name. Rückert studied the lampbrush chromosomes of elasmobranchs at the Stazione Zoologica towards the end of the 19th Century, and it was in this same institute that I first saw the germinal vesicles of newt oocytes in 1947, and came to realize the particular advantages for study that are offered by these enormous cell nuclei, and their equally gigantic chromosomes.

H. G. CALLAN

Contents

CHAPTER 1

Historical Introduction

Lampbrush chromosomes were first seen by the pioneer cytologist Walther Flemming in 1878. Flemming and his student Wiebe, in the course of studies on the development of ovarial eggs (oocytes) of Amphibia and fish, found „merkwürdige und zierliche Anordnungen" (strange and delicate structures) in stained sections through the nuclei of young oocytes of the axolotl *"Siredon pisciformis"* (*Ambystoma mexicanum*). Flemming published an account of these observations in 1882, and Fig. 1.1 is a copy of his single illustration. It shows elongate objects each made up of thin strands arranged normal to the long axis, and transverse views of these same objects where the thin strands are displayed radiating in all

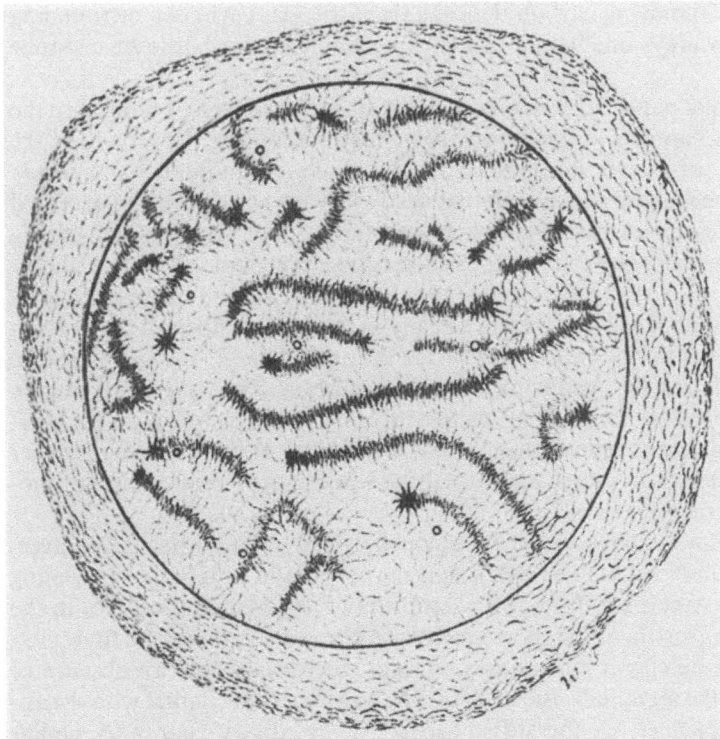

Fig. 1.1. Flemming's (1882) drawing of a stained section through a young oocyte nucleus of *Ambystoma mexicanum*

directions from the axis. Flemming reported seeing similar objects in the nuclei of young oocytes of salamanders, and also in those of frogs, though in the latter the transverse strands were of smaller dimensions.

Flemming discussed the question as to whether these cross-stranded objects might be artefacts. He noted that in ovaries fixed in chromic, picric or osmic acids the objects were distributed uniformly through the volume of the nucleus, but that their distribution was less regular after alcohol fixation. He came to the tentative conclusion that even though the objects that he saw in fixed preparations might be artefacts, they were more likely to represent some natural structure, perhaps something comparable to the cross-banded objects present in the salivary gland cell nuclei of *Chironomus*. This is the nearest Flemming came to relating his structures to chromosomes, though another 50 yrs were to pass before the cross-banded (polytene) structures in dipteran salivary glands were themselves recognized as such, long after the cross-stranded objects had secured their recognition already in 1892.

Structures similar to those found by Flemming were seen by Rabl (1885) in oocyte nuclei of the neotenic urodele *Proteus* and by Holl (1890) in oocyte nuclei of the chicken, but it was Rückert (1892) who first gave evidence that these structures are chromosomes. Rückert described them by making analogy to a „Lampencylinderputzer", or lampbrush, distorted through use, and this well-chosen term has since become generally accepted. In Rückert's day a lampbrush meant a brush used for removing carbon deposits from the glass cylinder surrounding an oil lamp; nowadays one would call such an implement a bottle or test-tube brush.

Rückert's 1892 paper deserves special attention. He addressed himself to the problem of what happens to the chromosome material during the growth of the germinal vesicle, or oocyte nucleus, in ovaries of the elasmobranchs *Scyllium, Torpedo* and especially *Pristiurus*. By the early 1890s it was becoming accepted that chromosomes are the carriers of hereditary factors, but what happens during the two divisions just prior to the production of mature germ cells was a matter for lively debate. Boveri (1890), from studies on a wide variety of invertebrates, held that the chromosomes are present already in reduced number on the division spindle which leads to the production of the first polar body, arguing therefore that in the female sex the "reduction" must have occurred, at the latest, within the germinal vesicle. Hertwig (1890) on the contrary, from his work on the spermatogenesis of *Ascaris,* claimed that there is a doubling of chromosome number prior to the reduction divisions, and that these two divisions reduce the chromosome number to one-quarter.

The problem was compounded by statements from many authoritative investigators, most notably Schultze (1887), that the chromatin present in very young oocyte nuclei of Amphibia disappears completely from germinal vesicles in the course of oocyte growth, and only reappears, as chromosomes, on the first polar body spindle of the egg at maturation. Rückert recognized that an absence of chromatin from the germinal vesicle, if correct, was clearly in conflict with the notion that chromosomes are the stable carriers of hereditary factors. A similar proposition, on similar evidence, was to be made nearly 50 yrs later regarding DNA!

By studying sections of *Pristiurus* ovaries Rückert established that the nuclei of the smallest (18 μm diam) oocytes contain readily stainable chromosomes which wind throughout the volume of the nucleus, and generally resemble the prophase of mitosis. The nuclei of these smallest oocytes do not contain a network of chromatin, like a typical "resting" nucleus, and neither do they come to do so in later development. As an oocyte grows to reach about 2 mm diam, so its germinal vesicle enlarges to reach a diameter of about 350 μm; thereafter the oocyte continues to grow but its germinal vesicle stops enlarging.

As the germinal vesicle in oocytes of *Pristiurus* enlarges up to about 350 μm diam, so the chromosomes grow longer and thicker and become distributed further apart from one another, showing up as rows of „Mikrosomen" (chromomeres) which appear like little rods arranged transverse to the chromosome's long axis. All the while this is occurring the chromomeres progressively increase in number and decrease in size, and lose their stainability. Rückert remarks that this loss of stainability cannot alone be ascribed to progressive loosening of the chromosome's texture, for individual chromomeres become less stainable; rather he states: „... die färbbare Substanz eines Mikrosoma auf eine grössere Masse einer nicht färbbaren Grundsubstanz sich verteilt". (... the stainable material of a chromomere becomes dispersed through a greater mass of unstainable ground substance).

There follows a stage of oocyte development when the chromosomes are exceedingly difficult to see in sections, the stage at which several previous authors had claimed they disappear entirely. Rückert describes the germinal vesicle at this stage as having a marbled appearance, with slightly denser regions occupied by the chromosomes, separated from one another, though without sharp boundaries, by less dense regions of about the same width and occupied by nuclear sap. By careful observation of osmium-fixed material Rückert was still able to make out tiny rod-shaped chromomeres, though by this stage of oocyte development he described them as being no longer arranged in neat parallel rows transverse to the long axis of the chromosome, but rather projecting irregularly and laterally from this axis. According to Rückert it is owing to this irregular distribution of the chromomeres, coupled with their poor stainability, that the chromosomes are so difficult to follow.

A little later on in oocyte development, when the oocytes have reached some 0.75 to 1 mm diam, it becomes easier to differentiate between the chromosomes and the nuclear sap because the former stain more intensely. Rückert describes this stage as being the high-point of the developmental process so far as the chromosomes are concerned. Rückert's drawing of a piece of such a chromosome is reproduced in Fig. 1.2, and he writes as follows: „Es besteht hier das Chromosoma aus einem Knäuel kompliziert gewundener und dicht verschlungener Fädchen, deren Hauptrichtung meist eine Quere ist. Die Mikrosomen, die schon in früheren Stadien zu kurzen Stäbchen sich verlängert hatten, sind jetzt zu gewundenen Fädchen ausgewachsen. Auf Querschnitten der Schleife tritt, wie Flemming für Siredon schon beschrieben hat, ‚das Bild eines Sternes mit dunkler Mitte auf', d. h. es sind die Fäden radiär zur Längsachse des Chromosoma gestellt. Man kann sich, ganz im Groben, eine plastische Vorstellung von dem Bau eines Chromosomenstückes entwerfen, wenn man an einen Lampencylinderputzer denkt,

Fig. 1.2. Rückert's (1892) drawing of part of a lamp-brush chromosome of *Pristiurus*

dessen Fäden nach dem Gebrauch stark verbogen und untereinander verfilzt sind." Freely translated, this passage reads: "The chromosome here consists of complexly wound and densely intertwined threads whose general direction lies normal to the chromosome's long axis. The chromomeres, which already in earlier stages had lengthened into short rods, by now have grown out laterally into the sinuous threads. Seen in cross-section a portion of the chromosome (Rückert here unfortunately uses the word Schleife – loop – to designate a portion of a chromosome, not a lateral loop) appears, just as Flemming described for the axolotl, like a star with a dark centre, i.e. the threads are disposed radial to the long axis of the chromosome. To take a rough analogy, one can visualize the structure of such a chromosome by thinking of a lampbrush whose bristles, through use, have become sharply bent over and intermingled with one another."

Rückert goes on to consider the nature of the main chromosome axis, and concludes that its density can be in part accounted for by the radial distribution of the thin threads which enter and leave it. But he makes the further point that the threads themselves are thicker and more strongly stained where they lie in the chromosome axis than where they run peripherally. Rückert's drawing shows the majority of the thin threads as loops projecting laterally from a chromosome axis without chromomeres, and although he was unable to observe with certainty that the loops continue one into another, this was clearly his general impression. But he leaves no doubt that in his opinion the thin lateral loops take their developmental origin from chromomeres, single chromomeres generating single loops, or several.

Now comes Rückert's positive demonstration that the objects just described are indeed chromosomes. As the end of the period of growth of the germinal vesicle approaches, in oocytes of 2 mm diam, the lampbrush structures become more readily distinguishable from nuclear sap. However, because of the great size of the germinal vesicle and the great lengths of the lampbrush structures, stained sections only provide views of chopped pieces of these objects. Rückert wished to see these structures in their entirety, and count them. He therefore isolated germinal vesicles free-hand under a dissecting lens, fixed them in acetic-sublimate, stained them for several days in borax carmine, and differentiated in acid alcohol until the nuclear sap was just colourless. Figure 1.3 is a reproduction of his drawing of the entire complement of lampbrush structures in such a preparation from a *Pristiurus* oocyte of 3 mm diam. Rückert described the structures as chromosome pairs, most about 80 µm long, but the longest up to 120 µm, each chromosome connected to its partner by one or more „Ueberkreuzungen" (cross-overs!)

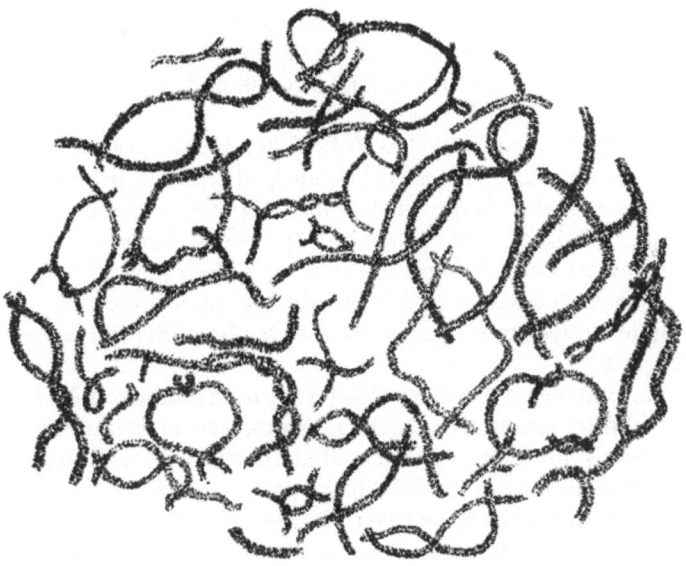

Fig. 1.3. Rückert's (1892) drawing of the entire lampbrush chromosome complement of *Pristiurus* from an isolated, fixed and stained germinal vesicle, when the chromosomes have reached their maximum size

to give the shapes now recognized to be so characteristic of diplotene bivalents. Rückert counted between 30 and 36 pairs of lampbrush structures per germinal vesicle, being in uncertainty over the exact number because a few of the structures looked to be unpaired. He then examined the germinal vesicles of larger oocytes over the size range 3 to 13 mm, and found that the structures progressively diminished in size and increased in staining intensity, but remained constant in number and continued to manifest their paired nature. He describes the reduction in size and increase in staining intensity as being the reverse of what happens at the beginning of oogenesis, retraction of the lateral fibres going hand-in-hand with the reconstitution of chromomeres. In oocytes of 12 mm diam all trace of the lateral fibres has gone, and thereafter the chromomeres amalgamate to form the chromosomes typical of a cell during division.

It remains to add that in one sense Rückert was unlucky in his choice of material, for during cell division the chromosomes of elasmobranchs are numerous, small and closely packed together. Rückert must have miscounted them, for his mitotic counts for *Pristiurus,* both germ line and somatic, are given as about 36, while his meiotic metaphase counts are given as about 18. Yet knowing what we now do about meiosis, the haploid number of *Pristiurus* must be the same as the number of pairs of lampbrush chromosomes, i.e. on Rückert's own evidence some 30 to 36. Because of these incorrect counts of metaphase chromosomes Rückert was led to the erroneous conclusion (though in keeping with that of Hertwig) that chromosome number doubles early in oocyte development, and is reduced to one-quarter during the elimination of the polar bodies. However this is a minor blemish in an otherwise remarkable study, and it is fitting that the cytologist who first showed that the lampbrush structures are indeed chromosomes, and that these chromosomes retain their individuality right through a stage in oogenesis where many other observers had claimed that they disintegrate, should have given them a name which endures.

Rückert's claim was speedily confirmed by Born (1892, 1894) working on the urodele *"Triton taeniatus"* (*Triturus vulgaris*). The illustration from Born's preliminary note, which was published in the same volume of the Anatomischer Anzeiger as that containing Rückert's paper, shows all the lateral strands as loops leaving from and returning to the chromosome axis, and Born makes the prescient point when describing these objects that the whole chromosome may well consist of a single, uninterrupted strand, many times woven back on itself to form the lateral loops.

In his later and more detailed paper Born (1894) returns to consider these two features of lampbrush chromosomes, that the lateral strands are invariably loops and that the chromosome is not multistranded, and he therefore remarks that to draw strict analogy with a lampbrush is misleading, for a lampbrush is certainly multistranded, and its lateral bristles have free ends. However Rückert must be defended, for his intention was clearly stated; he merely drew a rough and ready analogy to a lampbrush. Just like Rückert, Born observed that early in oogenesis the lateral loops appear to develop as extensions from axial chromomeres, that there follows a stage when no chromomeres are evident (the disappearance of stainable axial "chromatin" being accounted for by its extreme lateral dispersion) and that chromomeres reappear later in oogenesis as the lateral loops regress.

At the turn of the century Carnoy and Lebrun published three papers on the development of amphibian oocytes, two of which must be included in a survey of early work on lampbrush chromosomes. The first paper (1897) deals with *Salamandra* and *Pleurodeles,* the second (1898) with *Ambystoma* and *Triturus.* The texts of these two papers are hard to evaluate, for the authors do not recognize chromosomes as such. They speak of the disappearance of a "filament nucleinien primitif" early in oocyte growth, and they claim that the remarkable structures found in oocyte nuclei midway through the period of growth sprout out as filaments from nucleoli, these filaments in turn sprout lateral filaments which at the peak of their development take the form of granular loops, giving rise to "goupillons" (bottle-brushes), and that thereafter the loops and filaments disintegrate. This description is so bizarre, and so much at variance with the observations of the more authoritative German cytologists of the time, that Carnoy and Lebrun's papers could be disregarded were it not for the beauty and accuracy of their illustrations, one of which is reproduced in Fig. 1.4. Clearly they worked with excellent preparations, and drew what they saw with precision. What they saw evidently included chromomeres, lateral loops of great morphological variety, and recognizable bivalents with chiasmata. The third paper of the series (Carnoy and Lebrun 1899), which is concerned with the origin of the polar bodies in *Triturus,* starts off by denying continuity of the "element nucleinien" during oogenesis, claims de novo origin for the division chromosomes (this is the first occasion when Carnoy and Lebrun use this term) from a coalescence of nucleolar products, is equally well illustrated, and the haploid chromosome number is correctly given as 12!

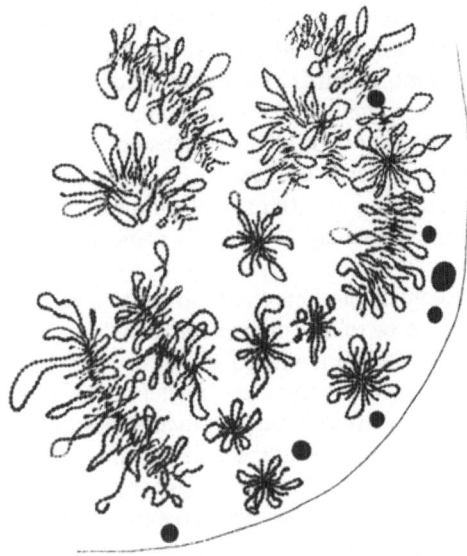

Fig. 1.4. Carnoy and Lebrun's (1897) drawing of parts of lampbrush chromosomes in a stained section through the germinal vesicle of an oocyte of *Salamandra*

Evidence for and against the structural continuity of chromosomes during oocyte development was actively sought during the first few years of the 20th century, and supporters of Carnoy and Lebrun's views included Goldschmidt (1902) and Hartmann (1902), biologists who were later to make important positive contributions to the study of heredity and its material basis. An early and formidable opponent was Janssens, soon after to achieve fame by his "chiasmatype" interpretation of bivalent chromosome structure. In a brief but cogent paper, without illustrations, Janssens (1904) emphasized the identity of the synaptic process in spermatocytes and oocytes of *Triturus,* and stated that he was able to follow the metamorphoses of the 12 bivalent chromosomes throughout oocyte growth and right up to the first maturation division.

The arguments of the time were reviewed by Maréchal (1907) who also came down firmly in favour of continuity despite his having worked at the Institut Carnoy in Louvain! Embarrassment is evident in the following passage taken from his introduction: «On pardonnera au bénéficiaire d'une hospitalité genereuse et cordiale l'espoir, un moment caressé, de confirmer les vues du fondateur de l'Institut Carnoy. Bientôt cet espoir dut en partie céder devant les faits».

Maréchal worked with stained sections through the ovaries of the elasmobranch *Scyllium canicula* and *Pristiurus melanostomus,* and his observations and elegant illustrations gave full support to Rückert's account of the genesis of lampbrush chromosomes during the early stages of oocyte growth, their appearance at full development and the retraction of lateral loops and contraction of chromosome axes as these chromosomes give origin to normal condensed meiotic bivalents; the nucleoli do not intervene directly in any of these metamorphoses. Maréchal also found «chromosomes barbelés» in oocytes of the teleosts *Trigla hirundo* and *Gasterosteus aculeatus;* he states that they are fully comparable to those of elasmobranchs, though his drawings show that their lateral filaments are much shorter.

Further evidence for chromosomal continuity through a lampbrush stage of oogenesis was provided about this time by Stevens (1903, 1904) for the chaetognath *Sagitta,* by Loyez (1906) for various reptiles, birds and cephalopod molluscs, and by King (1908) for the anuran *Bufo lentiginosus.* Moreover although King stated that the lateral filaments (not loops) grow out at the expense of chromomeric material, and augment chromomeric material as they later regress, a chromomeric organization of the chromosome axis, i.e. with discontinuous dense aggregates of chromatin in the axis, persists even when the laterally projecting filaments are maximally developed.

Around the turn of the century cytologists paid much attention to a certain "duality" of the constituents of cell nuclei, how to distinguish between them, and their functional significance. Heidenhain (1894) introduced the terms basichromatin and oxychromatin to designate materials in fixed cell nuclei which stain with basic and acidic dyes respectively, the former being relatively resistant to peptic digestion, the latter less so. A few years later Lubosch (1902) introduced the terms idiochromatin and trophochromatin, the former designating "stable" hereditary material, the latter being "metabolic" or "trophic". To a first approximation basichromatin = idiochromatin, and can be equated with DNA/histone in modern terminology, while oxychromatin = trophochromatin, and can be equated with nuclear proteins, some of which are associated with RNA. In the controversies surrounding the use of these terms the nature and behaviour of the constituents of germinal vesicles figured largely.

The transformation of an oocyte's chromosomes from basiphil (prior to oocyte growth) to acidophil (during oocyte growth) and back to basiphil (towards the end of oocyte growth in Amphibia, but much earlier in elasmobranchs and birds) led Rückert and most of his successors to the view that the predominantly acidophil lampbrush stage is a period when the chromosomes are metabolically active, i.e. generating trophochromatin according to the terminology of Lubosch, actively synthesizing RNA in modern parlance. Thus Jörgensen (1913), in the course of a morphological study of oogenesis in the urodele *Proteus anguineus,* the cave-dwelling olm of Carpathia, showed that if sections of ovary are digested in the presence of trypsin or pepsin, the chromosomes when basiphil resist digestion, whereas the acidophil lampbrush chromosomes are completely destroyed. Jörgensen concluded from these observations that the protein backbone of the chromosomes is protected from hydrolysis by the nucleic acid present during the basiphil stages, and that when the chromosomes are acidophil and digestible no nucleic acid is present. This view was upheld by Stieve (1921), the third cytologist to study oogenesis in such an unusual and hard to come by animal as *Proteus.*

The technique for specifically and unambiguously staining DNA, or thymonucleic acid as this substance was called at the time, was described by Feulgen and Rossenbeck in 1924, and a few years later Brachet (1929) applied the technique to sections of oocytes of various animals including *Rana, Triturus* and *Salamandra.* Brachet came to the conclusion that when they are maximally extended the lampbrush chromosomes of all three species are Feulgen-negative. He considered but rejected the possibility that the lampbrush chromosomes might appear to be Feulgen-negative merely because of an extreme dispersion of thymonucleic acid within the enormous germinal vesicle, though he made the qualification that thy-

monucleic acid might conceivably remain present but in a "masked" form. Brachet's observations were at the time generally taken as evidence that thymonucleic
acid cannot be the primary and persistent genetic material of chromosomes.

Brachet's observations were supported by those of several authors cited by
him in a later paper (1940), including in particular Koltzoff (1938) who studied
oocytes of *Triturus,* the hen and the pigeon by various techniques, including
Feulgen. Koltzoff made a contrast between the stable genotype of the chromosome and its phenotype, essentially a resuscitation of Lubosch's distinction between idiochromatin and trophochromatin. Koltzoff visualized the stable material in a lampbrush chromosome as a gigantic chain molecule or "genonema"
linking and including elementary chromomeres, possibly to be considered as
genes, which reproduce by a conservative template copying mechanism and
which, by a similar mechanism, induce the formation of maternal gene products
in the form of granular lampbrush strands and loops; the transversely disposed
lampbrush material ultimately detaches from the chromosomes' axes, and enters
the egg cytoplasm at maturation. In the English summary of his paper Koltzoff
states without qualification that "chromatin, i.e. thymonucleic acid completely
disappears from the chromosomes during the second period of oogenesis and reappears again only in the fifth period; the chromatin should, therefore, by no
means be included into the genotype of the chromosome and, hence, it cannot be
regarded as an ingredient of the genes. Chromatin is primarily a solid protective
sheath of the chromosome which at some stages isolates the genonema from the
caryoplasm and determines the definite form of the chromosomes which is convenient for caryokinetic movements."

Surprising though this may now appear, Koltzoff's view of chromosome organization was a very fair statement of cytological opinion prevailing at the time.
Thus Caspersson (1936), on the basis of microphotometric observations with ul-
traviolet light of the changing nucleic acid content of cell nuclei and chromosomes
during mitosis, and supported by the apparently total disorganization of *Chironomus* salivary gland chromosomes produced by tryptic digestion, held the view
that chromosomes have a protein framework to which nucleic acids become attached in preparation for mitosis, the nucleic acids dispersing once more to the
nuclear sap as nuclei pass from telophase to interphase. Caspersson's view was
championed by Darlington and La Cour (1940) in a paper where chromosomes
or parts of chromosomes were stated to be charged or over-charged or starved
of nucleic acid according to their staining behaviour, and this "nucleic acid charging" hypothesis was maintained by Darlington in several subsequent papers until
as late as 1947.

The question as to whether chromosomes really contain no thymonucleic
acid when in the lampbrush phase was evidently of crucial import in the late 1930s
because of its bearing on the discussions then in progress concerning the nature
of the primary genetic material; the problem was taken up again by Brachet.
Working with oocytes of *Rana fusca* and *R. esculenta,* and of *Triturus alpestris,*
T. cristatus and *T. pyrrhogaster* Brachet (1940) now showed that, despite previous
statements to the contrary, the chromomeres of the lampbrush chromosomes of
Triturus remain Feulgen-positive throughout oogenesis. He attributed his earlier
failure to demonstrate this property to the use of a fixative, corrosive sublimate/

acetic acid, which preserves cytological detail poorly and which leads to acute tissue swelling during the acid hydrolysis preceding exposure to Schiff's reagent; in
his later work Brachet used as fixatives Bouin-Allen, Helly and Flemming. Brachet was still unable to demonstrate Feulgen-positive chromomeres in *Rana* oocytes at the stage when their lampbrush chromosomes are maximally extended,
but he argued that anuran chromosomes are unlikely to be constructed in a fun-
damentally different way from those of urodeles, they are merely much smaller,
and that consequently the chromomeres of *Rana* are too small for the Feulgen
reaction to be detectable in the light microscope.

In his 1940 paper Brachet also described how Feulgen-positive granules are
associated with free nucleoli in the small oocytes of *Rana fusca*. This finding was
independently confirmed by Painter and Taylor (1942) in a study of the distribution of nucleic acids in oocytes of the toad *Bufo valliceps,* their paper including
the first description of nucleolar DNA amplification, a phenomenon that was to
be overlooked for another 25 yrs. Painter and Taylor (1942) were also able to establish that the chromomeres of toad lampbrush chromosomes are Feulgen-positive throughout oocyte development. Earlier Painter (1940), evidently with
thoughts of *Drosophila* polytene chromosomes in mind, had speculated that
lampbrush chromosomes are "... the result of some sort of reduplication process." However Painter and Taylor (1942) corrected this error with the statement
"... that the germinal vesicle of the toad is highly polyploid in nucleolar organizers but otherwise lampbrush chromosomes are normal meiotic structures."
Painter and Taylor's 1942 paper is important in another respect. Previous observers had generally viewed oogenesis as a process whereby material accumulates
within the germinal vesicle, under some kind of chromosomal control, and that
this "excess nuclear material" only becomes available to the developing embryo
when the germinal vesicle wall breaks down. On the contrary Painter and Taylor
were able to show, by methyl green/pyronine staining with and without prior ribonuclease digestion, and in further confirmation of Brachet (1940), that a more
dynamic situation exists, "... that there is a specific cytological mechanism in the
toad (and frog) which begins to deposit ribonucleic acid in the cytoplasm of the
oocyte soon after it is differentiated and continues to function through the
months required to build up the mature egg. While large amounts of nuclear material are set free in the cytoplasm when the germinal vesicle breaks down, this
appears to contain very little nucleic acid of either the ribose or desoxyribose
type."

The last paper to consider in this general context is one by Dodson (1948),
who made a morphological and biochemical study of the lampbrush chromosomes of the large American salamander ("Congo eel") *Amphiuma means* and to
a lesser extent those of the shark *Squalus suckleyi*. Ovaries were fixed and sectioned, and stained in various ways, with or without prior digestion by the enzymes nuclease ("thymo-nuclease") pepsin or trypsin. Dodson's morphological
observations on lampbrush chromosomes agreed with "... the classic picture as
reported by Rückert and most subsequent writers". Using the Feulgen technique
on *Amphiuma* Dodson confirmed once again that the chromomeres remain
Feulgen-positive throughout oocyte development, but he emphasized that the
smaller chromomeres are close to the limit of resolution of the light microscope

(he counted eight in a row only 7.2 µm long overall, their widths being about half their lengths). He found the lampbrush filaments to be Feulgen-negative, except occasionally at their bases. After prolonged nuclease digestion the chromomeres were found to be no longer Feulgen-positive, though the morphology of the chromosomes remained unchanged. Staining with Unna's methyl green/pyronine mixture resulted in faint pyronine colouration of the lateral loops, but after digestion with nuclease all capacity to stain was lost; from this Dodson concluded that his nuclease was contaminated with ribonuclease, and that the lateral loops must contain RNA.

When Dodson stained slides which had first been incubated with pepsin, the chromosomes showed up virtually intact in haematoxylin preparations, lateral loops as well as main axes, and Feulgen-stained preparations still showed the chromomeres to be Feulgen-positive. However, digestion with trypsin resulted in the destruction of both lateral loops and main chromosome axes; in the early stages of tryptic digestion Feulgen stainability was not restricted to chromosome axes, but instead was diffusely spread throughout oocyte nuclei, and from this observation Dodson deduced that "... the nucleic acid is bound to the chromosome by a protein which is susceptible to trypsin digestion, but not to pepsin digestion." From the results of his studies taken overall, Dodson reached the following conclusion: "The structural skeleton of the chromosome appears to be histone, while the nucleic acids are attached to this skeleton through protamine. In the main axis the nucleic acid is of the thymonucleic type, while in the loops this is rapidly converted to the ribonucleic type ... Functionally, the lampbrush chromosomes are regarded as agents for the synthesis of ribonucleic acid and enzymes, or other type of developmental agent, for use in the cytoplasm. ... In these chromosomes there is a functional separation of idiochromatin (the main axis) and trophochromatin (the side loops)."

It is appropriate at this point to turn to another approach to the study of lampbrush chromosomes, a method that was destined to lead to novel and far-reaching conclusions. Beginning his work in 1936 in Prof. Baltzer's laboratory in Bern, the American Duryee discovered how easy it is to isolate the germinal vesicles of amphibian and fish oocytes in saline media, and to remove the nuclear membrane, both these operations being done free-hand, and thereafter to handle the lampbrush chromosomes with the aid of a micromanipulator. Duryee published the first account of his observations in 1937, and one of his photographs (of a bivalent lampbrush chromosome of *Triturus pyrrhogaster*) was reproduced in Waddington's (1939) textbook on genetics. Duryee's crucial contribution was his discovery that lampbrush chromosomes must not be exposed to Ringer's amphibian saline if they are to be handled and examined in a lifelike state, and that for this deleterious action the calcium ions at the concentration present in Ringer's saline (CaCl$_2$ 1 mM) are responsible. Moreover because the germinal vesicle membrane is permeable to calcium, oocyte nuclei must from the start be isolated and cleaned of cytoplasm in calcium-free Ringer. In Ringer containing calcium the chromosomes quickly become brittle and contracted, loops as well as main axes, whereas in calcium-free Ringer they remain extended as in life, and furthermore show remarkable elastic extensibility when stretched by micromanipulation. When the chromosomes are stretched, the bases of the loops do not

Fig. 1.5. Duryee's (1941) schematic diagram of the structure of a lamp-brush bivalent

open up, as they would if the loops were merely the lateral projections of a loose coil; instead the lateral projections remain as loops, with their bases close together and associated with chromomeres visible in the stretched main chromosome axis.

Figure 1.5 shows Duryee's schematic diagram of the morphology of a lamp-brush bivalent as published in 1941. It is erroneous in several respects, and in part these errors must be attributable to diffraction artefacts; unfixed and unstained lampbrush chromosomes are exceedingly difficult to see, let alone to observe criti-cally, under a light microscope using ordinary illumination, and indeed Duryee states that in order to take his photographs he was forced to increase the refrac-tility of the chromosomes by adjusting saline pH to 4.5.

Duryee's view of lampbrush chromosome organization was that the main axis consists of a "single plastic cylinder" with granules, chromomeres, embedded in this cylinder. The chromomeres are present in pairs, or are dumbbell-shaped, *transverse* to the chromosome axis, and where loops are present these are *single,* running from a granule on one side to its partner on the other. Duryee's concep-tion of the loops was that they are the products of "molecular templates" which remain in the chromosome's axis cylinder, the loops starting as transverse bars between a chromomere pair (so giving rise to the dumbbell shape), then intermit-tently growing and buckling out from the main axis, then fragmenting into gran-ules which come to lie free in the nuclear sap and which ultimately enter the egg cytoplasm at maturation. Duryee identified the junctions between homologous chromosomes as chiasmata, and he was unable to break these junctions by micro-manipulation.

Considered in detail, there are a host of mistakes in Duryee's account of the structure of lampbrush chromosomes (and of the successive production of nu-cleoli at "... definite loci on 3 or 4 pairs of chromosomes ... during early and middle ovogenesis," and of the "eversion" of nucleolar contents through the ger-minal vesicle membrane into the cytoplasm), but all this should not distract the reader from appreciation of his main achievement, which was to show that lamp-brush chromosomes can be easily isolated free-hand provided calcium-free saline is used as the medium for isolation.

A year or so later, and independent of Duryee, another investigator, Gersch (1940) made numerous experiments on germinal vesicles that had been isolated free-hand. Gersch worked with oocytes of *Rana temporaria,* and he was primarily interested in the effects of various simple salines, also of pepsin and trypsin, on nucleoli. He made his observations and photographs using low magnification and dark-field illumination and made incidental notes on the reactions of lampbrush chromosomes, e.g. whether or not they remained visible, in the various media employed. Gersch found that *Rana* lampbrush chromosomes remain detectable in distilled water, in 0.5%, 10% and saturated NaCl, and also in pepsin/HCl, but that they are destroyed by 0.5% KOH, also by 0.5% Na_2CO_3 with or without trypsin.

In a paper primarily concerned with the structural organization of meiotic prophase chromosomes of grasshoppers, Ris (1945) considered the evidence for a "chromonemal" (i.e. continuous thread) as opposed to a "chromomeric" (i.e. beads on a string) type of organization, and came down strongly in favour of the former. Ris claimed that the chromomeres seen at leptotene by so many of the earlier cytologists are in reality regions of a coiled chromonema of uniform thickness where the gyres of the spiral are more tightly packed together than they are in neighbouring "interchromeric" regions. One reason for this claim was that at leptotene in grasshopper spermatocytes "... with Feulgen the chromosome stains evenly throughout its length and there are no Feulgen-negative interchromomeric fibrils." As will become apparent later, there is an underlying measure of truth in this claim, though it is not based on the evidence supplied by Ris. The reason for referring to Ris's 1945 paper here is that he included a photograph of a portion of a lampbrush chromosome from "... a frog oocyte, smeared in aceto-orcein," and stated that: "In contrast to Duryee, Koltzoff and Painter, it is here suggested that 'lamp-brush chromosomes' are typical diplotene chromosomes which differ from other diplotene chromosomes only in the tremendous longitudinal growth of the chromonemata. The loops are then the major coils of the laterally separated chromonemata, the chromomeres are simply overlaps of the strands just as in diplotene chromosomes of the grasshopper." In the photograph Ris claimed that the "... large loops of the major coil and the minor coil are easily visible." It is certainly true that no chromomeres are evident in Ris's photograph, but his claim that the loops are simply major coils of the laterally separated chromonemata (and Ris assumed there to be eight of these per chromosome) was based on supposition only. There is however another statement made in Ris's 1945 paper, for which the evidence was provided several years later. Unlike Duryee and Koltzoff, both of whom looked upon the lateral loops as gene products which are periodically sloughed off in their entirety from a primary chromosome axis, Ris claimed that "... the reduction in (lampbrush) chromosome size, just before the meiotic divisions, is accomplished ... not by throwing off parts of the chromosome ... but by the elimination of material on a submicroscopic level."

Ris's coiled chromonemal interpretation of lampbrush chromosome structure was challenged by Duryee (1950). In this paper Duryee made the certainly incorrect claim that the bivalents of *Rana temporaria* and four other species of *Rana* can be identified on the basis of the number and disposition of chiasmata (said to be constant for each individual bivalent) together with overall size; however he again demonstrated by micromanipulation that the bases of lateral loops do not open out when lampbrush chromosomes are stretched, and that dense chromomeres lie in the chromosome axis at the base of loops, some of the chromomeres bearing clusters of loops, others two or one loop, and some no loops at all. Furthermore he drew attention to the fact "... that loops are not all of the same construction," which he argued "... contradicts the assumptions of Ris and

Painter that the lateral loops are portions of the chromonemata." Duryee's view that the lateral loops are of quite a different nature from that of the chromosome axis was backed up by his observation that short exposure to 200 mM $NaHCO_3$ followed by mild acidification, or to 10 mM KCl, or to various other reagents, or indeed to heavy dosages of X-rays, appeared to remove the lateral loops but left the chromosome axis, complete with chromomeres, intact. Duryee continued to assert that in the normal course of oogenesis the lateral loops fragment, and that the loop fragments accumulate inside the germinal vesicle, being at their most numerous just before oocyte maturation.

It was in 1947 that I first came to realize the potentialities which amphibian oocytes present for experimentation, but this realization did not stem from my having read Duryee's papers. Instead it arose from a misconception of the process of oogenesis! I wished to compare first meiotic metaphase chromosomes of newt (*Triturus cristatus carnifex*) oocytes with those of newt spermatocytes, being already familiar with the latter (Callan 1942). For this purpose I fixed newt ovaries in Zenker's fluid, and used the Feulgen technique on the largest oocytes with the expectation of finding, under a dissecting binocular, a small purple object – the group of meiotic chromosomes – lying in an unstained mass of yolky cytoplasm. Instead I found that entire oocytes stained bright purple, a consequence of the plasmal reaction, and when broken open these oocytes each revealed a large white sphere inside, the germinal vesicle. It was only thereafter that Duryee's photograph of a lampbrush bivalent in Waddington's (1939) textbook of genetics came to my attention and I then read Duryee's paper. I did not at once turn to a study of lampbrush chromosomes, but rather took up the question of the structure and properties of the germinal vesicle nuclear membrane (Callan 1948, Callan and Tomlin 1950), and made some preliminary observations on the chemical and physical nature of its nuclear sap (G. L. Brown et al. 1950, Callan 1952).

With the help of Dr. S. G. Tomlin, who operated an early model of the Siemens electron microscope (EM) in the Physics Department of King's College, London, I then attempted the free-hand isolation of *T.c. carnifex* lampbrush chromosomes under a dissecting binocular, attaching parts of moderately stretched chromosomes to a film of metallized formvar, the film thereafter being mounted on a copper EM grid, shadowed and examined. These preparations showed that the axis of a newt lampbrush chromosome consists of a single filament of relatively even diameter, about 20 nm, to which lateral loops are attached at intervals. The preservation of the lateral loops left much to be desired, but the finding of a single axial filament of so little width (Fig. 1.6) came as something of a surprise, for even in the light microscope the dual nature of the axes of chromosomes at diplotene is readily observable in suitable material, such as grasshopper spermatocytes. This preliminary work was published by Tomlin and Callan in 1951.

While manipulating the chromosomes of *T.c. carnifex* for EM study it soon became apparent that there is considerable morphological variety amongst the objects attached to lampbrush chromosome axes; this was evident even under a dissecting binocular. I therefore attempted to examine in phase contrast unfixed preparations of lampbrush chromosomes which had been isolated in saline and then placed between a conventional coverslip and slide. This latter operation

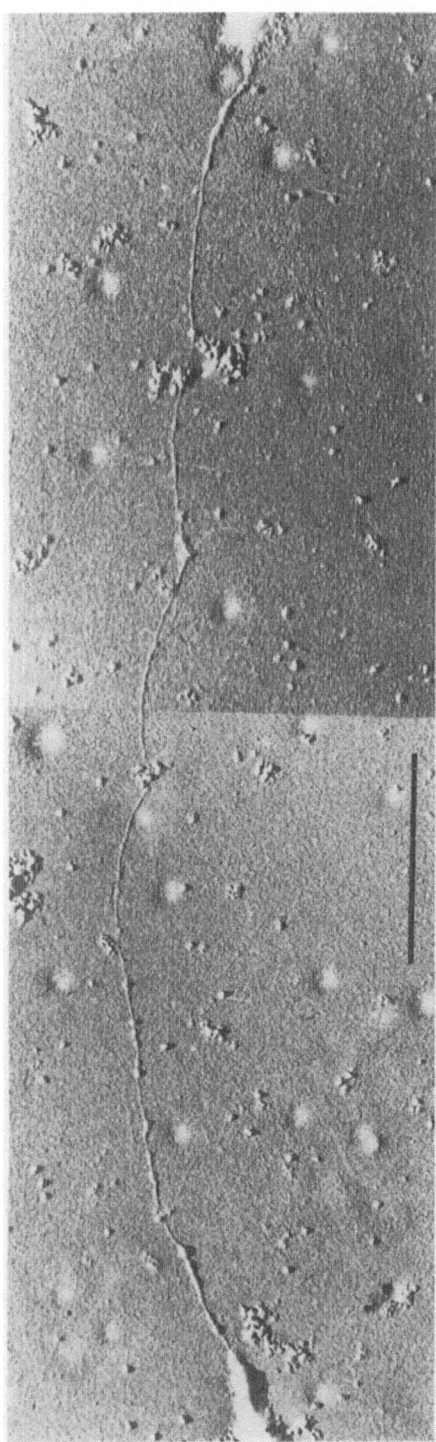

Fig. 1.6. Electron micrograph showing the single axial filament running between two chromomeres in part of a stretched lamp-brush chromosome of *Triturus cristatus*. *Bar* = 2 μm. (From Tomlin and Callan 1951)

proved to be fraught with difficulty; indeed it was at once clear that complete un-damaged chromosome complements could not be prepared in this manner. I started to experiment using depression slides, and it was at this juncture that Dr. J. G. Gall, then a postgraduate student, first wrote to me from Yale University, giving at once invaluable advice on how to proceed.

Gall had been studying the lampbrush chromosomes of *Triturus* (*Notophthalmus*) *viridescens,* and his first account of this work was published in 1952. In it he described how preparations can be made which permit protracted study under phase contrast, and with immersion objectives resolving to the limit achievable by the light microscope. A germinal vesicle is isolated and cleaned of adhering yolk in calcium-free saline, then transferred to a large drop of saline on a coverslip, where its membrane is removed. As the nuclear sap disperses, nucleoli and chromosomes both become more and more clearly visible, and they fall to the coverslip surface. The coverslip preparation is placed for an hour in a moist chamber saturated with osmic or formaldehyde vapour, after which time the chromosomes generally come to adhere firmly to the glass. The coverslip is now inverted over a thin depression slide filled with low viscosity immersion oil, and is finally ringed with paraffin wax. Gall had established that the method of fixation by vapour causes little morphological alteration of the chromosomes other than a measure of contraction which enhances their refractility, so that such preparations conserve the lifelike appearance yet are essentially permanent.

Gall observed and photographed specific parts of the *N. viridescens* complement that were regularly identifiable from preparation to preparation, he saw chromomeres of various dimensions, the patterns of which were similar, though not always identical, in homologous region of bivalents, and he identified the centromeres (kinetochores), which in *N. viridescens* are elongate chromomeres without lateral loops. Gall confirmed earlier demonstrations that the chromomeres, but not the lateral loops, are Feulgen-positive, and as regards the latter Gall stated that "... the lateral loops are dissolved away if the chromosomes remain an hour or more in distilled water." He further confirmed that the interchromomeric fibril, which he termed the chromonema, is a single strand.

Gall's detailed account of the lampbrush chromosomes of *N. viridescens* was published 2 yrs. later, in 1954. In the material and techniques section of this paper, which I will discuss in Chap. 2, he described a novel method for examining lampbrush chromosomes, the use of which was soon to lead to a better understanding of the organization of these objects. Gall identified all 11 bivalents of *N. viridescens* on the basis of their lengths (which at maximum extension range from about 350 to 800 µm), centromere positions, the positions of loops which can be regularly recognized because of their exceptional lengths (some up to 200 µm) or exceptional textures, and the positions of various other laterally attached objects, particularly "knobs" and nucleoli. As regards the latter, Gall compared the place where a nucleolus, morphologically similar to the free nucleoli, is attached to one lampbrush chromosome of *Amblystoma tigrinum,* with the position of the nucleolus organizer constriction in somatic chromosome complements of this organism, and demonstrated their correspondence. He thereby established that the hundreds of unattached nucleoli in amphibian germinal vesicles nuclei really are homologous to the nucleoli of somati cells, which up until this time had been merely an assumption.

Gall confirmed Dodson's observation, by toluidine blue staining of fixed chromosomes with and without prior digestion by ribonuclease, that the lateral loops and nucleoli (but not the "knobs") contain RNA. He also tested the reactions of unfixed chromosomes to digestion with trypsin and pepsin. Unlike Dodson, who had worked throughout with paraffin wax sections of oocytes, he found that trypsin, although it rapidly dissolves the lateral loops and causes the chromomeres to coalesce and lose optical contrast, does not destroy the linear integrity of lampbrush chromosomes. A few years later the full significance of this important observation was to become apparent.

Beginning their observations in 1949, Guyénot and Danon in Geneva carried out a parallel cytological study of oogenesis in *Triturus cristatus* and *Rana temporaria,* and published a long paper in 1953. Their work is not easy to evaluate, partly because several of the objects which they describe from stained sections are certainly misidentified. Thus for example their «macronucléoles» and «filaments nucléoplasmiques» are particularly large and peculiar lateral loops that are characteristic of *T. cristatus* chromosomes (see Chap. 3). Guyénot and Danon's (1953) theory concerning the origin and development of lateral loops is bizarre; these are said to grow out from the chromosome axis as long bristles («poils»), which are split lengthwise into two except where they are inserted into the chromosome axis (and presumably also at their distal extremities) thereby forming loops («boucles»). They assert that the DNA content of the lampbrush chromosomes is minimal when oocytes are growing most rapidly, when the lateral loops are maximally extended, and that the chromosomes reacquire a coating of DNA as maturity approaches; this is a revival of the nucleic acid charging hypothesis of Darlington. Their paper could be altogether disregarded were it not for an observation made in phase contrast of lampbrush chromosomes isolated in «eau salée acétifiée» (to enhance contrast). They noted «... la fissuration du chromosome en deux chromatides, dont les grains sombres (chromioles) se correspondent deux à deux. C'est sur ces chromioles jumelés que sont insérés les poils et boucles qui sont disposés par paires». This is the first mention in the literature of lateral loops occurring *in pairs.*

My own first studies of lampbrush chromosomes with the light microscope, using Gall's observation chamber and an inverted microscope, began in 1950. I started to work with oocytes of *Triturus vulgaris, T. helveticus* and *T. marmoratus,* and published a preliminary review of this work in 1955. An early and provocative observation was that if a freshly isolated lampbrush chromosome is stretched with a micromanipulator, not only does extension primarily occur in regions between chromomeres, without affecting the close juxtaposition of individual loops' two insertions in the chromosome axis (thus confirming Duryee's earlier observation), but further, occasional chromomeres split transversely when under tension; when they break in this way, the gap in the chromosome's axis *is invariably spanned by a pair of lateral loops* (Fig. 1.7). It later transpired that such "double-loop bridges" may be generated accidentally during the isolation of a germinal vesicle and the removal of its nuclear membrane (Fig. 1.8a), more frequently in some newt species than in others. Still later I found that one particular pair of loops (the "giant fusing loops") on the smallest chromosome of *Triturus cristatus cristatus* often occur in double-bridge form, i.e. with their axial insertions separate from one another, in the natural state (Fig. 1.8b).

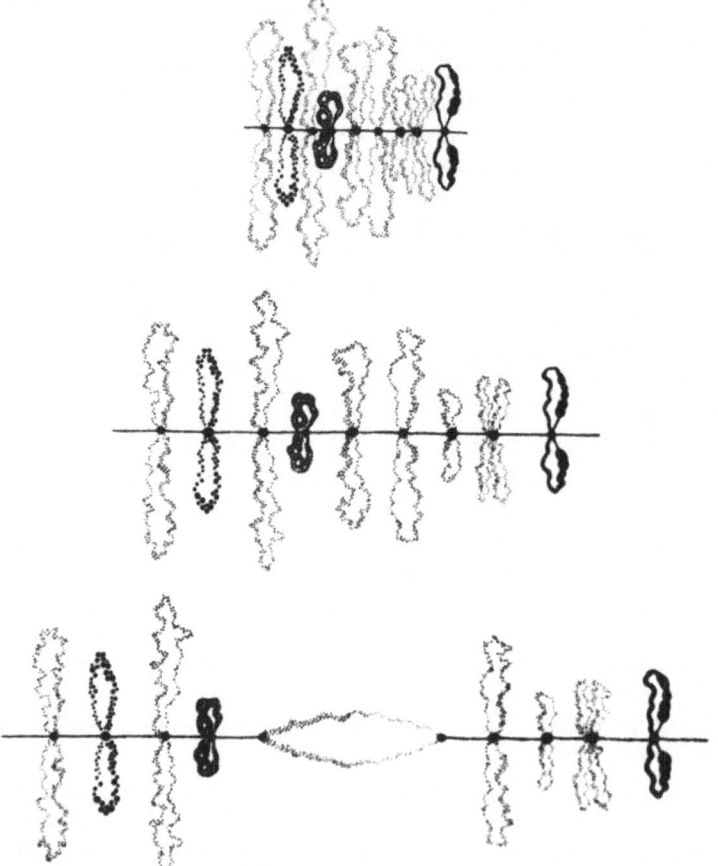

Fig. 1.7. Diagrams to illustrate what occurs when part of a lampbrush chromosome is stretched and a chromomere cleaves transversely, so producing a "double-loop bridge". (From Callan 1963)

The fact that mechanically produced or natural loop bridges occur at all, and are double, and that particularly large and easily recognizable loops on the smallest chromosome of *T. marmoratus* are always present in pairs (Fig. 1.8c), prompted the speculation that there is a general duality of lateral loops (Fig. 1.9), indicating that each lampbrush chromosome consists of two chromatids, that each loop contains a fibrillar "core" running throughout its length, and that these loop cores are reflected, extended portions of the chromatids' axis, each being surrounded, asymmetrically, by the accumulated products of synthesis. This view

Fig. 1.8 a–c. Phase contrast micrographs. **a** Double bridge accidentally produced by a pair of lampbrush loops of *Triturus cristatus carnifex. Bar* = 20 μm. **b** Naturally occurring double bridge formed by the pair of giant fusing loops on chromosome XII of *T.c. cristatus. Bar* = 20 μm. **c** Bivalent XII of *T. marmoratus* dispersed in dilute saline; *arrows* point to the two homologous pairs of giant loops which have unravelled from their naturally compact state, while *arrowheads* point to the double axis regions. *Bar* = 50 μm. **a** From Callan (unpublished); **b** from Callan and Lloyd (1960); **c** from Callan (1955)

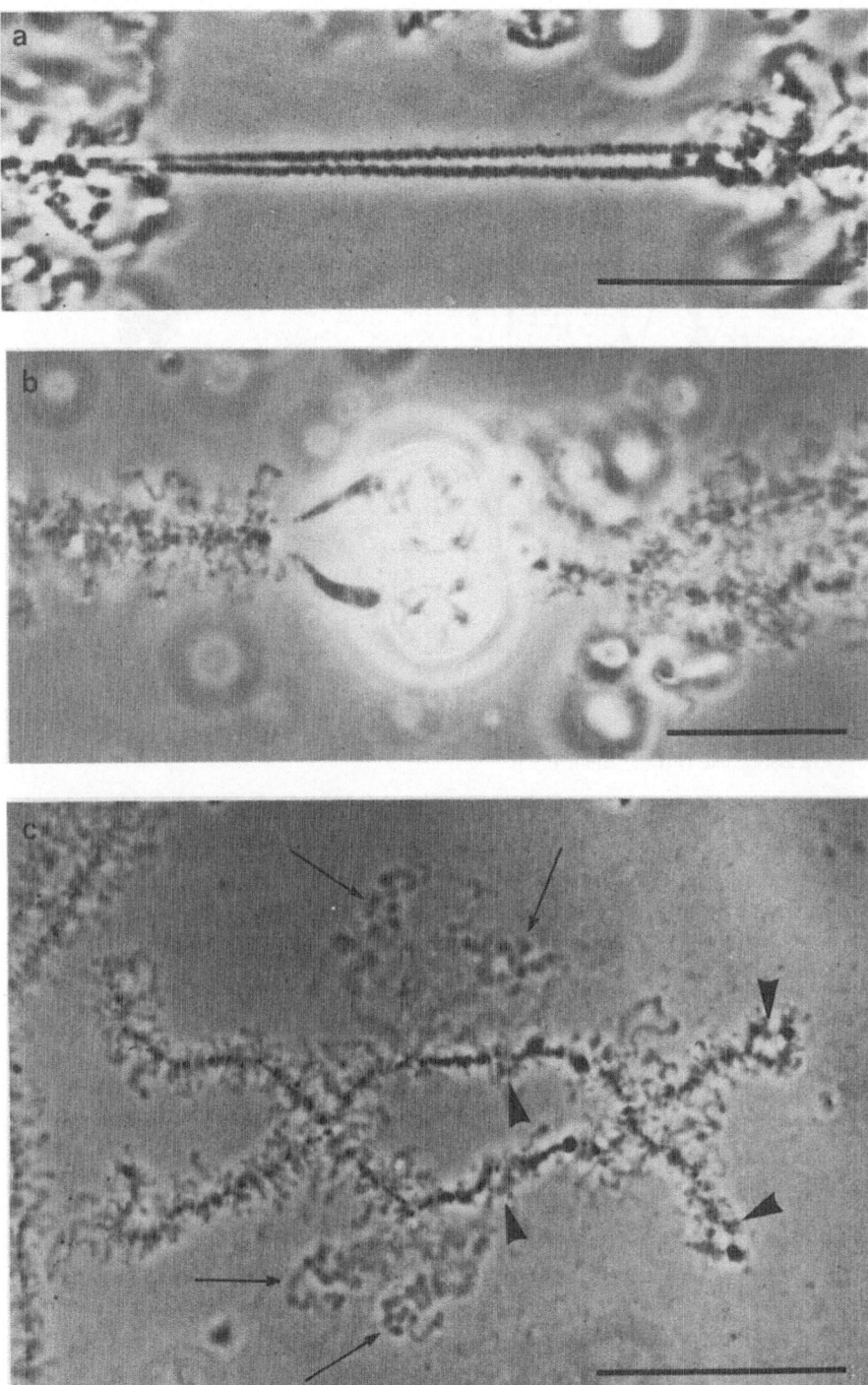

Fig. 1.8 a–c

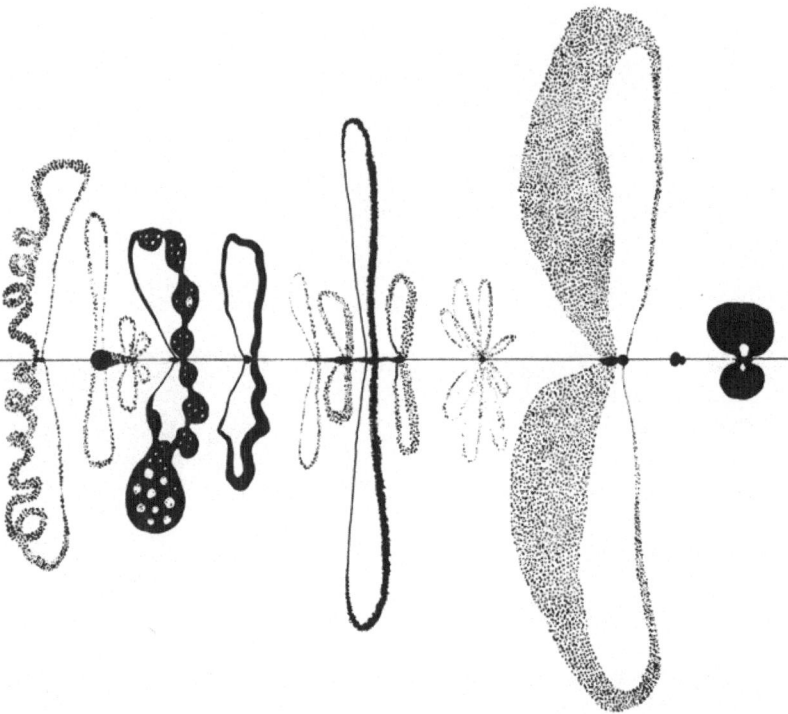

Fig. 1.9. Some characteristic types of newt lampbrush loops and the chromomeres with which they are associated. (From Callan 1955)

was reinforced by the observation that along the smallest chromosome of *T. marmoratus* there are two short regions, one near the middle and the other at one end, where the chromosome axis in its entirety is double (Fig. 1.8c), with matching pairs of chromomeres on each axis, and each axis generating single-loop bridges when mechanically broken. Thus the relationship between lateral loops and the chromosome axis was put in rather a new light, betwixt and between the Duryee and Ris interpretations. Chromomeres are a reality, but the lateral loops are not simply "gene products" periodically shed from chromomeres, nor are they merely portions of a continuous chromonema; instead, embedded within each lateral loop, is a portion of the chromonema.

This interpretation of the organization of lampbrush chromosomes was discussed by Gall in 1955, and within a year the existence of a loop axis, resistant to digestion with pepsin and in this respect unlike the bulky surrounding material, had been demonstrated by electron microscopy (Gall 1956). For this paper Gall drew a diagram showing the structure postulated, which is reproduced here in Fig. 1.10. At the close of his 1956 paper Gall calculated the total length of DNA present in a lampbrush chromosome complement having the 4C weight of DNA, and from this concluded that a single averaged chromatid of *Notophthalmus viridescens* contains about 90 cm of DNA. He also computed to a first approximation the total length of the lateral loops associated with such a lampbrush

Fig. 1.10. The organization of a chromomere and of its suggested relationship to a pair of lateral loops. (From Gall 1956)

chromatid as 5 cm. He then went on to consider the possibility that if DNA is found uniformly throughout the chromatid (i.e. in the interchromomeric fibril and in loop axis as well as in chromomeres) then one can predict an upper limit to the number of DNA strands per chromatid (18) and the further likelihood, because of the known compaction of DNA in chromomeres, that the number of strands per chromatid may be considerably less than this figure, perhaps only one. So far as I am aware, this was the first tentative step towards the proposition of uninemy.

By 1955 I had decided to concentrate my attention on lampbrush chromosomes of newts of the multiracial species *Triturus cristatus,* and for three reasons. One of these was that there is a remarkable wealth of morphological diversity amongst the objects, loops and other structures, projecting laterally from the chromosomes of these animals; this permits the rapid and precise identification of particular chromosome regions. A second reason lay in the fact that one of the races of *T. cristatus,* the Italian *T.c. carnifex,* is easy to maintain and breed in the laboratory, there being no obligatory terrestrial stage in its life history, and both sexes coming repeatedly into annual cycles of reproductive activity without any special treatment; the males of many other species of newt, including the American *N. viridescens* and the British race of *T. cristatus, T.c. cristatus,* are recalcitrant in this regard. The third reason was that I was already familiar with the male meiosis of four of the races of *T. cristatus,* and of several of its interracial hybrids, F_1 and other (Spurway and Callan 1950, Callan and Spurway 1951), I had many interracial hybrids still alive in the laboratory and I hoped by studying female interracial hybrids to find evidence for the Mendelian inheritance of "phenotypic" chromosomal characters, i.e. lateral loop morphologies, differentiating the races from one another.

The construction and examination of interracial hybrids in order to demonstrate the Mendelian inheritance of lateral loop characters later proved to be, at least in part, unnecessary, indeed already in 1956 Callan and Lloyd published a preliminary note in which it was shown that individual females of *T.c. carnifex* may differ from one another in respect of the morphologies of loops at specific chromosomal loci, including spectacular heterozygosities that are consistent in all the oocytes of certain individuals, therefore providing strong presumptive evidence that loop morphologies are directly determined by "local" genetic constitution.

For a short period in 1956 I turned my attention to the possibility of studying the oocyte chromosomes of other organisms, particularly invertebrates, by the techniques already applied to newts, and found (Callan 1957) that there are genuine lampbrush chromosomes, as defined by the criteria that the chromosomes bear lateral loops, and that axial breaks are regularly spanned by double-loop bridges, in the cephalopod mollusc *Sepia* and the isopod crustacean *Anilocra*. These findings, and a perusal of the already voluminous literature on oogenesis, prompted the speculation that all animal oocytes may contain lampbrush chromosomes at some stage during oogenesis. This generalization is not however valid, as will be discussed in Chap. 6.

Returning to newt lampbrush chromosomes, the possibility now presented itself that, by applying enzymes directly to unfixed lampbrush chromosomes floating freely in saline, the resolution of an urgent and contentious problem of the time might be in sight: how is DNA distributed along the length of a chromosome, is it interrupted, and if so, what material maintains a chromosomes's linear integrity? The great advantage of this experimental system lies in the fact that isolated lampbrush chromosomes surrounded by saline are in violent Brownian movement, yet they remain coherent entities; any disruption of their linear integrity at once becomes manifest because portions then move independently and drift apart.

Macgregor and I applied the enzymes trypsin, pepsin, ribonuclease and deoxyribonuclease to freshly isolated lampbrush chromosomes (Fig. 1.11). We found that: "Pepsin, trypsin and ribonuclease all bring about the solution of loop matrix materials, together with certain other structures attached laterally to the axes of the chromosomes, but these enzymes do not destroy the linear integrity of the chromosomes. The action of deoxyribonuclease, however, is dramatically different. The submicroscopic fibrils connecting adjacent chromomeres snap, and the lateral loops fragment into smaller and smaller pieces, each piece however conserving for a time at least the fine-scale morphological peculiarities of the loop from which it originated. A drastic disruption of the chromosomes is already visible within a few minutes of the beginning of enzyme treatment, and in half an hour Brownian movement has so scattered the fragments that no recognizable trace of the original forms of the chromosomes remains. In our experience no other chemical agent has been found capable of breaking lampbrush chromosomes in this fashion, and the conclusion is inescapable that deoxyribonucleic acid runs throughout the lengths of lampbrush chromosomes, including the axes of their lateral loops" (Callan and Macgregor 1958).

Following up these observations Gall (1963 a, b) studied the kinetics of breakage of unfixed lampbrush chromosomes when digested by pancreatic DNase I. This enzyme was already known to attack the two polynucleotide chains of DNA in solution independently and at random, and that scission of the double helical molecule occurs only when breaks in the two chains occur within a few nucleotide pairs of each other; thus breaks in DNA in solution accumulate as a function of time raised to the power two. Gall found that breaks of single loops (the giant loops near the centromeres of *Notophthalmus viridescens* were chosen because of their considerable lengths) accumulates as a function of time raised to the power 2.6 ± 0.2 (Fig. 1.12). Breaks of interchromomeric main axes accumulated as a function of time raised to the power 4.8 ± 0.4. Under the conditions of Gall's experiments, and bearing in mind that DNA is associated with histones in lampbrush chromosomes, both these figures are likely to be overestimates. Gall

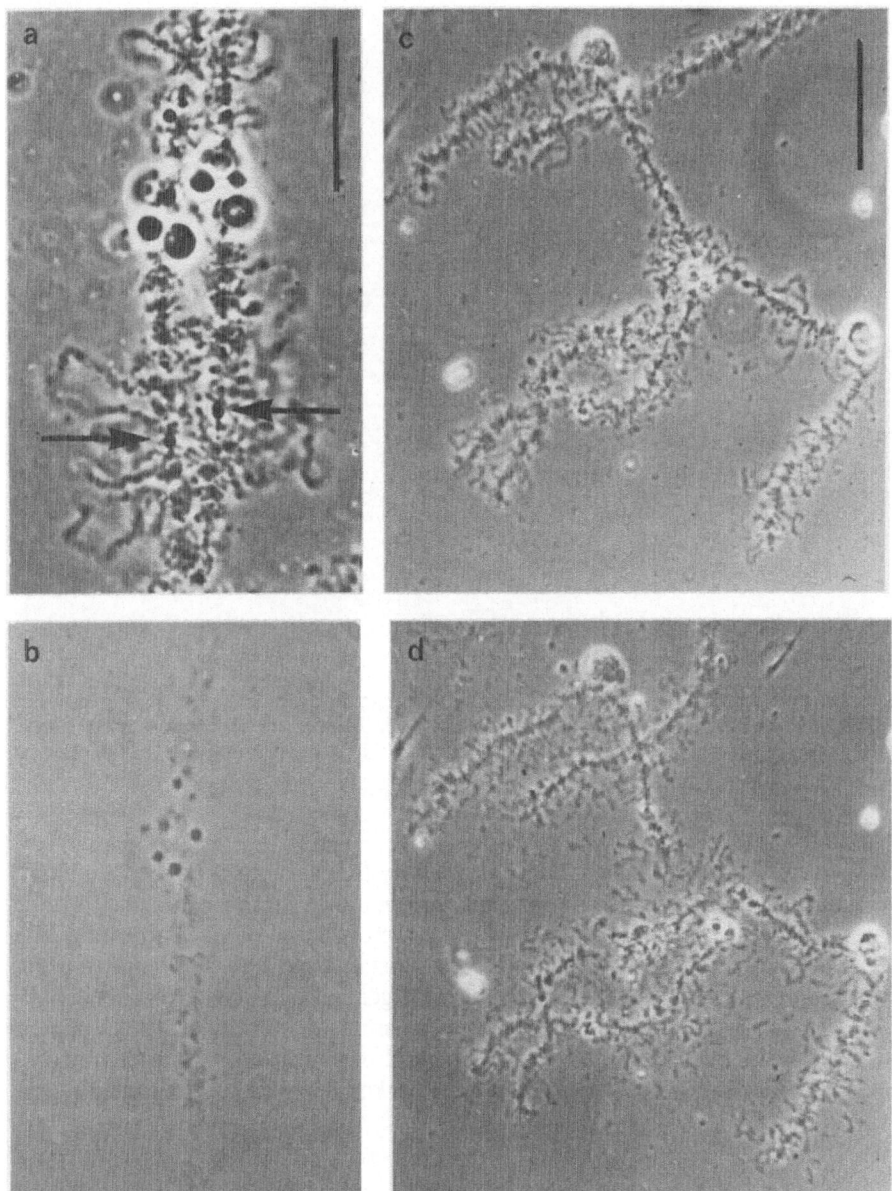

Fig. 1.11 a–d. Phase contrast micrographs. **a** The middle region of lampbrush bivalent II of *T.c. carnifex; arrows* point to the centromeres; the bulky refractile objects are lumpy loops, regular features near the centromeres of chromosome II. *Bar* = 20 μm. Photograph **b,** as **a,** but after 40 min digestion with trypsin at pH 7.8. **c** Bivalent X of *T.c. carnifex,* entire. *Bar* = 50 μm. Photograph **d,** as **c,** but after 3 min digestion with pancreatic DNase at pH 6.3. (From Macgregor and Callan 1962)

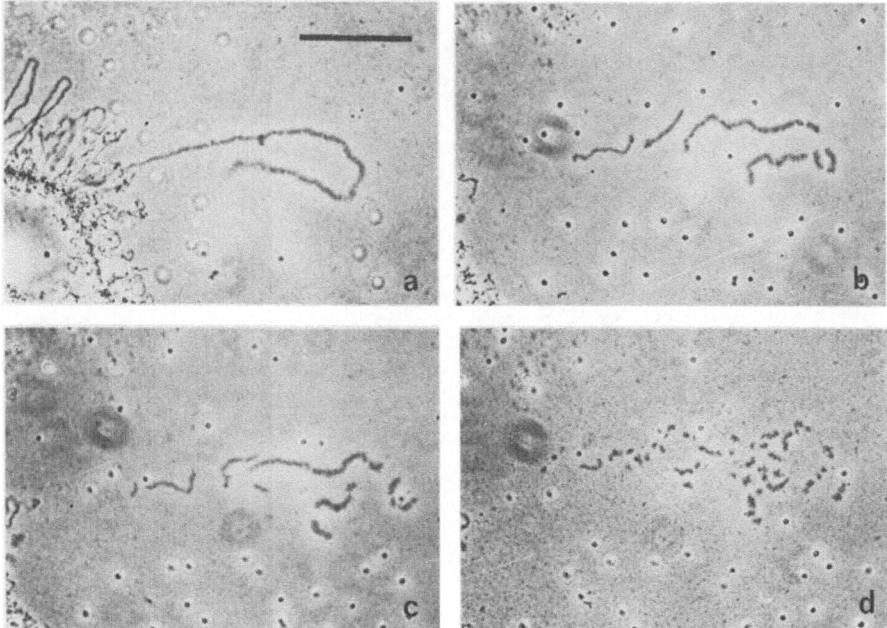

Fig. 1.12 a–d. Phase contrast micrographs. Lampbrush chromosomes of *Notophthalmus viridescens* were isolated and dispersed in saline containing DNase, and flash photographs of one of the giant loops on chromosome II were taken to determine the rate of accumulation of breaks with time. **a** First break 7 min after isolation; **b** after 18 min; **c** after 22 min; **d** after 27 min. *Bar* = 50 μm. (From Gall 1966)

concluded that the simplest model consistent with his results is that a chromatid, as seen in single lampbrush loops, contains one DNA double helix, and that the interchromomeric strand in a lampbrush chromosome contains two running side by side. These conclusions have now been substantiated by experiments of other kinds, and the unineme nature of chromatids generally thereby established.

   This is an appropriate moment to close an historical introductory chapter, for a reasonable understanding of the general organization of lampbrush chromosomes had by now been provided. It had taken 85 yrs. since Flemming's original discovery.

CHAPTER 2

Techniques for the Study of Lampbrush Chromosomes

This chapter will be primarily concerned with techniques that have been devised for isolating and studying the lampbrush chromosomes of Amphibia. No attempt will be made to cover conventional methods applied to whole oocytes for light or electron microscopy, though it is worth a mention that in my experience Sanfelice's (1918) fixative is by far the most reliable for morphological studies with the light microscope on fixed, sectioned and stained oocyte nuclei. Two accounts of the procedures appropriate for isolating and examining lampbrush chromosomes are available in the earlier literature (Callan and Lloyd 1960, Gall 1966) but they do not include recent technical developments; several of these are to be found in Macgregor and Varley's (1983) book on techniques for studying the chromosomes of animals.

Instruments

Scissors. Operations on small Amphibia, and the removal of their ovaries, or parts of their ovaries, are best performed using spring-loaded iridectomy scissors (scissor forceps).

Surgical Needles, Suture and Anaesthesia. Amphibia should be anaesthetized before operation, and a 0.1% solution of MS222 Sandoz (ethyl m-aminobenzoate) in tap water is a suitable anaesthetic. Only a small portion of ovary may be required at any one time, and it is convenient that most Amphibia will survive repeated operations if the cut body wall is sutured after laparotomy. In the United Kingdom when an amphibian is allowed to recover after such an operation a vivisection license issued by the Home Office is required.

An appropriate surgical needle is ½ circle triangular size 23, threaded with plain $^3/_0$ catgut suture. This suture lasts intact for a few days, long enough for the wound in the body wall to heal, but it does not need to be removed as it eventually disintegrates. The surgical needle can be held with large forceps, or with artery clamp forceps, while the suture is inserted.

Forceps. For general purposes, such as operating on newts and grasping oocytes prior to the isolation of germinal vesicles, Dumont No. 4 stainless steel watchmaker's forceps (jeweller's tweezers) are suitable, and two pairs should be at hand.

For the removal of the germinal vesicle membrane it is imperative to have available a pair of particularly well-sharpened Dumont No. 5 "original" *not stainless* steel watchmaker's forceps. Forceps made of stainless steel are less satisfactory, for they do not hold their points as well. These forceps need to be sharpened under a binocular dissecting microscope, and examined at the magnification which will be used for membrane removal. The stone recommended for sharpening is a small, hard Arkansas polishing slip. This stone should be soaked in molten vaseline, wiped clean and its surfaced smothered with graphite from a lead pencil prior to use, so that its cut is rendered as fine as possible. For sharpening, the forceps should be held with their points apposed to one another, and both points now sharpened by manipulating the stone, not the forceps. Both points should be made to terminate at the same length, and they should meet together precisely when the forceps are closed. Dr. J. G. Gall recommends the use of two such forceps; I prefer to use one, and a needle about to be described.

Needles. An ordinary steel needle mounted in a wooden handle serves for lacerating the follicle tissue surrounding an oocyte so as to release its contents. As this needle will become blunt through being pressed against the bottom of the container (embryo cup, or "solid" watchglass) housing the oocytes, it will need repeated sharpening on a carborundum stone, the needle being rotated as it is drawn across the stone.

A fine-pointed tungsten needle is required, if used in conjunction with a single pair of No. 5 Dumont forceps, for removing the germinal vesicle membrane. The virtue of tungsten needles was first brought to my attention by the late Professor C. F. A. Pantin, and they are mentioned in his book on microscopical technique (Pantin 1946). To make such a needle, break off a 3 cm length of tungsten wire, 0.5 mm diam. Take a small *nickel* crucible half filled with sodium nitrite ($NaNO_2$), place it in a pipeclay triangle, and heat over a Bunsen flame until the nitrite has just melted. Now dip one end of the tungsten wire into the molten nitrite, and allow it to etch clean. Take a 10 cm length of Pyrex glass capillary tubing, about 7 mm o.d., insert a few mm of the cleaned end of the tungsten wire into the tubing, and fuse the wire to the glass over a hot flame. Now increase the heat applied to the nitrite, and repeatedly dip the exposed tip of tungsten wire into the nitrite. At first the wire will continue to etch quietly, but when the nitrite has reached a critical temperature the tip of the wire will produce a bright flash of light as it is dipped under the surface. Allow this to occur three or four times, and then examine the tip under a binocular. It should terminate with a smooth polished surface and an extremely fine point; if the wire still ends bluntly, repeat the dipping procedure until the desired sharp point has been obtained. Wash off all traces of nitrite from the needle, and store ready for use. It is worth making several such needles at a time; they do not deteriorate with storage and they can be repeatedly re-pointed as necessary. Do not use a porcelain crucible for melting sodium nitrite, for molten nitrite bonds to porcelain, and when it cools and contracts it is liable to crack the crucible.

Pipettes. The pipette used for transferring germinal vesicles should be hand drawn from soft glass tubing, of the kind used commercially for disposable Pasteur pipettes; it is convenient to use a 15-cm-long disposable pipette as the starting material. Heat such a pipette over a narrow gas flame at a point about 7 cm from its 'teat' end, and rapidly draw the two ends a long way apart, about a metre, so that the drawn tubing is of fine bore. Nick and fracture the drawn region, using a writing diamond, at about 12 cm from the teat end, and examine the pipette's new orifice under a binocular with a calibrated eyepiece scale. The orifice should have an i.d. of about 0.6 to 0.7 mm, and it should now be fire-polished. Pipettes that happen to have too wide an orifice need not be thrown away; they can be used for transferring fluids.

The fire-polishing of the narrow orifice is best done at a tiny gas flame emanating from a 20 gauge hypodermic injection needle connected by tubing to a gas tap. Hold the pipette so that you can see its tip clearly against a bright background; before it has been fire-polished the tip will glisten, and it will go dull as the glass at the orifice melts. It is easy to overheat and occlude the tip, in which case cut off the old tip, and try again.

It is essential to have good control over fluid movement in a "nuclear" pipette, and it is therefore important to choose a teat which ensures such control; a flabby, thin-walled bulbous teat should not be used, instead choose a relatively thick-walled (2 mm) cylindrical teat, preferably with an even thicker (4 mm) wall at its open end which makes an absolutely airtight seal to the pipette. Teats must be kept scrupulously clean.

Binocular Dissecting Microscope. A stable instrument with good optical qualities is required. I use a Carl Zeiss Stereomicroscope II for manual operations, with a clear glass plate on its stage just (15 mm) above bench level, and a dull black background below. The binocular is fitted with $\times 10$ eyepieces, one of which contains a scale. When handling germinal vesicles it is important to be able to change magnification rapidly; I usually isolate and clean a germinal vesicle at a magnification of $\times 16$, and remove its membrane at $\times 25$. Higher magnifications are available on this instrument, but they are not generally needed. However the lowest magnification, $\times 6$, is useful for operations on newts.

Illumination for Manual Operations. Zeiss supply a low voltage (6 V, 15 W) epi-illuminator, including a heat filter, for use with the Stereomicroscope. A much more powerful and versatile illuminator, which makes use of a halogen lamp and fibre optics, is now available (Schott Cold Light Source KL150), and this is ideal for fine-scale manipulations.

Inverted Phase-Contrast Microscope. Gall's introduction in 1954 of a home-made inverted microscopic system for examining unfixed preparations of lampbrush chromosomes was a major advance. The Zeiss Plankton microscope, fitted with appropriate optics, now meets this need. Above the stage of this microscope the IS condenser, backed by annular stops, has sufficiently long working distance (about 7 mm) to admit micromanipulator needles yet still provides Köhler conditions of illumination. Above the condenser, and mounted on the same dovetail slide as that which carries the condenser, the Plankton microscope is fitted with a low voltage illuminator similar to that used as a light source for the dissecting binocular; however additionally the housing of this illuminator includes a slot to receive a xenon flash, and a field stop iris diaphragm. The illuminator should be mounted high on the pedestal above the condenser so that the wide annular stop that is used in conjunction with the ×100 phase objective is uniformly illuminated, and Köhler illuminating conditions achieved with this lens. The most generally useful phase objectives are the Neofluars ×16, ×25, ×40, and ×100 (oil).

Photography with the Light Microscope. As unfixed lampbrush chromosomes lying free in saline undergo Brownian motion, a flash is required for their photography. The Zeiss Ukatron 60 flash unit, the bulb of which fits into a slot in the illuminator housing (and which can be left permanently in place), has made photography simple. A splendid example of the outcome is shown in Fig. 2.1. Zeiss supply a special camera attachment for use with the Plankton microscope, but in my experience this system produces too great a magnification on the negative, and the flash does not provide sufficient light to produce well-exposed negatives with the ×100 phase objective when normal film is being used. I have instead used a single exposure camera constructed by Dr. C. Muir and mounted in place of the binocular eyepiece head that, with the ×100 phase objective, gives an overall magnification of ×450 and produces well-exposed negatives on cut lengths of Ilford Pan F film when the flash unit is operating at 60 W s^{-1}. With the ×40 phase objective a 30 W s^{-1} flash provides sufficient light, while with lower objective magnifications neutral density filters are necessary to reduce light to appropriate levels. The material to be photographed is first focussed on a ground glass screen incorporated in a dark slide, and after switching off the continuous illumination the carriage in the dark slide is pulled across so that the film lies in the position previously occupied by the ground glass screen. The flash is now triggered, and the exposed film returned to its original position in the dark slide. Developing and fixing these cut lengths of film is simple. Each piece of film is taken from its dark slide and inserted, sensitive side inwards, into a cylindrical glass specimen tube some 5 cm long and 2 cm wide. The developing, rinsing and fixing are done by dipping the tube into the various fluids housed in beakers in the dark room.

It remains to be added that, in a laboratory where expense is no restraint, the Zeiss IM 35 inverted microscope (Gundlach and Trendelenburg 1981) offers all the facilities required for observing, manipulating and photographing lampbrush chromosomes.

Centrifuge. For various purposes preparations of lampbrush chromosomes in saline, which thereafter are centrifuged and attached to slides, as will be later described, have played an increasingly important role in recent research. In passing, it is worth a mention that such centrifuged preparations do not require an inverted microscope for their examination, nor a flash for their photography.

--→

Fig. 2.1. Phase contrast photograph of a freshly isolated unfixed lampbrush bivalent of *Notophthalmus viridescens. Bar* = 50 μm. The *arrow* points to fused centromeres, the three *arrowheads* to chiasmata. (From Gall 1966)

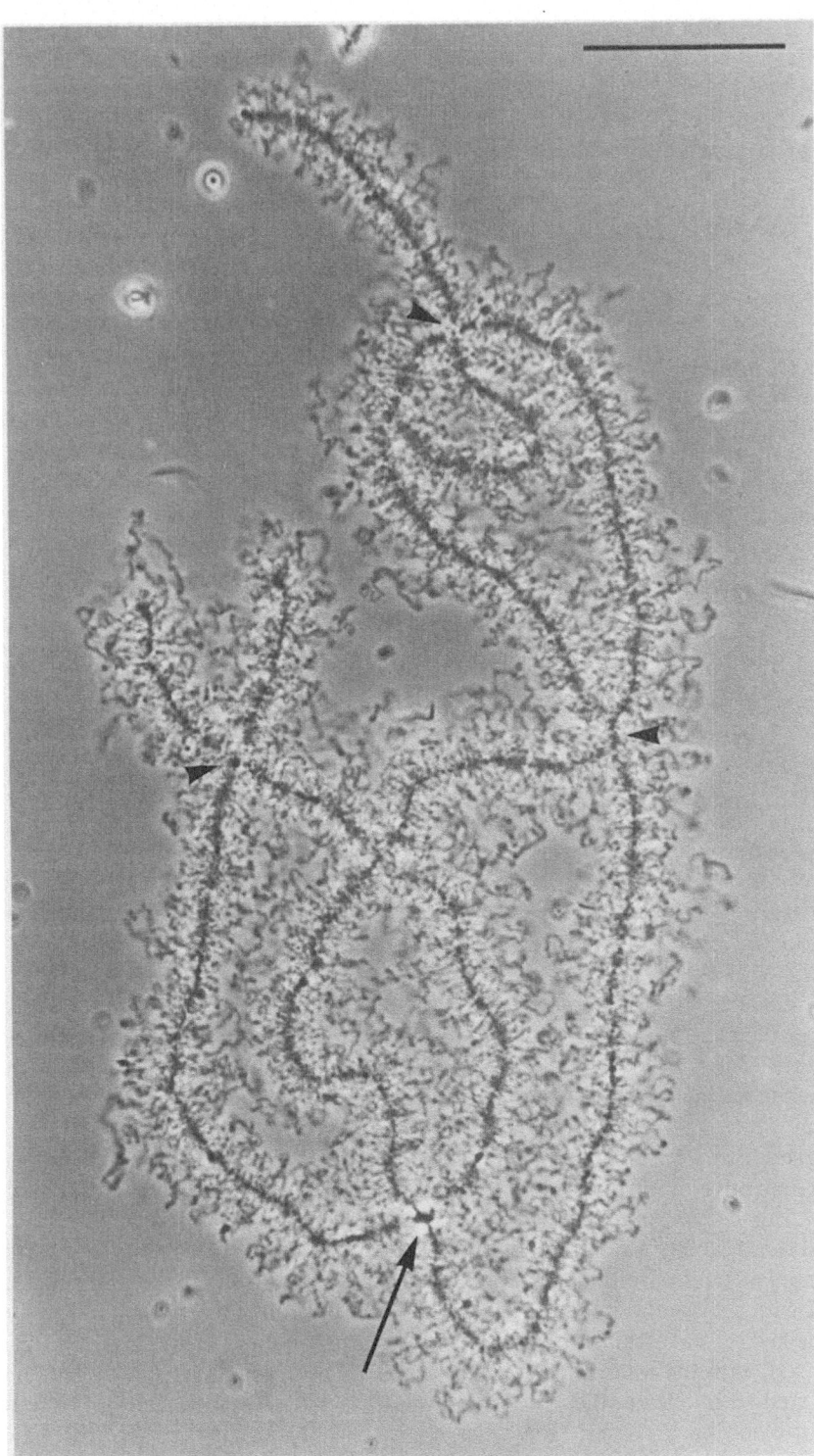

Fig. 2.1

The centrifugation does not involve speeds in excess of 5000 rpm so a costly centrifuge is not essential. An MSE Minor centrifuge fitted with an HB-4 swinging bucket rotor has given me good service. Several other suitable centrifuges are available, such as the Sorvall GLC-1 with an HS-4 rotor. Macgregor and Varley (1983) recommend in particular the Sorvall RT-6000 bench-top refrigerated centrifuge, for it requires no specially constructed baskets for slide preparations and takes up to 16 preparations at a time.

Other Materials

Embryo Cups ("Solid" Watchglasses). These substantial square glass containers, each with a circular well about 1 cm deep, are convenient for housing pieces of amphibian ovary in saline when germinal vesicles are to be isolated. The surface of the well should be clear, not ground. Embryo cups also provide convenient receptacles for the short-term storage of amphibian ovaries "dry", i.e. bathed in coelomic fluid only. When used for this purpose, the top surface of the embryo cup should be ground flat if it is not already flat as supplied by the manufacturer; this surface should be smeared with vaseline and covered with a small glass plate to prevent the ovary from desiccation.

Covered embryo cups also serve for the short-term (up to 4 h) incorporation of radioactive precursors by oocytes, and at its simplest an appropriate quantity of precursor in solution is placed in the middle of the cup and allowed to dry out. The precursor passes into solution in the coelomic fluid carried over with the batch of oocytes, and provided the oocytes are moved about periodically, incorporation is tolerably uniform throughout the sample. For more stringently controlled in vitro incorporation special oocyte or oocyte nuclear incubation media are needed.

Expanded Polystyrene "Igloo" Containing Crushed Ice. This will be required to keep oocytes cool until they are used for making preparations.

Moist Chambers. These serve several needs that will be mentioned later in this chapter. Petri dishes, 10 cm square, are convenient because each can accommodate three slides. A 9-cm-diam filter paper is placed on the floor of the dish, with two L-shaped glass rods above to keep the slides from contact with the paper.

Observation or Dispersal Chambers (Flat-Bottomed Well Slides). These chambers were first introduced by Gall in 1954, and each is prepared from a standard 3 in. × 1 in. microscope slide with a ¼ in. diam hole bored through its centre. Bored slides can be made by cementing ten slides together with an ethanol-soluble thermoplastic resin, so that they form a solid block; then bore through the entire block with a diamond-tipped drill. The top and bottom slides in the block may be cracked, or have holes with irregular walls, but the inner slides should be undamaged, with the walls of their holes smooth. The bored slides should now be separated from one another, thoroughly cleaned and dried, and laid out flat on a matt-black background. Now place two small dabs of molten paraffin wax

(M.P. 49 °C for preference) close to and on either side of the hole on each bored slide. Place a coverslip (No. 1 thickness, 13 mm square is a convenient size) on top of the now solid dabs of wax, and gently heat the coverslip with a gas flame from above until the wax has melted and sealed the coverslip over the hole. Try to avoid melting the wax so that it runs far across the hole and down its ground sides, for chambers with too much wax around their cavities tend to reject the saline later to be placed within; experience will quickly teach how small the dabs of paraffin wax need to be in order to make good chambers. The gas flame used for sealing coverslips to bored slides is conveniently one emanating from a hypodermic injection needle, the same as that used for fire-polishing nuclear pipettes.

Flat-Bottomed Well Slides for Centrifugation. These are similar to the observation chambers just described, but instead of a coverslip a standard microscope slide is cemented to a bored slide with paraffin wax to form the base of the chamber.

An ingenious method for making dispersal chambers on slides has recently been devised by Dr. E. De Robertis. A strip of Sellotape 25 mm wide, adhesive on both sides, is stuck to a strip of Teflon plastic sheet, similarly 25 mm wide and 0.5 mm thick or slightly thicker. The plastic plus Sellotape strip is cut into 25 mm squares, and in the middle of each square a hole of 6 mm is punched out, using a sharp paper punch. The other adhesive surface of the Sellotape is now exposed, and plastic square plus Sellotape pressed down on to the middle of a microscope slide.

Recycling Bored Slides. Provided observation chambers have not encountered noxious materials during their use (ribonuclease and detergents both rank high in this regard) bored slides can be cleaned by flushing them out under a stream of hot tap water, hot enough to melt the paraffin wax that was used to cement them to coverslips or slides when chambers were constructed. The bored slides are then rinsed in glass-distilled water and stored in 95% ethanol until they are required for use, when they should be dried with a paper tissue that leaves no fluffy fibres, such as a Kimwipe. It can be more troublesome to remove vaseline, and if forced to do so set the bored slides in a rack and place them in a solution of commercial detergent for 1 or 2 h before flushing them out very thoroughly indeed with running tap water.

If the bored slides are really dirty or newly made they should be set in a rack and placed in the powerful cleaning agent that consists of equal volumes of concentrated sulphuric acid and saturated potassium dichromate, and left for several days. After this treatment they must again be flushed out thoroughly with running tap water. Bored slides cleaned by this means suffer from a perverse defect, in that when they are first made up to form observation chambers, saline placed in the well runs out over the top glass surface instead of forming a convex-up meniscus and remaining confined to the well. Slight contamination with wax or vaseline prevents this from happening!

Carrier (Basket) for Centrifuging Lampbrush Chromosomes onto Slides. A special carrier is needed for centrifuging slide preparations when used in conjunction with a swinging bucket rotor. The brass carrier designed by Prof. H. C. Macgregor is illustrated and described in Macgregor and Varley (1983). Its upper part bears short trunnions that fit the rotor. The lower part is a rectangular plate connected to the upper part by rods at the four corners. The rods are so positioned that a microscope slide can be freely accommodated between them on the lower plate, and to prevent the slide from falling out as the rotor gathers speed a small screw is fitted at the "inside" end of the lower plate. The radius of rotation of a slide on a carrier with dimensions appropriate for an MSE minor centrifuge is 14 cm.

Syringes, Bacterial Filters and Containers for Salines. There is no need to work under strict aseptic conditions, but control against bacterial contamination of lampbrush preparations is desirable, particularly those that are to be centrifuged. Most salines can be prepared in bulk, and small quantities dispensed as required. Erlenmeyer flasks capped with metal foil provide suitable containers, 100 ml flasks for nuclear isolation salines, 25 ml flasks for dispersal salines. Erlenmeyer flasks must be kept free from all traces of noxious chemicals, for after they have been isolated lampbrush chromosomes are extremely sensitive to detergents, incorrect pH, and many potential contaminants of salines.

Before starting to make preparations, salines should be filtered. This is conveniently done by fitting a plastic Swinnex-25 holder containing a 25-mm-diam Millipore filter of 0.22-μm-pore size to a plastic disposable injection syringe. Saline is now poured into the syringe, and the plunger inserted. The first few drops of saline expelled from the syringe should be discarded, just in case the Swinnex holder or Millipore filter was contaminated with a trace of detergent; the rest can be dispensed into an Erlenmeyer flask, and the flask capped. The simpler saline is likely to be that used for nuclear isolation, and the dispersal saline a derivative thereof, so the same syringe and filter can often be used for both. Filter the isolation saline first, then remove the Swinnex holder, remove the plunger from the syringe, replace the Swinnex holder, pour the dispersal saline into the syringe and re-insert the plunger. Now proceed as before, discard the first few drops of dispersal saline, dispense the rest into an Erlenmeyer flask, and cap it.

The pipette used for filling flat-bottom well slides should be pushed through the metal foil cap, and kept there except when being used to dispense dispersal saline. It is all to easy to contaminate a saline from a pipette that was inadvertently laid flat on top of a laboratory bench.

Saline Solutions for Isolating and/or Incubating Oocyte Nuclei

All salines to be used for making preparations of lampbrush chromosomes should be made up with glass-distilled water. An unbuffered saline containing KCl and NaCl of overall concentration 100 mM is suitable for isolating oocyte nuclei. The proportion of K^+ to Na^+ does not appear to be critical. In early work I used a 5:1 mixture on advice given me by Prof. W. T. W. Potts, resulting from his K^+ and Na^+ determinations carried out on *T.c. carnifex* oocytes, and this mixture has been widely used since. Riemann et al. (1969) on the same material found a $K^+:Na^+$ ratio of about 3:1 for the nuclei of oocytes ranging in diameter from 0.3

to 1.5 mm, and on this account I changed to a 3:1 mixture. Still later determinations made by Whitley and Muir (1974) are more in accord with Professor Pott's original ratio, except for oocytes in the size range 0.5 to 0.8 mm diam.

Duryee was certainly correct in stating that the nuclear isolating saline should contain no calcium, for even a trace of this cation, apart from its effect on nuclear contents (see later), causes cytoplasm to adhere to the germinal vesicle membrane at the time of isolation, and impairs the cleaning operation.

Gall et al. (1981) have given details of a saline which greatly simplifies the making of lampbrush preparations of *Notophthalmus viridescens*. It contains 83 mM KCl, 17 mM NaCl, 6.5 mM Na_2HPO_4, 3.5 mM KH_2PO_4, and has a pH of 7.2. In this saline, called "5:1 plus PO_4", the nuclear contents remain as a gelatinous ball, which facilitates removal of the nuclear membrane without damaging the lampbrush chromosomes. This saline will probably serve well for other species of urodeles whose oocyte nuclear sap disperses inconveniently rapidly in unbuffered saline.

Schultz et al. (1981) have devised an incubation medium in which isolated oocyte nuclei, or "nuclear gels", i.e. nuclei from which the membranes have been removed, continue to synthesize RNA in vitro at an undiminished rate for at least 2 h. The nuclei are isolated, cleaned and stored (at 4 °C) in a saline consisting of 60 mM KCl, 15 mM NaCl, 11.2 mM Na_2HPO_4, 3.8 mM KH_2PO_4, 0.5 mM EGTA (ethyleneglycolbis(aminoethylether)-tetraacetic acid), 1.5 mM dithiothreitol, pH 7.2. When sufficient nuclei or nuclear gels have been collected, the volume of this saline is reduced to a convenient amount, and an equal volume of a mixture of the appropriate reaction components added to give an incubation medium containing 60 mM KCl, 7.5 mM NaCl, 16 mM Na_2HPO_4, 4 mM KH_2PO_4, 5 mM $MgCl_2$, 0.25 mM EGTA, 0.75 mM dithiothreitol, 1 mM cytidine 5'-monophosphate, pH 7.4, and the four ribonucleotide triphosphates each at 0.5 mM. One or more of the ribonucleotide triphosphates is [32]P- or ^{3}H-labelled so that RNA synthesis can be monitored, the incubation being carried out at 20 °C. In both of these salines the lampbrush chromosomes of *N. viridescens* remain compact within a gelatinous ball of nuclear sap, and it is immaterial whether the nuclear membranes are removed before, or after, incubation.

Free-Hand Manipulative Procedures

An ovary or part of an ovary should be excised from a newt that has been anaesthetized in 0.1% MS222 Sandoz. Anaesthesia takes 5 to 10 min. The newt should be rinsed in tap water after it has ceased movement, and before operating, because a solution of MS222 is acid (pH about 4). It should be placed on a damp pad on the stage of a dissecting binocular, and the laparotomy carried out at low magnification. An entire ovary can be excised, or a portion only, preferably from the ventral region, so that no major blood vessels need be cut. The ovary should be placed just as it is (i.e. no saline added) in an embryo cup, the top surface of the cup smeared with vaseline, and a glass cover pressed down on the vaselined surface to prevent desiccation. If preparations are to be made throughout a day the embryo cup containing an ovary should be placed on crushed ice in a covered, expanded polystyrene "igloo". The ovary will remain without detectable deterioration for at least 6 h.

After excision of the ovary the small cut in the newt's abdomen (to the left or right of the mid-ventral line) should be closed with two or three stitches of $^3/_0$ plain catgut suture previously soaked in water, using a ½ circle triangular surgical needle, size about 23, held by forceps. The stitches should include both skin and muscle, and also peritoneal lining. When working with anurans such as *Xenopus,* which have large ventral lymph sacs, it is advisable to make two separate cuts, laterally displaced from one another, through skin and abdominal musculature, and stitch these tissues independently. The stitches should enter the tissues well away from the cut, they should not be drawn tight and should be tied with reef knots. With most newts, though not with *Pleurodeles,* (Lacroix 1968 a), sterile working conditions and antiseptic precautions are unnecessary. With axolotls it is particularly important to keep the skin well wetted throughout the operation. After laparotomy the animal should be returned at once to fresh water in a covered container. If kept in clean conditions, *Triturus cristatus* recovers remarkably quickly after a part of its ovary had been excised. The wound has generally healed and the stitches have disintegrated within a week of operation, and the animal can later be used for breeding.

Newts must be well fed and in good condition at the time of operation if they are to provide oocytes from which good lampbrush chromosome preparations can be made; in this regard temperature is also an important factor. A temperature of 18 °C is satisfactory for *T. cristatus* and its near relatives, also for *N. viridescens.* Other species have other optimum temperatures, for example 22 °C or thereabouts is recommended for the axolotl and *Pleurodeles.* If for one reason or another optimum temperature conditions cannot be met, use a lower rather than a higher temperature. The same consideration also applies to the temperature of the room in which preparations of lampbrush chromosomes are to be made.

It is convenient to have available within a single newt ovary a wide range of oocyte sizes from which to choose, and both *T. cristatus* and *N. viridescens* are in this respect ideal experimental organisms, as also are the axolotl and *Pleurodeles.* Amphibia with short breeding seasons tend to mature their oocytes in synchronous waves, and their ovaries may therefore lack oocytes of the desired size except for a relatively restricted period during the year. This is a disadvantageous feature of plethodontid salamanders, and of many anurans, though not *Xenopus.*

Isolation of Oocyte Nuclei. I will first describe the technique that has served me for many years. Take a small piece of excised ovary and place it in 100 mM KCl plus NaCl saline (a 5:1 or 3:1 mixture are equally suitable) in an embryo cup. Each oocyte is surrounded by a sheet of flat follicle cells, the sheet also enclosing blood capillaries, and it projects from the ovary wall, anchored by a follicular stalk, into the cavity inside the ovary. Gently tear the ovary wall with two pairs of forceps so that individual oocytes become well exposed. Accidentally broken oocytes are likely to have clouded the saline with yolk, so shake the piece of ovary and transfer it to a fresh embryo cup containing saline. Place the embryo cup on the stage of a dissecting binocular and over a matt-black background. Arrange the light source so that the ovary is well illuminated from above; I like to have the light beam directly in front of the binocular and angled down at about 45°.

Select an oocyte of suitable size, and grasp the neighbouring ovary wall with forceps. For a start choose an oocyte of roughly half the diameter of one that is mature, say 0.9 mm for *T. cristatus,* a trifle less for *N. viridescens.* Press the forceps lightly against the floor of the embryo cup, so that the selected oocyte is held rigid, and gently puncture the oocyte, at the pole opposite to that held by the forceps, with a steel mounted needle. The needle should be directed nearly tangential to the oocyte's surface, and should be a "slash" rather than a prick, so that a relatively large opening is made in the follicular envelope. Sometimes it pays to press the needle point, which has engaged the follicular envelope, right down on to the floor of the embryo cup, and then tear the envelope with a lateral movement. The object of this operation is to make a large opening quickly, without putting more pressure than necessary on to the oocyte. If you fail in this aim, as soon as a tear has been made in the follicular envelope the germinal vesicle will move towards the hole and may impale itself on the steel needle. It is important not to make too small a hole, for if this happens the germinal vesicle will move towards the hole as cytoplasm flows out, will come to block the hole, and any pressure now applied to the oocyte will burst the germinal vesicle membrane. In the case of species whose oocyte cytoplasm is relatively fluid, such as *T. cristatus* and *N. viridescens,* once the follicular envelope has been slashed the needle can be lifted and pressed against the oocyte, forcing the cytoplasm and germinal vesicle to extrude. However in species whose oocyte cytoplasm is relatively stiff, such as *Xenopus,* it is best not to press on the oocyte (doing so is likely to burst the germinal vesicle), but instead allow the contents to emerge slowly.

It is usually possible to see the germinal vesicle, as a dark region within the light-scattering opaque yolk, while the oocyte contents extrude. Unless its membrane was previously damaged, the germinal vesicle will start to swell as soon as it comes to lie in saline; once it becomes visible the needle should be put down, a nuclear pipette taken up instead, and used to clean the germinal vesicle of adhering cytoplasm.

One needs to have particularly good control of fluid movement in the nuclear pipette, and I will give my own prescription for achieving this. Pick up the pipette between forefinger and thumb, with the hand loosely clenched, the pipette resting across the terminal joint of the middle finger and the thumb controlling pressure on the teat. Draw up some saline into the pipette, keeping the end of the pipette all the while submerged so that no air bubbles form in the constricted portion, and squirt saline at the germinal vesicle. Generally, if it was not previously punctured, the membrane will quickly lose all trace of yolk; if there is any residual yolk left attached, this will detach if the germinal vesicle is drawn up and expelled from the pipette once or twice. When clean, the germinal vesicle should be taken up, then released into a region of saline free from yolk, the pipette removed from the saline, its contents ejected outside the embryo cup, clean saline drawn up (again without air bubbles), and the germinal vesicle picked up towards the end of this operation; it is now ready to be transferred to an observation chamber. Throughout the cleaning operation take pains to ensure that the isolated germinal vesicle does not come into contact with an air bubble, either in the embryo cup or pipette, for if it does the membrane will adhere to the bubble and cannot thereafter be detached. For this reason do not work with saline which has been stored in a refrigerator and just previously taken out, because as the saline warms up to room temperature air bubbles will form on the surface of the embryo cup. Also for this reason take care to use a scrupulously cleaned (preferably unused) nuclear pipette, for a trace of fat on the pipette wall will tend to trap air bubbles.

Removal of the Nuclear Membrane. Once a germinal vesicle has been cleaned of cytoplasm it should be transferred as quickly as possible and with a minimum of saline to dispersal saline in an observation chamber. The chamber should have been filled with dispersal saline just prior to isolating the germinal vesicle, sufficient saline to give a convex-up meniscus. To start with, use a saline that does not cause rapid sap dispersal, as this will permit more leisurely removal of the germinal vesicle membrane, this being the operation which, to be done successfully, requires most experience.

With one hand lift the pipette holding the germinal vesicle out of the embryo cup, and take care not to alter the pressure on its teat. Now with the other hand remove the embryo cup from the stage of the binocular, put the observation chamber in its place, refocus the binocular, plunge the end of the pipette under the saline meniscus and gently press on the teat to drive out the germinal vesicle. Aim to place the germinal vesicle a little off centre in the chamber, for its membrane inevitably carries traces of cytoplasm which will contaminate the glass bottom of the chamber, and one wishes later to observe the lampbrush chromosomes lying on a clean surface.

Now change to a higher magnification; I do the nuclear cleaning and transfer at a magnification of ×16, but membrane removal at ×25. Focus on the upper surface of the germinal vesicle, pick up the Dumont No. 5 forceps in one hand and the tungsten needle in the other. Approach and touch the equator of the germinal vesicle with both forcep tips, with the forceps partly closed so that the tips are about as far apart as one-quarter of the germinal vesicle's diameter, and gently press the germinal vesicle towards the forceps using the side of the tungsten needle. Now close the forceps completely, raise the germinal vesicle just off the bottom of the chamber, move it to the middle of the chamber, poke the tip of the tungsten needle into the membrane where this has been puckered by the grasp of the forceps, and steadily pull the forceps and needle apart. This should produce a large tear in the germinal vesicle membrane, through which the contents will spill out. The whole operation can be monitored precisely by watching through the binocular provided its plane of focus was appropriately arranged beforehand, and the chamber well illuminated. The germinal vesicle membrane glistens because of light scattered from the many nucleoli attached to it, and the nuclear contents are faintly opalescent before they have dispersed.

Best control can be achieved if the membrane remains adherent to both forceps and needle. The tear in the membrane will initially lie above, but as the nuclear contents begin to spill out the whole nucleus will tilt over so that the tear soon comes to lie below. Slowly raise forceps and needle as the nuclear contents can be seen slipping down towards the bottom of the chamber, and after they are free from the membrane (which is indicated when slight movement of the forceps and needle no longer shifts the nuclear contents en masse), lift the membrane to the saline/air interface, on which it will spread and disappear from view.

Because the hands need to be steady when manipulating forceps and needle to tear the membrane, and have to be held motionless for a few seconds while the nuclear contents slip way from the torn membrane, the binocular should stand well in from the edge of the work bench so that the whole lengths of the forearms are supported by the bench. This is also why the binocular stage should be only just above bench level; at no time is there need to use light transmitted through a glass stage from below.

The most common source of trouble during these manipulations occurs when the forcep tips grasp too small a region of the germinal vesicle membrane, i.e. when the tips are held too close together just prior to closure. There is a natural reluctance to take too big a 'bite' at the membrane for fear of grasping some of the chromosomes as well. Naturally enough too big a bite will do just this but if the forceps are used as described earlier, i.e. with their tips apart about one-quarter of the diameter of the germinal vesicle, when the membrane is grasped the entire nuclear contents are displaced from this region. If too small a bite is taken the needle often produces too small a tear, and the nuclear contents then remain hung up at the opening, partly extruded and partly still inside the membrane; alternatively when the needle is used to tear the membrane, the membrane may become detached from the forceps, again leaving too small a hole through which the nuclear contents must pass. The end result is that the chromosomes remain trapped by the membrane, and when the membrane is brought to the interface they are stretched and broken. It is best to abandon the preparation when this occurs, just as it is best to discard a germinal vesicle which has accidentally been punctured (recognizable because it remains flabby, undistended) during isolation and cleaning.

Another source of trouble arises when isolated germinal vesicles stick to clean glass surfaces. I have often encountered this conditions in the axolotl (Callan 1966), occasionally in specimens of *N. viridescens,* much more rarely in *T. cristatus.* Adhesion can occur while the germinal vesicle is being isolated in an embryo cup, or after its transfer to an observation chamber. It is more likely to occur immediately after isolation, before much swelling has occurred, than later, and consequently one can to some extent overcome the problem by deliberately delaying transfer, keeping the germinal vesicle continually on the move by pipetting, until it has lost its adhesive tendency. If the germinal vesicle does stick to the embryo cup, the chances are that when it is forcibly driven free by pipetting it will have left an area of membrane stuck to the glass, and its contents will therefore start to exude before transfer, in which case further manipulation should be abandoned.

It was mentioned earlier that as soon as an unbroken germinal vesicle comes to lie in saline it starts to inbibe fluid and expands (see Macgregor 1962). This swelling aids the cleaning operation, and in some measure aids the removal of the germinal vesicle membrane, for in all amphibian oocyte nuclei there are coextensive colloidal phases in the nuclear sap, one relatively rigid, the other fluid (Callan

1952). Initially it is only the fluid phase which expands, exerting colloidal osmotic pressure on the membrane, and so increasing the distance separating the membrane from the chromosomes, the latter being held in place by the stiffer colloidal phase.

In species such as *N. viridescens* and *T. cristatus,* where the gel component of the oocyte nuclear sap generally starts to disperse in 100 mM KCl + NaCl isolating saline, it is important not to delay too long between isolation of the germinal vesicle and rupture of its membrane. If the lampbrush chromosomes start to spread out inside the isolated germinal vesicle while the membrane is still intact, then as time passes it becomes increasingly more difficult to avoid picking up the chromosomes as well as the membrane when the latter is grasped by forceps. Moreover the more distended the germinal vesicle becomes, the more difficult it is to grasp the membrane with forceps without simultaneous puncture. Consequently it is a matter of judgment to decide just how long one can afford to delay the transfer to an observation chamber when dealing with germinal vesicles that have "sticky" membranes.

As soon as possible after the nuclear membrane has been removed a clean coverslip, 18 mm square is a suitable size, should be dropped from above on to the saline meniscus, without trapping an air bubble underneath. The coverslip should not be slowly lowered at an angle, using a needle to prop it up, as one would cover a stained preparation being mounted in balsam. If the preparation is to be examined without further treatment, the edges of the top coverslip should be sealed to the bored slide with molten vaseline, applied with a bent copper wire gently heated beforehand in a Bunsen flame, to prevent evaporation from the chamber. This too should be done as quickly as possible, before there has been much dispersal of the chromosomes, because the colloidal phase of the nuclear sap temporarily cushions the chromosomes from tangling with one another while movement of fluid occurs within the chamber.

If the preparation is to be put through further treatment, when the top coverslip has been dropped in position the preparation should be placed in a previously prepared moist chamber, the filter paper of which had been saturated with isolation saline. When not being examined or otherwise treated, sealed preparations and those in moist chambers should be kept cool.

Once a preparation has been made the tips of the nuclear forceps and tungsten needle should be wiped clean with paper tissue, saline ejected from the nuclear pipette, glass-distilled water taken up and likewise ejected, and all three instruments place on a clean surface, with their tips projecting beyond the edge of the bench, to avoid contamination.

There are two unsatisfactory features about the technique that has just been described. One is the need to hurry over removal of the nuclear membrane once the nucleus has been transferred to the saline in the observation chamber, for this saline will have been chosen because it promotes dispersal of the nuclear sap. The other defect is that some of the saline in which the nucleus was isolated and cleaned will be transferred with the nucleus to the dispersing saline in the observation chamber, thereby altering the dispersing saline's composition in an uncontrolled way.

Both these snags have been circumvented by Gall et al. (1981). As mentioned earlier in this chapter, if oocyte nuclei of *N. viridescens* are isolated and cleaned in the saline referred to as "5:1 plus PO_4" (83 mM KCl, 17 mM NaCl, 6.5 mM Na_2HPO_4, 3.5 mM KH_2PO_4, pH 7.2), the gel component of the nuclear sap remains compact and the nuclear membrane can be peeled off at leisure. Thereafter the nuclear gel, including the lampbrush chromosomes, can be pipetted from one saline to another, finally to the dispersing saline that it will encounter in the observation chamber and then, quickly, into the latter.

The success of this method probably depends on the intrinsic rigidity of the oocyte nuclear gel. In *T. cristatus* this varies according to oocyte size, season and hormonal state, being more rigid in large oocytes than in small, and in individual newts with relatively fluid oocyte nuclear sap rigidity increases following injection of gonadotropin (Macgregor 1963). In my limited experience the method is not successful with *T. cristatus* oocytes whose nuclear gels are intrinsically fluid, but succeeds if they are rigid, in other words the "5:1 plus PO_4" maintains rather than confers rigidity.

In the oocytes of a great many amphibian species, for example *Xenopus, Necturus, Cynops sinensis, Ambystoma mexicanum, A. macrodactylum* and all plethodontid salamanders, the rigid gel phase within the germinal vesicle remains compact even after isolation in 100 mM KCl plus NaCl, so also with these species the nuclear membrane can be removed at leisure prior to the transfer of its contents to a dispersing saline. The problem with these species is not how to avoid gel dispersal, but how to achieve it. Amphibian germinal vesicles contain high concentrations of actin, of the order 4 mg ml^{-1} in *Xenopus* (Clark and Rosenbaum 1979) and *Pleurodeles* (Gounon and Karsenti 1981). Nuclear actin exists in two forms, globular and filamentous. The filamentous form is more abundant in *Xenopus* (37%) than in *Pleurodeles* ($<10\%$) and its relative abundance presumably determines the degree of rigidity of the nuclear gel and the ease or otherwise of achieving sap dispersal.

Salines for Dispersing the Nuclear Sap

Salines that can be employed for dispersing oocyte nuclear sap, thereby releasing the lampbrush chromosomes, depend on the species of amphibian being studied. In early work on *T. cristatus* and *T. marmoratus* Callan and Lloyd (1960) found that sap dispersal could be induced or accelerated by using salines of lower molarity than the 100 mM KCl plus NaCl saline used for isolating germinal vesicles. Two salines proved useful, 58 mM KCl, 12 mM NaCl, 0.3 mM KH_2PO_4, and 30 mM KCl, 6 mM NaCl, 0.3 mM KH_2PO_4. In the stronger of these two salines the various landmark structures and the general run of lateral loops retain a lifelike appearance for several days if preparations are stored at 2 °C when not being examined. In the more dilute of these two salines sap dispersal is more rapid, but most of the landmark loops swell and partially disintegrate. In so doing, the underlying loop organization of most "lumpy" and "fusing" loops is revealed, regions where the chromosomes' main axes are double can be seen clearly, and complicated fusions involving chromosomes' axes can be analyzed. However preparations dispersed in the more dilute saline deteriorate rapidly in a few hours.

In his paper on technique Gall (1966) described an improved saline for dispersing the gel component of the nuclear sap of *N. viridescens,* one which also works well for *T. cristatus.* $CaCl_2$ is added to a 5:1 or 3:1 mixture of 100 mM KCl and NaCl but its concentration must not exceed 0.1 mM. Above this concentration, $CaCl_2$ not only liquifies the nuclear sap, but causes the lampbrush chromosomes to contract and become rigid, as though they had been denatured at low pH. At 0.1 mM $CaCl_2$ nuclear sap dispersal is rapid, sometimes inconveniently so because of the need to hurry over removal of the nuclear membrane. A good compromise is achieved by using $CaCl_2$ at 0.05 mM, and with "easy" individual newts $CaCl_2$ at 0.01 mM is a sufficiently high concentration. The particular virtue of the dispersing salines just described is that they include the full 100 mM concentration of KCl plus NaCl, which ensures that both the lampbrush chromosomes and nucleoli retain the morphologies that they had prior to nuclear isolation.

As already mentioned the nuclear gels of the oocytes of many amphibian species will not disperse in 100 mM saline and this is so even if the highest permissible concentration of $CaCl_2$ is present. I encountered a typical problem of this kind when working with the oocytes of *Ambystoma mexicanum* (Callan 1966). In this species the nuclear sap starts to disperse in 45 mM KCl, 15 mM NaCl, 0.03 mM $CaCl_2$, but dispersal ends abruptly when a meshwork of fibres builds up amongst and around the lampbrush chromosomes. These fibres are probably aggregates of filamentous actin. This difficulty was overcome by accident. In some of our earliest studies both Gall and I had placed freshly made, uncovered preparations of *N. viridescens* and *T. cristatus* lampbrush chromosomes for 1 or 2 min in a chamber saturated with the fumes of neutralized formaldehyde in the expectation that this would control bacterial contamination and would also stabilize loop matrix RNP. Thereafter the preparations were covered and left for further dispersal to occur. When this formaldehyde vapour treatment was applied to freshly made preparations of *A. mexicanum* the meshwork of fibres did not form, and dispersal proceeded satisfactorily.

The use of formaldehyde (prepared from paraformaldehyde) to aid in nuclear sap dispersal was taken a step further by Macgregor and Kezer (1970) when they added neutralized formaldehyde in solution to both isolating and dispersing salines while working with oocytes of the anuran *Ascaphus truei.* The isolating saline was a mixture of 100 mM KCl plus NaCl containing 0.2 to 2% formaldehyde, while the dispersing saline was the very dilute mixture of 15 mM KCl plus NaCl containing 0.5% formaldehyde. Essentially the same procedure was also used successfully on oocyte nuclei of *Plethodon cinereus* (Kezer et al. 1971, Kezer and Macgregor 1973), on *P. vehiculum* and *P. dunni* (Vlad and Macgregor 1975), on *Siren intermedia* (León and Kezer 1974) and on *Ambystoma macrodactylum* (Kezer et al. 1980) while Müller (1974) was able to disperse the nuclear sap and analyze the lampbrush chromosomes of *Xenopus laevis,* until more recently considered a most refractory species, in a phosphate-buffered (pH 6.9) saline of 60 mM overall concentration containing 0.03% formaldehyde.

Salines containing formaldehyde have made possible the examination of the lampbrush chromosomes of "difficult" organisms, where without its use little information would have been obtainable; the investigator can to some extent balance the preserving action of formaldehyde against the disadvantageous dispersive action of low molarity saline.

Gall et al. (1981) and Schultz et al. (1981) dispersed nuclear gels of *N. viridescens* in $^1/_4$ strength "5:1 plus PO_4" containing 0.1% paraformaldehyde, deliberately using this dilute saline to obtain centrifuged preparations with well-extended loop matrix, thereby improving the resolution of autoradiographs. To my great surprise this dispersing saline also serves for *Xenopus* (Dr. J.G. Gall, personal communication 1984), an important discovery because the lampbrush chromosomes of *Xenopus,* from the standpoint of molecular biology by far the best known amphibian, have now become accessible to in situ hybridization (see Jamrich et al. 1983). Nuclear gels of *T. cristatus,* prepared after isolation in "5:1 plus PO_4", can similarly be dispersed in this dilute saline, but they can be dispersed equally well in 100 mM KCl plus NaCl containing 0.1 mM $CaCl_2$, which gives better preservation of the landmark loops of this species.

Salines for Observing the Actions of Enzymes on Unfixed Lampbrush Chromosomes

Significant information about their composition has come from experiments where lampbrush chromosomes were dispersed in media containing proteases or nucleases and observed while digestion proceeded. In the early experiments (Callan and Macgregor 1958, Macgregor and Callan 1962, and Gall 1963a) simple unbuffered dispersing salines were used, but they suffered from one drawback; in the absence of a reducing agent chromosomal proteins in lampbrush preparations progressively denature by oxidation and this undesired "fixation" can obscure the action, in particular, of DNases. Sufficient co-factors such as the magnesium ions required by pancreatic deoxyribonuclease were presumably provided by the nuclear sap accompanying the chromosomes.

Ideal ionic conditions for various enzymes are by no means ideal for dispersing and observing lampbrush chromosomes; however Gould et al. (1976) devised a dispersion medium which is a reasonable compromise for studies involving several enzymes. It consists of 100 mM KCl plus NaCl; 10 mM Tris buffer, pH 7.2; 1 mM $MgCl_2$; 0.05 mM $CaCl_2$; and 1 mM mercaptoethanol. The mercaptoethanol must be diluted from a refrigerated stock and added fresh each day; it prevents the oxidation of chromosomal proteins, but as on its own it breaks DNA in the presence of oxygen, the dispersing medium should be degassed under vacuum for a few minutes just before it is used.

Micromanipulation of Lampbrush Chromosomes

As long ago as 1941 Duryee published a paper that included two photographs showing lampbrush chromosomes of *Triturus pyrrhogaster* being stretched by needles controlled by a micromanipulator. In the early 1950s I made some similar observations using a de Fonbrune micromanipulator, and found that gross fluid movements in the observation chamber when the needles are inserted, which can prove troublesome, can be avoided if the saline in the chamber is covered with medicinal paraffin before the needles are inserted. At the time I used a simple un-buffered saline, but if this approach were to be repeated I would now advocate the use of the Tris-buffered saline containing mercaptoethanol (or dithiothreitol) specified in the preceeding section.

Centrifugation of Lampbrush Chromosomes

In some of the early studies on lampbrush chromosomes where autoradiography was involved (for example Gall and Callan 1962) it was necessary to attach the isolated chromosomes to a glass surface. This was done by making preparations in the usual way in observations chambers, but with a slide instead of a coverslip sealed by wax to the bored slide. The preparations were left uncovered and placed in a moist chamber for 1 or 2 h to allow nuclear sap dispersal to occur. The un-covered preparations were then transferred as delicately as possible (for lacking a top coverslip any jolt will seriously entangle the dispersed chromosomes), to a large desiccator vessel containing formalin below its tray. The length of time such preparations required to be left in the formalin chamber for chromosome attach-ment was critical, and the technique altogether too capricious. Attachment turned out to be determined not by the formaldehyde vapour per se, but rather by formic acid, a regular contaminant of commercial formalin; and if preparations were left overlong in the formalin chamber a precipitate of coagulated nuclear sap descended on the chromosomes, which were by now attached to the floor of the depression slide.

In retrospect it is surprising that several years passed before centrifugation was first employed to attach lampbrush chromosomes to glass (R.J. Hill et al. 1973); the idea of doing so stemmed from Miller and Beatty's (1969 a) use of cen-trifugation for spreading rDNA transcription complexes of *Notophthalmus* and *Xenopus* oocytes for examination by electron microscopy.

In the method that I came to adopt for *T. cristatus* the chromosomes were attached to coverslips by centrifugation, using as observation chambers specially constructed "ring cells" that could be spun in plugged, 50 ml polypropylene cen-trifuge tubes. The technique was described in detail by Old et al. (1977) and has served for various experimental studies, including autoradiography, transmission and scanning electron microscopy, immunolocalization visualized by light or electron microscopy, as will be discussed in this and later chapters. However for autoradiography in particular it has been largely superseded by the use, as disper-sal chambers, of bored slides cemented by wax to conventional slides. Unsup-ported coverslips bearing chromosomes are fragile and awkward to pass through a series of reagents, whereas chromosome preparations attached to slides can be handled with much greater ease and assurance that they will not break at some crucial stage in the procedure. For details specifically relating to the use of ring cells, where the chromosomes are attached to coverslips, reference should be made to Old et al. (1977); here I will assume that the chromosomes are to be attached to slides.

As soon as the nuclear membrane has been removed from an oocyte nucleus lying in saline in a dispersal chamber, the preparation is covered with a coverslip, sealed with vaseline, and put aside for a time to allow sap dispersal to occur. For *T. cristatus* and using simple unbuffered media 1 h generally suffices. It is imperative that complete sap dispersal should have taken place before proceeding to the next step, and only when the chromosomes are seen to be lying flat in one plane should the preparation be centrifuged. For examining a coverslip preparation in a ring cell the normal phase objectives can be used to view the preparation from below, but for a slide preparation an objective with a sufficiently long working distance is required, whether the preparation is to be viewed from above or below. Zeiss manufacture a × 40 plan phase lens, with a correction collar for 1.1 to 1.5 mm of glass between the specimen and the lens, that gives excellent image resolution when used for this purpose (Dr. J. G. Gall, personal communication).

Slide preparations are placed in the carrier baskets that were mentioned earlier in this chapter. Centrifugation should be carried out in a cool room; a refrigerated centrifuge is not needed unless long spins are involved. Rotation must be started as gently as possible, without a jerk, and speed gradually increased during about 3 min to 3000 or 4000 rpm. I have found that 5 min centrifugation at 4000 rpm will generally attach the lampbrush chromosomes of *T. cristatus* to the glass at the bottom of a dispersal chamber; with 14 cm radius of rotation they experience a centrifugal force of about 2500 g. Other amphibians and/or other dispersing media may require more or less centrifugation to ensure chromosome attachment. Thus for example Gall et al. (1981) needed to spin preparations of *Notophthalmus viridescens* for 30 to 45 min at 2500 g, whereas Scheer et al. (1984) found 10 min at 800 g to suffice for *Pleurodeles waltlii*.

After centrifugation a good preparation will have the lateral loops distributed evenly on either side of the chromosome axis, with the chromomeres clearly evident, and the loops motionless. Preparations of poorer quality have the loops distributed asymmetrically with respect to the chromosome axis, which may have been caused by a jerky stark to the centrifugation, particularly if sap dispersal was incomplete.

Each dispersal chamber, whether it be a ring cell with a coverslip bottom or a bored slide with a conventional slide bottom, should now be fully immersed in nuclear isolation or dispersing saline in a deep Petri dish and its top coverslip pushed off, releasing any air bubbles that may have accumulated during centrifugation. The next step depends on what one intends to do with the preparation.

I have no personal experience of the immunostaining of centrifuged lampbrush chromosomes for study by light microscopy; for technical information on this topic reference should be made to Sommerville et al. (1978a), Scheer et al. (1979b) and Lacroix et al. (1985).

For permanent stained mounts of centrifuged preparations, and for autoradiography, each dispersal chamber should be lifted from its saline bath and the bottom slide pried apart from the bored slide with a razor blade. While the bottom slide remains wet and there is still wax on it indicating on which side the chromosomes are attached, this side should be appropriately scribed with a diamond pencil.

Permanent Stained Mounts of Centrifuged Preparations. Place slides carrying lampbrush chromosomes in a rack in 4% commercial formalin. To ensure adequate fixation they should remain for at least 1 h in formalin; there is no harm in leaving them for longer, and this may be a convenient stage to accumulate preparations. Wash the preparations standing in a rack in running tap water to remove the formalin, and then run them up an ethanol series and into xylene, where they should be left for several minutes to allow the wax to dissolve. Rinse the preparations in a second xylene bath, and then run them back down an ethanol series to water.

The preparations may now be stained in one of several ways. The schedule which I have found to give the most reliable information on morphology is the following, based on Heidenhain's (1896) iron haematoxylin method. Mordant in 3% iron alum overnight. Rinse in distilled water. Stain in Heidenhain's haematoxylin for 6 h. Rinse in distilled water. Differentiate in 0.5% iron alum very briefly, for not more than 1 to 2 min. Rinse the preparations thoroughly in distilled water, take them up an ethanol series, clear briefly in xylene, and mount in Canada balsam. The differentiation should aim at leaving sufficient stain in the generality of lateral loops for their detailed morphology to be apparent, without obscuring their relationship to the chromomeres which, except for some landmark loops and a few other laterally attached objects, are the last chromosomal components to lose their stain during differentiation.

It may be thought irrational to take the preparations up and down an ethanol series merely to remove wax, but if the wax is left in place during mordanting and staining it becomes stained, and residues of this stained material persist even after a prolonged final xylene bath. It is in any case essential not to leave these stained preparations overlong in xylene, for the loops destain while in contact with this fluid; once they have been mounted in Canada balsam and dried on a hot plate, however, the staining is permanent.

At no stage during the technique described above do the chromosomes come into contact with air, and indeed it is imperative that they should not be allowed to do so if one is aiming at the best possible preservation of morphological detail. These centrifuged preparations stained in iron haematoxylin can give remarkably clear views of the morphology of lampbrush chromosomes, and they do not require an inverted microscope for examination. An example is shown in Fig. 2.2. With a little practice it is also possible to dispense with an inverted microscope while making them.

Centrifuged Preparations for Autoradiography. Lampbrush chromosomes attached to coverslips or slides by centrifugation can be autoradiographed if they have incorporated radioactive substances in the natural course of synthesis, or if they have been labelled by in situ hybridization. In the former case the appropriate procedure (details are given by Hartley and Callan 1978, and by Schultz et al. 1981) is identical to that just described for making permanent mounts, except

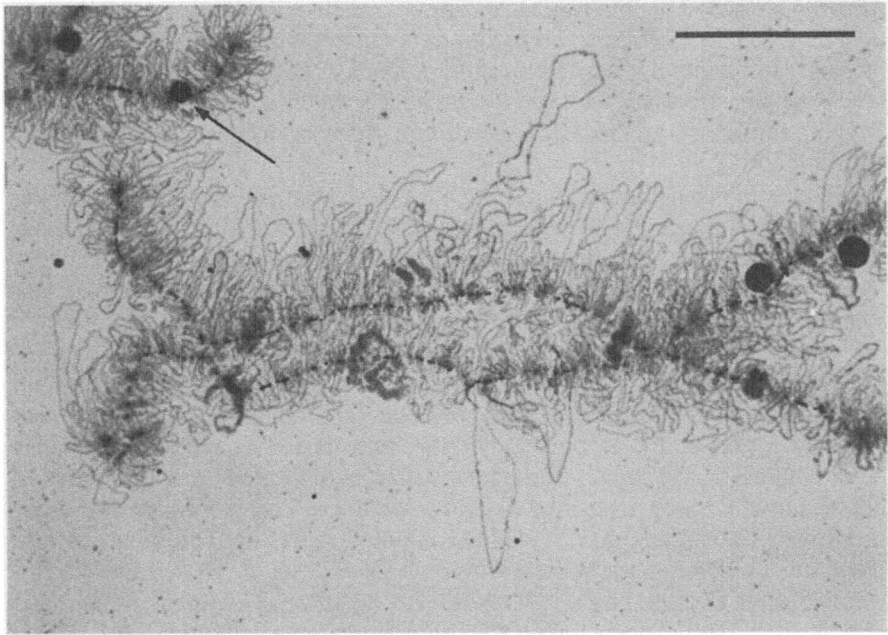

Fig. 2.2. About two-thirds of lampbrush bivalent X of *Triturus cristatus carnifex* after centrifugation, fixation in 4% formalin, and staining by Heidenhain's iron alum/haematoxylin technique as described in the text. The *arrow* points to an attached sphere on chromosome VIII. The two round, black objects on the *right* are free spheres. *Bar* = 50 μm. (From Callan 1982)

for the insertion of a 5 min bath in ice-cold 5% trichloroacetic acid after fixation and prior to washing to extract unincorporated labelled precursors, as far as the point where preparations are left in xylene for removal of wax. When the wax has dissolved the preparations are rinsed in clean xylene, then passed through two or three changes of acetone, and air dried. The dry lampbrush chromosomes can be readily seen with the naked eye; dilute subbing fluid (0.1% gelatin and 0.01% chrome alum in water, Gall and Pardue 1971) is now painted over the slide, except for the area immediately around the chromosomes, with a camel hair brush, and the slide allowed to dry. In this state the preparations are ready for filming, and they can be stored in boxes with slotted sides, protected from dust.

For in situ hybridization of labelled nucleic acid probes to centrifuged lampbrush chromosomes, the preparations are made as just described except for the omission of formalin fixation and trichloroacetic acid extraction after centrifugation. Details of procedures for in situ hybridization to lampbrush chromosomes have been given by Pukkila (1975), Macgregor and Andrews (1977), Old et al. (1977), Gall et al. (1981) and they will not be repeated here. Gall et al. (1981) used a dispersing saline containing 0.1% paraformaldehyde, which at this concentration did not interfere with hybridization. For a detailed account of in situ hybridization techniques in general reference should be made to Chap. 8 of Macgregor and Varley (1983).

Autoradiography, Staining and Mounting. It is not advisable to use AR10 stripping film to cover lampbrush preparations because it is difficult to stain the chromosomes differentially from the thick film; Kodak NTB-2 dipping emulsion, diluted one part emulsion with one part distilled water, is much to be preferred. Dipping emulsion is applied by the procedure which has by now become standard (Prescott 1964).

The completed autoradiographs, after development, fixing and washing, must now be stained. Of those methods which I have tried, Giemsa, though somewhat erratic, is the best. A well-made Giemsa preparation shows the chromomeres densely stained purple, the generality of lateral loops pale mauve-pink, and some exceptional "landmark" structures blue.

After the photographic fixer has been thoroughly washed out, the preparations are rinsed in distilled water and placed in Coplin jars, each jar containing 50 ml of 10 mM phosphate buffer, pH 7.0. Two ml of Giemsa stock stain (B.D.H.) is now added by pipette, and the fluids mixed by pipetting in the jar. After 2 h the Giemsa stain is flushed out with a jet of filtered tap water, the slides are then briefly rinsed in distilled water, and air dried. This procedure was recommended by Gall and Pardue (1971) to avoid the preparations becoming contaminated by the surface scum which forms on a Giemsa staining solution. More recently Gall et al. (1981) have advocated staining autoradiographs of lampbrush chromosomes with Coomassie blue because the staining is more intense and less capricious than Giemsa. Prior to staining, the developed, fixed and washed preparations are air dried in order to minimize the danger of film reticulation. They are then passed through 50% methanol into 0.1% Coomassie Brilliant Blue R250 dissolved in 50% methanol, 10% acetic acid and stained for 10 min. After staining, the preparations are dipped in 50% methanol, washed in running water for 10 min to get rid of acetic acid, and air dried from distilled water. Once the preparations are dry they can be mounted in Canada Balsam.

Centrifuged Preparations for Electron Microscopy

Sections. A procedure for studying centrifuged preparations of lampbrush chromosomes in section in the electron microscope was described by Mott and Callan (1975). The first part of the technique made use of "ring cells" with circular coverslip bases, and centrifugation in plugged centrifuge tubes, as described by Old et al. (1977). Dispersal chambers with a standard microscope slide instead of a coverslip base should serve just as well. The slide which makes the floor of the chamber in which the chromosomes are to be centrifuged is glow-discharged and carbon-coated before attachment to its bored slide. In all other respects the operations leading to the attachment of centrifuged chromosomes to slides follow the procedures given earlier in this chapter. When the slide bearing chromosomes has been pried away from the bored slide it is placed in 3% glutaraldehyde in 0.1 M Sørensen's phosphate buffer, pH 7.2, for 1 h. After this first-stage fixation it is washed in several changes of buffer, and then post-fixed in 1% osmium tetroxide in the same buffer for 30 min. After second-stage fixation the slide is washed in several more changes of buffer, and then dehydrated by passage through an ethanol series. Preparations in absolute ethanol are passed through propylene oxide if they are to be embedded in Araldite; this step is omitted if they are to be embedded in Spurr.

Slides bearing chromosomes, while still bathed in alcohol or propylene oxide, are now placed "butter side up" on metal foil, flooded with embedding medium, and incubated overnight at 60 °C for polymerization. The chromosome group is now covered by a capsule containing embedding medium, and the assembly left to polymerize at 60 °C, again overnight. A capsule with end-embedded chromosomes is now split from the slide by placing the assembly in a bowl containing solid CO_2 and liquid nitrogen. The capsule is removed, and the back of the plastic block inside filed down to a height of about 5 mm.

The surface of the block which had previously been in contact with the slide is examined under a dissecting binocular, and when the position occupied by the chromosome group has been identified the block can be trimmed down to a surface area appropriate for thin sectioning, which includes the objects of primary interest. As the chromosome group only occupies some 2 μm depth from the block face, at the most only forty 50 nm sections can be taken from it. Thin sections are mounted and stained using conventional EM techniques. Thick sections can alternatively be taken, and examined by HVEM.

Dr. U. Scheer (personal communication 1985) has recently adopted this technique for immunolocalization studies. After centrifugation the lampbrush chromosomes of *T. cristatus* were treated with antibodies, the secondary antibodies being coupled to colloidal gold of 5 nm diam. The chromosomes were then fixed, end-embedded and ultra-thin sectioned. His observations are discussed in Chap. 9.

Whole Mounts. Angelier et al. (1984) have studied centrifuged lampbrush chromosomes of *Pleurodeles waltlii* by scanning electron microscopy (SEM).

They used ring cells with bottom coverslips, 12 mm diam, and a physiologically balanced dispersing medium Tris-buffered to pH 7.3. Chromosome attachment required 30 min centrifugation at 1500 g. The preparations were first photographed in phase contrast, then processed for SEM. Several methods of fixation were assessed, and of these brief exposure to 1% glutaraldehyde buffered to pH 7.2 with 0.1 M phosphate proved most satisfactory. Notably paraformaldehyde, a good fixative for Miller-spreads (see next section) was found to be inadequate because it failed to prevent lateral loops from collapsing. Angelier et al.'s (1984) observations are discussed in Chap. 4.

Miller-Spreads. The first attempts to prepare whole mounts of lampbrush chromosomes for electron microscopy were carried out on chromosomes isolated in salines of near-physiological concentration (Tomlin and Callan 1951, Gall 1952, 1956, 1958). Apart from providing information on the width of the interchromomeric chromonema such preparations were not very informative because of the high electron opacity and poor state of preservation of the lateral loops.

Significant progress was made by Miller (1965) who introduced a technique whereby lampbrush chromosomes, that had been isolated and allowed to disperse in saline, were thereafter centrifuged through several millimetres of fluid in a com-

pound chamber and driven on to an EM grid. By means of this technique the fibrillar components of lateral loops matrix RNP were revealed, and an accurate
estimate was obtained of the width of lateral loop axis remaining after tryptic digestion, 2 to 3 nm, which supported the contention that the loop axis contains
only one DNA duplex.

KCl plus NaCl solutions at concentrations significantly less than 50 mM
cause the RNP matrix of the lateral loops of freshly isolated lampbrush chromosomes to spread, and in distilled water the degree of dispersal is so great that
scarcely any objects, or none at all, can be resolved in the phase microscope (Gall
1954, Macgregor and Callan 1962). Miller and co-workers took advantage of this
extreme degree of dispersal in distilled water, and devised a remarkably simple
but effective technique for visualizing RNA transcription, first in amphibian oocyte nuclei (Miller and Beatty 1969 a, b) and later in other cells, in the electron
microscope. The technique, as applied to amphibian oocyte nuclei, has been described in detail by Miller and Bakken (1972).

An oocyte nucleus is isolated and freed from cytoplasm in 100 mM KCl. It
is quickly transferred to clean saline, where the nuclear membrane is removed.
The nuclear contents, which remain gelled for longer or shorter times according
to the amphibian species, are now transferred by fine-bore pipette and with a
minimum volume of saline to a 5 to 7-mm-diam drop of double glass-distilled
water, pH adjusted to 9 with 0.1 mM borate buffer, lying on a plastic surface. The
nuclear contents are allowed to disperse for several minutes before being transferred to a centrifugation chamber.

The centrifugation chamber used for EM studies is a ring cell. It consists of
a 25-mm-diam disc of 6-mm-thick polymethylmethacrylate (Perspex, Lucite),
with a centrally bored hole 4 to 5 mm wide, and a coverslip sealed across the hole
to make a floor. The chamber is prepared by filling the well with a Millipore-filtered (0.22 μm pore size) 10% solution of formalin in 100 mM sucrose, pH adjusted to 8.5 with borate buffer. An EM grid, covered with a thin carbon support
film and freshly glow-discharged to make the surface hydrophilic, is placed in the
bottom of the well, carbon film up, and liquid removed until the well is only one-
half to one-quarter full.

The drop with dispersed nuclear contents is now transferred by pipette to the
sucrose/formalin "cushion" in the centrifuge chamber, a round coverslip is placed
on top, and excess fluid absorbed by filter paper. The chamber is placed on a flat
rubber pad lying at the bottom of a 100 ml swinging centrifuge bucket, and centrifuged for 3 to 5 min at 2000 to 2500 g. The chamber is now taken from the
bucket, its top coverslip removed, and the fluid meniscus rounded up with su-
crose/formalin.

With a fine-pointed needle the edge of the EM grid is touched to ensure that
it is free to leave the floor of the chamber when the latter is inverted; on inversion,
the grid falls to the air/fluid interface, from which it is picked up by forceps with
sharply bent tips. The forcep tips are bent so that no liquid is trapped between
them during this and later operations. The grid is gently rinsed in dilute Kodak
Photo-flo (three drops in 50 ml water, pH 7), its edge then touched against lens
tissue to absorb liquid, and thereafter air dried. The use of Photo-flow is a crucial
step in this procedure, for it much improves the spreading of the preparation by

decreasing surface tension as the air/water interface crosses the specimen on the grid.

The grid is stained for 30 s in a freshly prepared mixture of six drops 95% ethanol and two drops of Millipore-filtered 4% aqueous phosphotungstic acid, rinsed for several seconds in 95% ethanol, touched against lens tissue to absorb liquid, and air dried. It is now ready for examination in an electron microscope.

Miller and Bakken (1972) introduced one variant on this procedure which has proved to be generally advantageous, and for some biological materials essential; the use of a detergent instead of mechanical means for disrupting the nuclear membrane. Alone of several detergents tried by Miller and Bakken (1972), a dilute solution of a commercial product "Joy" (Proctor and Gamble, formula not divulged) in the dispersing water proved to be suitable in that it disrupts the nuclear membrane without adversely affecting the morphology of transcription complexes. This modification in procedure enables an intact oocyte nucleus in pH 9 water to be transferred direct to the centrifugation chamber if the upper part of the well of the chamber has been filled with pH 9 water containing "Joy". Dispersal can then occur in the chamber instead of prior to transfer, thereby avoiding mechanical damage to the nuclear contents before centrifugation. R.S. Hill (1979) has used this method, with "Joy" at 0.25% in pH 9 water, for studying lampbrush transcription complexes in *Xenopus* oocytes including early diplotene, when the oocyte nuclei are much too small for the free-hand removal of their membranes.

"Miller-spreads" of transcription complexes in a wide range of organisms have now been made in several laboratories. Another advantageous variation in procedure has been the use of shadow casting to improve specimen contrast in the electron microscope (Angelier and Lacroix 1975, Franke et al. 1976). A recent review of the techniques in use for the visualization of transcription is given in Chap. 7 of Macgregor and Varley (1983).

The lampbrush chromosomes of the oocytes of other animals, such as birds, reptiles, fish and species belonging to several invertebrate groups, can be studied by techniques similar to those described in this chapter, provided their nuclei are sufficiently large to manipulate free-hand. Micromanipulators have not been employed with success for the removal of the membranes from nuclei too small to be dealt with free-hand. Chemical methods for the removal of nuclear membranes, which is the trickiest manual operation, may prove to be feasible, but the extreme sensitivity of loop matrix to non-physiological saline concentration and to a wide variety of chemical agents could prove a serious stumbling block if one wishes to examine lampbrush chromosomes of small nuclei in a reasonably life-like state.

The lampbrush Y-loops of *Drosophila* spermatocytes have been studied by different, and for the most part simpler, techniques; references to these will be given in Chap. 7.

CHAPTER 3

Chromosome Identification

Except where stated otherwise, this chapter will be concerned only with the Amphibia.

The identification of chromosomes in lampbrush preparations depends on determining their axial lengths relative to one another, and on the recognition of various "landmark" structures that occur at particular sites. Despite the differences in structural organization between lampbrush chromosomes, with their intermittently distributed compact chromomeres and lateral structures loaded with RNA/protein (RNP), and mitotic chromosomes where, apart from the centromeres and secondary constrictions (when present), DNA/histone (DNP) complexes are uniformly compacted, the lengths of chromosomes relative to one another within the lampbrush and mitotic complements of individual species are in good accord. This was established by Gall (1954) for *Notophthalmus viridescens*, where in regard to absolute lengths the axes of the lampbrush chromosomes, when maximally extended, are about 50 times longer than the mitotic, and by Callan and Lloyd (1960) for *Triturus cristatus carnifex*, where the lampbrush chromosomes are some 25 times longer. The same holds for *Ambystoma mexicanum* (Callan 1966), though in this species the lampbrush chromosomes are about 80 times longer.

Although these figures might appear to imply that lampbrush chromosomes extend (i.e. decompact) to different degrees in the three species being compared, this is not confirmed when axial extension is expressed in absolute terms, as shown by measurements based on a comparison between the bivalents shown in Fig. 3.1. The C-value of *T.c. cristatus* is 19 pg (Olmo 1983). Chromosome XI represents 0.056 of the *T.c. cristatus* genome (computed from Callan and Lloyd 1960) and so contains 2.13 pg of DNA in its two chromatids. Its axial length at full extension measures 307 μm (Fig. 3.1 a), so 2 pg of DNA in this lampbrush chromosome extends to 288 μm. The C-value of *A. mexicanum* is twice as large, 38.4 pg (Bachmann 1970). Chromosome XIII represents 0.039 of the *A. mexicanum* genome (computed from Callan 1966) and therefore contains 2.30 pg of DNA in its two chromatids. Its axial length at full extension measures 408 μm (Fig. 3.1 b) so 2 pg of DNA in this lampbrush chromosome extends to 355 μm. Measurements made on two other lampbrush chromosomes of these species gave similar figures, 280 μm per 2 pg of DNA for *T.c. cristatus*, 350 μm for *A. mexicanum*. Thus the apparent diversity in the decompaction of the lampbrush chromosomes in these two species is in the main attributable to different degrees of compaction

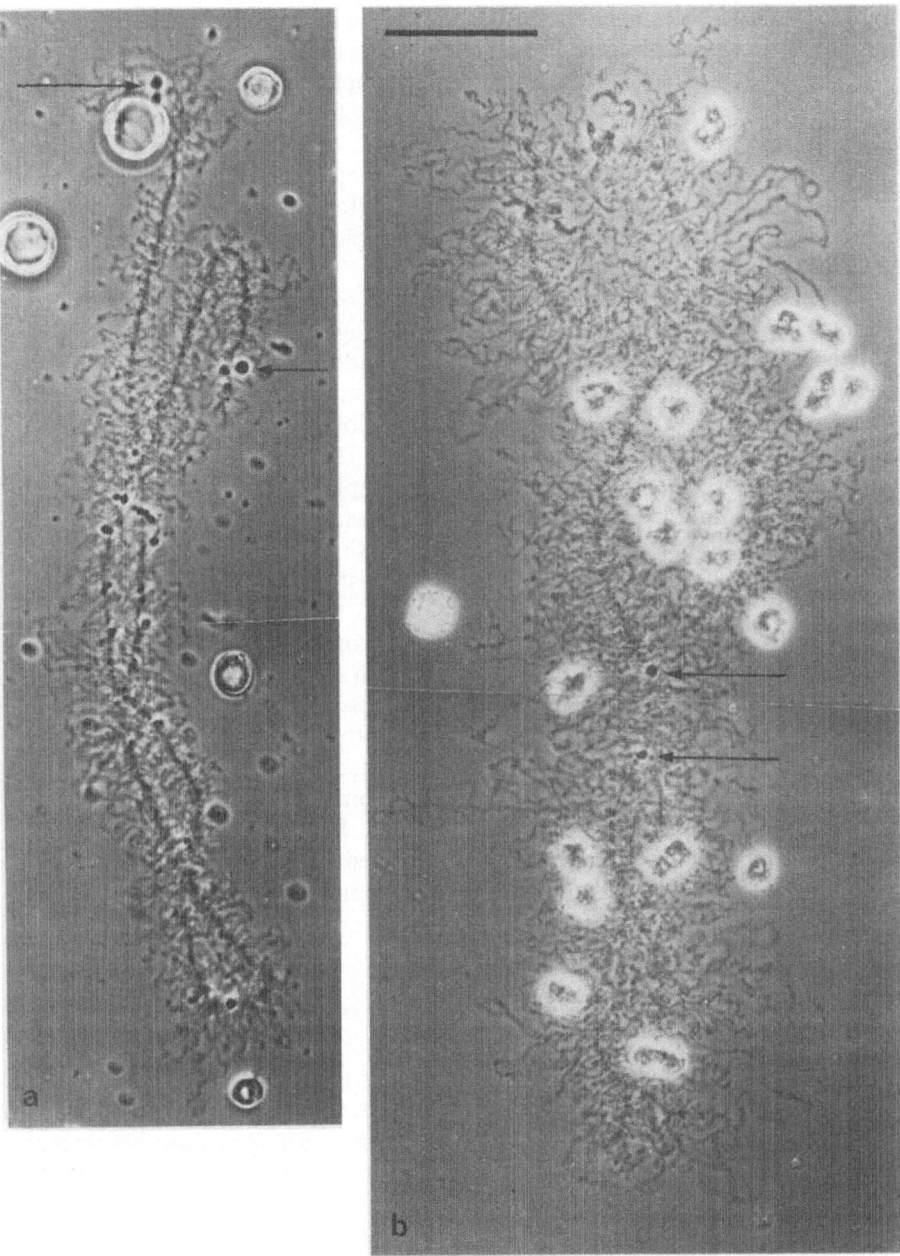

Fig. 3.1 a, b. Phase contrast photographs of unfixed lampbrush chromosomes from oocytes that have reached comparable stages of oogenesis. **a** Bivalent XI of *Triturus cristatus cristatus* from an oocyte of 1.0 mm diam; *arrows* point to small subterminal fusing loops in the longer chromosome arms; the telomeres of the shorter arms are united with one another. **b** Bivalent XIII of *Ambystoma mexicanum* from an oocyte of 1.4 mm diam; *arrows* point to fused spheres. The large refractile objects are free nucleoli. Both are printed at the same magnification; *bar* = 50 μm. **a** From Callan and Lloyd (1960); **b** from Callan (unpublished)

of the mitotic chromosomes with which they are compared; when related to DNA content, both are decompacted to much the same degree.

There is, however, a substantial difference between the lampbrush chromosomes of *T.c. cristatus* and *A. mexicanum* that is apparent in Fig. 3.1; the lateral loops of *Ambystoma* are much longer than those of *Triturus*. The two bivalents came from oocytes at roughly comparable stages of oogenesis, and were dispersed in salines of comparable concentrations. The difference between them is for the most part attributable to C-value diversity, a topic that will be considered again in Chaps. 4 and 5.

The accordance between the relative lengths of lampbrush and mitotic chromosomes within complements implies that although there are some axial regions of lampbrush chromosomes that are more extended than others, and also some regions containing unusually long stretches of compacted chromomeric material, when considered overall the degrees of extension of different lampbrush chromosomes within a complement are much the same. This is borne out by the general impression to be gathered when scanning a complement of lampbrush chromosomes, for although their chromomeres vary in size (within limits characteristic of the species), and although there are the exceptional regions mentioned above, the density of chromomeres per unit length of unstretched chromosome is generally much the same throughout the complement.

When measuring the axes of lampbrush chromosomes in order to establish their relative lengths it is advisable to work with fresh preparations, as soon as nuclear sap dispersal is complete and the chromosomes have come to lie on a flat plane, and to disregard the lengths of regions that are under tension brought about by some mechanical accident. Centrifuged preparations should not be used, because the attachment of lampbrush chromosomes to a glass surface sets up irregularly distributed mechanical tensions. In fresh preparations regions under tension can be recognized by the restricted Brownian movement of their axes, and a control is often provided by the homologous region of the partner chromosome.

Centromeres

Having established the lengths of chromosomes relative to one another, a first concern of a cytologist trying to identify mitotic or meiotic chromosomes at metaphase is to locate the positions of their centromeres. In compact mitotic chromosomes, unless they are acrocentric, this generally presents no great problem, but it can be otherwise with lampbrush chromosomes.

In some amphibian species relatively extended regions of chromosome axis alongside and including the centromeres are conspicuous features in that they are devoid of lateral loops. Thus when the lampbrush chromosomes of these species are viewed at low magnification the centric regions are "landmarks" and appear as gaps. This holds for *N. viridescens,* many ambystomid salamanders, many plenthodontid salamanders and one subspecies of *T. cristatus, T.c. karelinii.*

In other amphibian species the centromeres of the lampbrush chromosomes are not at all conspicuous, and their identification has presented difficulty. In *T.c.*

cristatus and *T.c. carnifex* (Callan and Lloyd 1960), also *T. alpestris apuanus* (Mancino and Barsacchi 1965, Ragghianti et al. 1972) and *T. italicus* (Mancino and Barsacchi 1969), urodeles where in female meiosis chiasmata tend to be pro-centrically localized, the centromeres have been identified by examining the positions of chiasmata. In these species the centromeres are compact spherical or el-lipsoidal chromomeres, often somewhat larger than the generality, smooth in out-line and without associated lateral loops (Figs. 1.11 a and 3.2 a). Determination of their exact location depends on the fact that although the centromeres lie within chromosome regions where chiasmata are formed at particularly high fre-quency, chiasmata never occur precisely at the centromeres. However because the centromeres are compact, the absence of associated loops can only be established by observing them at high magnification.

There are several urodele species whose centromeres must be inconspicuous, for although their lampbrush chromosomes have been studied with care, the cen-tromeres have not been identified. This was true of *T. marmoratus* (Nardi et al. 1972) despite the fact that its chromosome complement closely resembles that of *T. cristatus,* with which species it can hybridize (White 1946, Lantz and Callan 1954); however in a later paper on *T. marmoratus,* Batistoni et al. (1974) have shown that Giemsa staining of preparations after exposure to C-banding treat-ment (denaturation followed by renaturation, Gall and Pardue 1971) reveals the centromeres as particularly dark-stained chromomeres each a trifle under 3 µm long, lacking lateral loops and with a median constriction. It remains true of *Pleurodeles waltlii* and *P. poireti* (Lacroix 1968 a), *Siren intermedia nettingi* (León and Kezer 1974), *Taricha granulosa* (Dr. James Kezer, personal communication), *T. helveticus helveticus* (Mancino and Barsacchi 1966), *T. vulgaris meridionalis* (Barsacchi et al. 1970) and *Salamandra salamandra* (Mancino et al. 1969). Like-wise centromeres have not been identified in lampbrush chromosomes of the anurans *Rana esculenta* (Morescalchi and Filosa 1965, Giorgi and Galleni 1972) and *Xenopus laevis* (Müller 1974). Srivastava and Bhatnagar (1962) claimed to have identified the centromeres of the Indian frog *Rana cyanophlyctis,* but gave no supporting evidence. Ohtani (1975) could not distinguish the centromeres of the Japanese frogs *R. nigromaculata* and *R. brevipoda,* but in a later paper by Nishioka et al. (1980) Ohtani supplied evidence (from the distribution of chias-mata and comparisons between lampbrush and mitotic karyotypes) that the cen-tromeres of both species are minute axial chromomeres lying for the most part in regions amongst, or close to, conspicuous groups of landmark loops. This jux-taposition is reminiscent of the centromeric loci in *T. cristatus.*

In the urodele *N. viridescens,* where the centric regions (kinetochores) of lampbrush chromosomes were identified for the first time (Gall 1952, 1954) the regions appear as elongate chromomeres some 10 µm long and 2 µm wide, with irregular outlines and without associated lateral loops. These regions are uni-formly Feulgen-positive. Occasionally the centric region includes a discrete small sphere, either embedded within or displaced to one side. In some instances fusions between homologous centric regions were observed by Gall (1952, 1954), and where they occurred these fusions always involved the embedded granules, not the entire centric region. Gall found the centric regions of *Ambystoma tigrinum* chromosomes similar to those of *N. viridescens.*

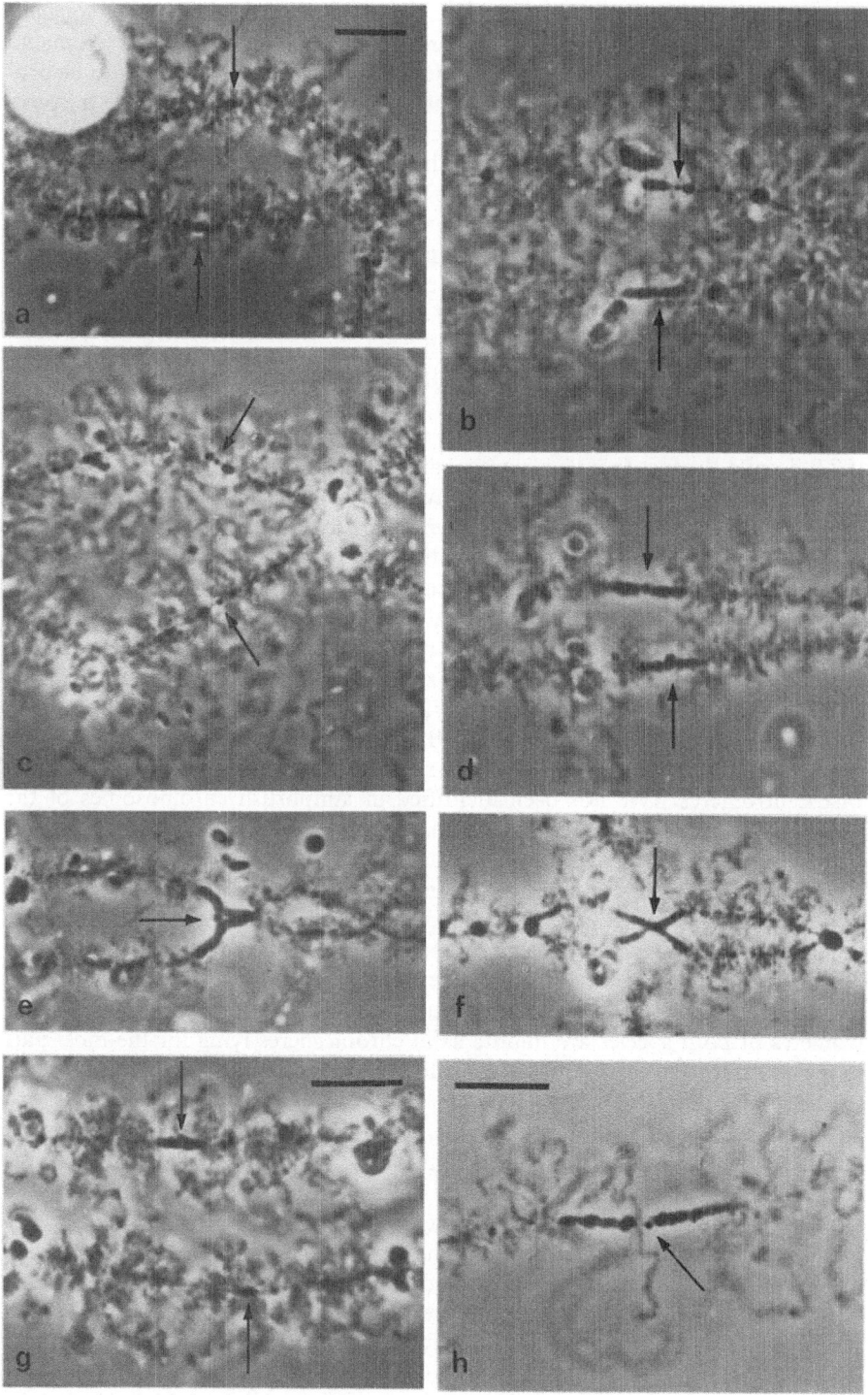

Fig. 3.2 a–h

In the axolotl *A. mexicanum* (Callan 1966) the centric regions of the lamp-brush chromosomes resemble those of *N. viridescens* in being elongate chromomeres of irregular outline and lacking lateral loops, but they are more slender, about 10 μm long and not more than 1 μm wide (Fig. 3.3 a). There is no sign of a discrete granule within the centric regions, and centric fusions do not occur. In *A. macrodactylum* (Kezer et al. 1980) the centric regions are similarly identifiable as elongate chromomeres without lateral loops, but in this species they are thicker, frequently with an associated 1.5 μm granule lying within a depression in the loop-free region (Fig. 3.3 c), or adjacent to it but displaced laterally, or several microns away but connected to the loop-free region by strands that look like transcriptionally active lateral loops (Fig. 3.3 b). Very occasionally the loop-free region includes a pair of 1 μm granules embedded in, and lying transverse to, the chromosome axis (Fig. 3.3 d), similar to the centromere granules of diplotene spermatocytes of the salamander *Plethodon cinereus* described by Kezer and Macgregor (1971). Centric fusions were not recorded by Kezer et al. (1980). The regions around and including the centromeres of the mitotic and male meiotic chromosomes of *A. macrodactylum* stain heavily with Giemsa when preparations are subjected to the C-banding technique. Kezer et al. (1980) conclude that the centric loop-free regions of *A. macrodactylum* lampbrush chromosomes consist for the most part of constitutive heterochromatin, and that the centromeres proper, i.e. the foci to which spindle fibres attach at metaphase, are the granules embedded in the heterochromatin. In *Plethodon cinereus, P. vehiculum* and *P. dunni* lamp-brush chromosomes (Kezer and Macgregor 1973, Vlad and Macgregor 1975) each centric region consists of a variable number of tiny loop-free chromomeres occupying some 5 to 10 μm of chromosome axis. In mature oocytes of *P. dorsalis* at a stage when the lateral loops have fully retracted and the chromosome axes are for the most part compact, the centric regions are easily recognizable as stretches of diffuse axis several tens of microns long, smothered in granules (Dr. J. Kezer, personal communication; Fig. 3.3 f). Presumably here again what can be seen in the phase microscope is primarily constitutive heterochromatin, and in view of its unusually diffuse appearance it may be engaged in transcription. However this has not been demonstrated, and its significance is in any case obscure.

Fig. 3.2 a–h. Phase contrast photographs of regions including the centromeres in unfixed lamp-brush chromosomes of *Triturus cristatus*. Photographs **a, b, c** and **h** are from oocytes of about half the mature diameter, while **d, e, f** and **g** are from somewhat larger oocytes. *Arrows* point to centric granules. **a** In bivalent I of *T.c. carnifex;* the *white spherical object* represents the phase halo around a free nucleolus. **b** In bivalent XII of *T.c. karelinii;* in the upper chromosome the axial bar to the right of the centric granule is less compact than the homologous bar in the lower chromosome, accounting for the different distances separating the two centric granules from the landmark axial granules to their right. **c** In bivalent X of *T.c. karelinii*, with unusually short axial bars. **d** In bivalent IX of *T.c. karelinii*, showing the centric granule displaced from the axis of the lower chromosome. **e** Fused centric granules in bivalent X of *T.c. karelinii*. **f** Fused centric granules in bivalent V of *T.c. karelinii;* the other connections between the two chromosomes are axial granule fusions. **g** In bivalent VII of an F_1 hybrid between *karelinii* ♀ × *cristatus* ♂; the upper chromosome originated from the ♀ parent. **h** The *karelinii*-derived centric region of chromosome X of a backcross hybrid (*karelinii* ♀ × *carnifex* ♂) ♀ × *carnifex* ♂. A pair of loops extend from the constriction between the centric granule, marked with an *arrow,* and the axial bar to its left. *Bars* = 10 μm, **a** to **f** being of the same magnification. **a–f** From Callan and Lloyd (1960); **g** and **h** from Callan (1982)

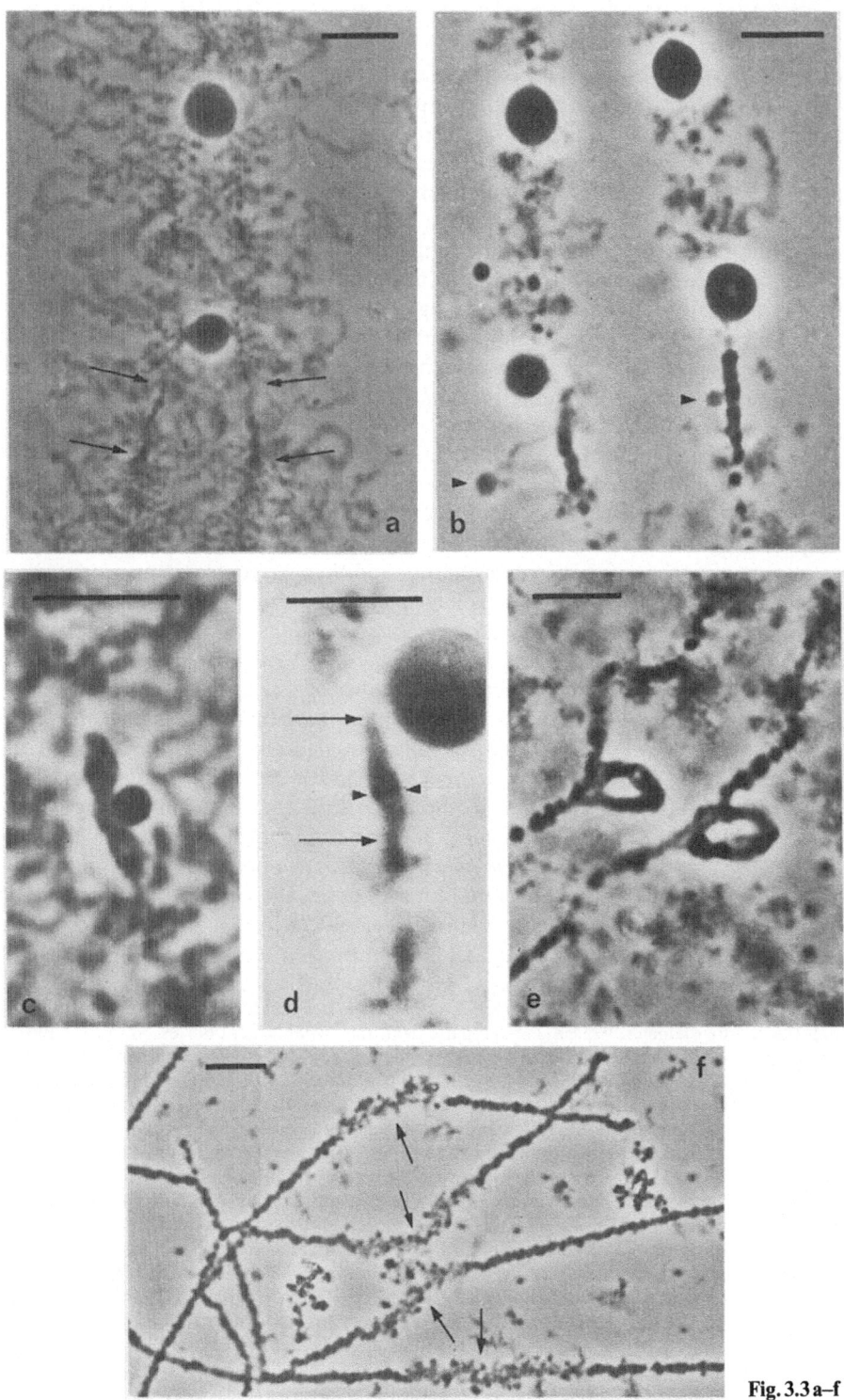

Fig. 3.3 a–f

The centric regions of another plethodontid salamander, *Eurycea bislineata,* are dense and massive structures 10 to 15 µm long and 3 to 5 µm wide, while those of *Chiropterotriton bromeliacea,* a plethodontid salamander that lives in bromeliads in the cloud forests of Guatemala, are bizarre in the extreme (Dr. J. Kezer, personal communication); they are long and dense objects with attached lumps or rings of material of comparable density (Fig. 3.3 e) and similar lumps or rings lie free in the nuclear sap as though they had been shed from these centric regions; whether these free objects contain DNA is not known.

Unlike the other races of *Triturus cristatus,* the centric regions of *T.c. karelinii* (Callan and Lloyd 1960) are easily recognized in lampbrush preparations. In oocytes of up to 0.8 mm diam, each centric region consists of a smooth round granule of less than 1 µm diam flanked by "axial bars", elongate chromomeres with smooth outlines about 1 µm wide, on either side (Fig. 3.2 b). Constrictions separate the granule from the two flanking axial bars, and lateral loops are absent from all three components of the centric complex. Although the axial bars are usually conspicuous, each about 4 µm long, those of certain chromosomes are regularly shorter in particular individuals, only 1 to 2 µm long (Fig. 3.2 c). As oocytes increase in size so too do the centric granules, and the constrictions between each granule and its adjacent axial bars are either eliminated (Fig. 3.13 d) or remain and the granule is displaced to one side of the chromosome axis (Figs. 3.2 d and 3.13 c).

Fusions between homologous centric granules, though not between the neighbouring axial bars, occur in about one *T.c. karelinii* lampbrush bivalent out of every five (Fig. 3.2 e, f). Furthermore non-homologous centric granules are occasionally fused together in this subspecies and more than two granules may be included in a fusion. Fused centric granules are not dissociated by prolonged digestion with pepsin, trypsin or pronase (Macgregor and Callan 1962), suggesting that the associations depend on interaction between homologous regions of DNA. However the fusions do not involve recombination between the DNA components of centric granules that persist as chiasmata, for there is nothing untoward about *T.c. karelinii* bivalents at first meiotic metaphase; just as in the bivalents of *T.c. cristatus* and *T.c. carnifex,* the centromeres of each bivalent are sep-

---

Fig. 3.3 a–f. Phase contrast photographs of regions including the centromeres in unfixed lampbrush chromosomes. **a** In bivalent VI of *Ambystoma mexicanum* from an oocyte of 1.5 mm diam; the limits of the centric regions are marked by *arrows;* above them are two homologously-fused spheres. **b** In bivalent V of *A. macrodactylum* from an oocyte of 1.8 mm; the spheres and centric regions are distributed as in **a**, but the spheres are not fused, and the centric regions are more massive; *arrowheads* point to granules associated with the centric regions, that on the left being attached by diffuse fibres. **c** In a chromosome of *A. macrodactylum* from an oocyte of 1 mm diam, where a dense granule is associated with the centric region. **d** In a chromosome of *A. macrodactylum* from an oocyte of 1.8 mm diam; the limits of the centric region are marked by *arrows,* and *arrowheads* point to a pair of dense granules embedded within it; the *round object* above and to the right is a sphere. **e** Part of a bivalent of *Chiropterotriton bromeliacea* showing dense rings connected to the centric regions, characteristic of this plethodontid salamander. **f** From a mature oocyte of *Plethodon dorsalis;* the chromosome axes are for the most part compact and dense, the lateral loops having retracted, whereas the centric regions are diffuse and smothered in granules. *Bars* = 10 µm. **a** From Callan (1966); **b, c** and **d** from Kezer et al. (1980); **e** and **f** courtesy of Dr. J. Kezer

arate from one another and they co-orientate normally on either side of the spindle equator (Watson and Callan 1963).

In *T.c. karelinii* it seems reasonable to assume that the centric granule, not the flanking axial bars, represents or includes the centromere. I will return to consider the nature of the flanking material, but one point is worth emphasizing here; in some manner the fusions between the centromeres of lampbrush chromosomes, which in the normal course of meiosis come to lie close together during synapsis, are positively influenced by the presence of neighbouring axial bars; for it was precisely the lack of association between the centromeres of *T.c. cristatus* and *T.c. carnifex* lampbrush bivalents that played a crucial part in their identification.

Whether the centric granules in the lampbrush chromosomes of *T.c. cristatus* and *T.c. carnifex* consist of centromeric material only is debatable, for C-banded mitotic chromosomes of both these races also show signs of constitutive heterochromatin at their centromeres (Rudak and Callan 1976), and Baldwin and Macgregor (1985) have demonstrated that short satellite DNA sequences present in abundance around the centromeres of *T.c. karelinii* are also present, though much less abundant, in the centric regions of *T.c. cristatus, T.c. carnifex* and *T. marmoratus*. I mentioned earlier in this chapter that the centromeres in the lampbrush chromosomes of several urodeles, including some other species of *Triturus* and also *Pleurodeles*, have not yet been identified; this may be because there is little or no constitutive heterochromatin in their vicinity.

Nucleolus Organizers

In the mitotic complements of many plants and animals, unless one or other of the "banding" preparative technique has been employed, the nucleolus organizer loci are amongst the few reliable characters available for chromosome identification. At mitotic metaphase the centromeres are generally evident as places where sister chromatids are jointly constricted; the nucleolus organizer loci, the secondary constrictions, are by contrast independent constrictions in sister chromatids, often subterminal in one or more chromosomes of the complement (Heitz 1931). The value of nucleolus organizer loci as recognition characters in the lampbrush chromosome complements of Amphibia varies a great deal from species to species.

The several hundred free nucleoli in the oocytes of many Amphibia, which during most of oocyte development lie just inside the membrane of the germinal vesicle, arise as a result of rDNA amplification, a process which in *Xenopus laevis* begins in oogonial stages but continues on a much more extensive scale in pachytene, prior to the chromosomes assuming the lampbrush form (Gall 1968, Macgregor 1968, Gall and Pardue 1969, Bird and Birnstiel 1971, Watson-Coggins and Gall 1972). This topic is peripheral to the main concern of this monograph; it has been covered by several reviews (Gall 1969, 1978, Macgregor 1972, Tobler 1975) which give references to the already extensive literature.

In Gall's paper (1954) describing the lampbrush chromosomes of *N. viridescens* he identified a structure attached subterminally on the short arm of chromosome III (Gall's 7) as being the only structure attached to the chromosomes that is at all comparable in shape and refractility to the peripheral, free nucleoli (Fig. 3.5). He noted that in some preparations the attached nucleoli were very small, and occasionally absent altogether. These observations led Gall to the tentative conclusion that "... some or all of the larger peripheral nucleoli are produced at this locus by successive growth and detachment". As the time I shared Gall's opinion; it seemed reasonable to take this view, for objects formed at several loci on the lampbrush chromosomes of *T. cristatus* are certainly released into the nuclear sap during oogenesis, and if these, why not nucleoli too?

In his 1954 paper Gall went on to consider what other evidence might be obtained to prove that the assumed nucleolus organizer locus on the lampbrush chromosomes corresponds with the nucleolus organizer locus in mitotic chromosomes. In *N. viridescens* there are five secondary constrictions in haploid mitotic chromosome complements prepared from embryos, and one of these lies subterminally in the short arm of the third longest chromosome. However there was no assurance that this particular constriction out of the five does indeed represent the nucleolus organizer locus. The matter was substantially settled by Gall, in the same paper, by observations made on a different urodele, *Ambystoma tigrinum*. In this salamander the site of origin of the nucleolus in epithelial cells at telophase had been established by Dearing (1934) as a subterminal constriction on the short arm of one of the longest chromosomes. Gall found that an object similar in morphology to the peripheral nucleoli was attached subterminally in the short arm of a long lampbrush chromosome having a similar arm length ratio, that there was only one such object within the haploid lampbrush complement, and on morphological grounds the relationship thereby established.

The situation in *T. cristatus* proved to be different, and for several years it gave rise to confusion. Callan and Lloyd (1960) were unable to find any structures attached to the lampbrush chromosomes of these newts whose morphology corresponded with that of the peripheral bodies. As a result, and because we were convinced, on the information thought to be then available, that nucleoli must originate from the chromosomes, we speculated that the peripheral bodies might not be nucleoli at all. Had we been aware of the observations of King (1908), and of Painter and Taylor (1942) on young oocytes of *Bufo*, the latter authors having shown that nucleolar DNA has already separated from chromosomal DNA by pachytene, we would have been less sceptical. In passing it may be mentioned that in *T. cristatus*, unlike *Xenopus* and *Bufo*, the amplification of rDNA is completed well before the onset of meiosis (Callan and Ross, unpublished).

Having failed to find the nucleolus organizer loci in *T. cristatus*, and being still unsure of the nature of the free peripheral bodies in its oocytes, I turned my attention to the axolotl *A. mexicanum* (Callan 1966). The diploid somatic nuclei of this urodele contain two nucleoli, or one only if they happen to fuse. In mitotic complements a pair of large chromosomes with submedian centric constrictions have single subterminal secondary constrictions in their short arms. These are the only secondary constrictions present, unless the cells have undergone low temperature treatment, and in the light of Dearing's (1934) study of the related *A. tigrinum* I assumed them to be the somatic nucleolus organizers; this assumption proved to be correct. During oogenesis in *A. mexicanum* the peripheral free bodies undergo a remarkable series of transformations. They start as roughly rounded solid objects attached to the nuclear membrane in pre-vitellogenic oocytes; each then enlarges, and hollows out to form a ring connecting together a few lumpy objects, which still lie at the periphery of the germinal vesicle. Similar rings develop from solid nucleoli of *T. pyrrhogaster* if preparations are made in saline of low molarity, and Miller (1964) showed that such rings are cut into fragments by DNase, though not by RNase; in the same year Kezer demonstrated that this also holds for ring nucleoli of *Plethodon cinereus* (cited in Macgregor 1965).

In *A. mexicanum* as oogenesis proceeds the rings at the periphery now stretch radially inwards towards the contracting chromosome group, increasing in contour length and becoming beaded as they do so; then the distal ends of most of the rings detach from the nuclear membrane, allowing these rings to contract and concentrate around the chromosomes. Finally all the rings revert to solid round objects, most of them closely investing the chromosome group, the rest remaining at the nuclear periphery.

These transformations are precisely matched, in accordance with the stage of oogenesis, by transformations undergone by objects attached at a single subterminal site in the short arm of lampbrush chromosome III (Fig. 3.9), whose relative size and arm length ratio corresponds with that of the mitotic chromosome which has a subterminal secondary constriction in its short arm. These observations demonstrated that the peripheral objects really are nucleoli, and that comparable objects on lampbrush chromosomes III lie at the nucleolus organizer locus.

The next point established by this study was that there is no increase in the number of free nucleoli during oogenesis, at least between the stage prior to ring formation and the stage when rings revert to solid round objects. Consequently one cannot ascribe variations in morphology at the nucleolus organizer loci on the lampbrush chromosomes to cycles of formation and detachment of nucleoli; this deduction was important, because the scale of such variation is great. Exceptionally there may be no nucleoli at all attached to the organizer locus, which can then be seen as a chromomere somewhat larger than most of the chromomeres of *A. mexicanum* oocytes at a stage when the lateral loops are fully extended (Fig. 3.4a). When there are nucleolar rings attached to the organizer locus (Fig. 3.4b, d) the number can vary from one to six, possibly even more; their great refractility and consequent phase haloes make accurate counting difficult. The attached nucleoli can be so bulky that they stretch over more than 50 μm of the chromosome axis.

Fig. 3.4 a–e. Phase contrast photographs of regions including the nucleolus organizers in unfixed lampbrush chromosomes. **a–d** Ends of the short arms of bivalent III of *Ambystoma mexicanum* from oocytes of 1.5 mm diam. **a** *Arrows* point to the nucleolus organizers; there is a minute nucleolus attached to the organizer on the left, but no trace on that to the right; the three bright objects are free nucleoli. **b** The attached nucleoli are stretched rings, more attached to the organizer on the left than to that on the right. **c** The attached nucleoli are rings, as in **b**, but on the right one a double bridge has formed. **d** The attached nucleoli are lumpy rings, about the same number on both organizers. **e** The nucleolus organizers on the short arm of bivalent VII of *Plethodon c. cinereus* from an oocyte of 1.5 mm diam; the many nucleoli attached to the extended organizers are rings, and several free-ring nucleoli are also evident. *Bar* in **a** = 20 μm; **a–d** are at the same magnification. *Bar* in **e** = 50 μm. **a–d** From Callan (1966); **e** from Kezer and Macgregor (1973)

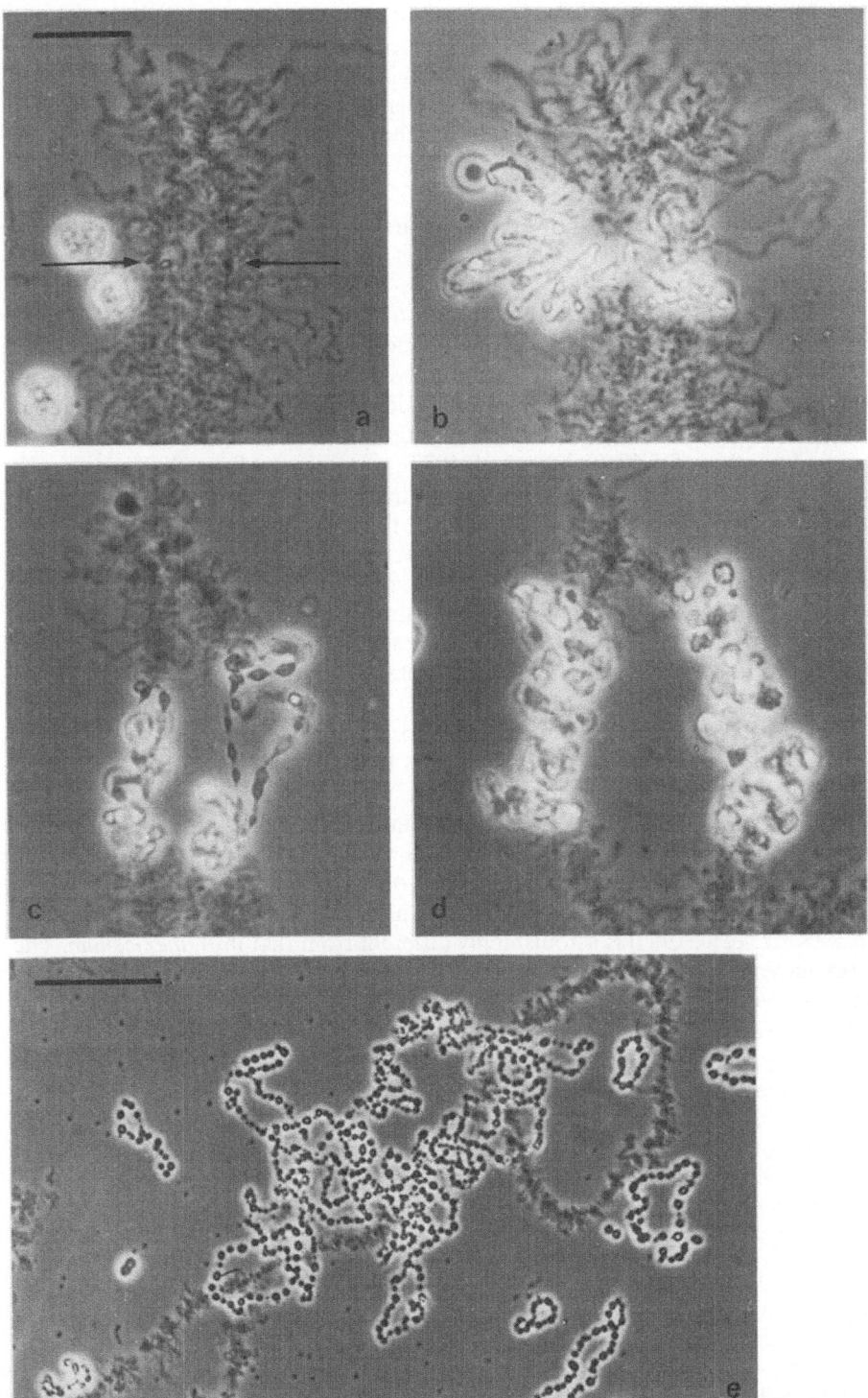

Fig. 3.4 a–e

Humphrey (1961) described a mutant of *A. mexicanum* in which the somatic nucleolus is smaller than normal, and in heterozygotes a size distinction between the two somatic nucleoli is evident. This was later shown to be reflected in degrees of rDNA sequence repetition, and in the lengths of the subterminal secondary constrictions in pre-gastrula stage mitotic chromosomes (Sinclair et al. 1974). In lampbrush bivalents III of *A. mexicanum* the nucleoli attached at the two organizer loci may be more or less equal in size (Fig. 3.4 d), or they may be grossly asymmetrical (Fig. 3.4 b); however comparison between "nucleolar" homozygotes and heterozygotes failed to reveal an association between heterozygosity and asymmetry.

That 18S + 28S rDNA is amplified by a "rolling circle" mechanism was established by Bird et al. (1973) and by Hourcade et al. (1973), but how (and when) the initial detached length of rDNA is derived from the chromosomal DNA remains unknown. On somewhat indirect evidence the rDNA that becomes amplified in *Xenopus* is thought to detach as single units (Wellauer et al. 1976) or as a small number of units (Scheer et al. 1977), certainly not from the entire length of genomic rDNA (Bird 1977). It might be tempting to suppose that the variable quantities of nucleolar material attached at the organizer locus of *A. mexicanum* in a consequence of incomplete "separation" of rDNA preparatory to amplification, varying in degree and for unknown reasons from one organizer locus to another. However one observation argues against such an assumption. It is that attached nucleoli, in their ring form, occasionally break open where their bases are connected to the chromosome axis, and thereby produce "double bridges" in the same manner as do lateral loops (Fig. 3.4 c). This indicates that the attached nucleolar rings bear the same relationship to the chromosome axis as do lateral loops, and that therefore the rRNA which they synthesize is being transcribed from the "parental" or "persistent" rDNA, i.e. that which is an integral part of the genome. Although this observation eliminates one possible explanation for the variable quantity of nucleolar material attached at the organizer locus, the source of that variability still remains unknown.

In regard to lampbrush nucleolus organizers the situation in *A. macrodactylum* (Kezer et al. 1980) is astonishingly different from that in *A. mexicanum*. The free nucleoli in *A. macrodactylum* oocytes do not transform into beaded rings during the course of oocyte development, but their morphology is nevertheless distinctive. Kezer et al. (1980) did not find a single example of a nucleolus attached at the organizer locus in any of their preparations of the lampbrush chromosomes of this species though, just as in *A. mexicanum* mitotic chromosomes, there is a subterminal secondary constriction in the right arm of chromosome III of *A. macrodactylum* which is almost certainly the site of the nucleolus organizer. This precisely parallels my own experience with *T. cristatus* lampbrush chromosomes. Similarly in a salamander whose phylogenetic relationship to ambystomids is remote, *Siren intermedia nettingi,* yet whose free oocyte nucleoli consist of beaded rings, León and Kezer (1974) found no nucleoli attached to its lampbrush chromosomes.

Macgregor (1965) made an investigation of peripheral nucleoli in *T. cristatus,* and paid particular attention to their number arguing that, other things being equal, if nucleoli periodically detach from the lampbrush chromosomes and add to those lying free, there should be more in older oocytes than in younger. In broad outline he found that the numbers remain nearly constant at about 1000 in *T.c. carnifex, T.c. karelinii* and *T.c. danubialis* oocytes throughout the lamp-brush phase, except for a minor increase thought to be brought about by the division of pre-existing free nucleoli, but that in *T.c. cristatus* there is a wide variation in the number of nucleoli from oocyte to oocyte, and a decided increase in their number as oocytes grow in size. Macgregor also demonstrated that hexaploid oocytes from triploid gynogenetic females of *A. jeffersonianum* have three times as many free nucleoli as diploid oocytes from diploid females of the same species, i.e. that there is a direct relationship between the number of chromosome complements and the number of free nucleoli per oocyte, though he found no objects attached to the lampbrush chromosomes of this urodele that resemble the free nucleoli. The study provided no conclusive answer to the original question; it proved to have been bedevilled by the complication that in *T.c. cristatus* and several other Amphibia, though not *T.c. carnifex*, many of the free nucleoli contain not one but several "cores" (the "fibrous component" of rDNA plus associated transcripts) within single "hulls" (the "granular component" of ribosome precursors). This complication was discussed by Macgregor in a later paper (1972) by which time he had shown that in *T.c. cristatus* the number of nucleolar cores, though not nucleoli as such, remains constant during the lampbrush stage.

The location of the nucleolus organizers on the lampbrush chromosomes of *T.c. carnifex* remained unknown until Mancino et al. (1972b) compared three semi-albino and some wild-type females of this newt. They failed to find any lampbrush feature that was associated with the semi-albino mutant, but in the course of this study they did find attached objects of variable size whose morphology matched that of the free nucleoli. These occurred at two loci in the haploid complement (Fig. 3.7) one subterminal in the short arm of chromosome VI, the other about halfway along the short arm of chromosome IX, though attached nucleoli were not invariably present at these loci.

The status of nucleolus organizers in *T. cristatus* is still not entirely clear. The various subspecies have different maximum numbers of nucleoli per somatic nucleus, uniformly two in *T.c. karelinii*, but from three to six according to individual in *T.c. cristatus* and *T.c. carnifex* (Callan and Morton, unpublished). Furthermore in situ hybridization of denatured ^{125}I-labelled *Xenopus* $18 + 28$S rDNA to denatured mitotic and lampbrush chromosomes of *T.c. carnifex* (Hennen et al. 1975) confirmed the presence of rDNA at the expected site on chromosome IX, but not that on chromosome VI. In a later study (Morgan et al. 1980) where ^{3}H-labelled *Xenopus* rRNA and denatured rDNA probes were hybridized in situ to denatured mitotic chromosomes, there was variation between individual newts, the most consistently labelled site being at the end of the long arm of chromosome VI (not subterminal in its short arm), with minor labelled sites about the middle of chromosome IX (as anticipated) and also elsewhere. Morgan et al. (1980) made many lampbrush preparations from several *T.c. carnifex* females for in situ hybridization and other purposes; in only two were nucleoli attached at the organizer loci on chromosome XI, and no chromosomes VI with attached nucleoli were seen. I will return to this vexing question in Chap. 5.

In *T. marmoratus* there is a single subterminal nucleolus organizer locus in the long arm of chromosome X (Nardi and Mancino 1971; Nardi et al. 1972). Although there are several secondary constrictions in the mitotic chromosomes of embryos, the only secondary constriction that is regularly evident in mitoses from adult tissues is that on chromosome X. In the lampbrush chromosomes of *T. marmoratus* this same locus is often, though not invariably, the site of attachment of an object whose morphology corresponds with that of the free nucleoli. When this site is unoccupied, the organizer can be seen as a short loopless bar in the chromosome axis, as in *A. mexicanum*. It is particularly evident, by virtue of its intense colouration, in Giemsa-stained preparations exposed to C-banding pretreatment (Batistoni et al. 1974).

The situation in *T. vulgaris meridionalis* is similar. Attached nucleoli were identified by Barsacchi et al. (1970) as being sometimes present at a subterminal site on lampbrush chromosome XI, close to a much more striking landmark, a sphere locus, in this species. In mitotic chromosomes a subterminal secondary constriction on chromosome XI is regularly present, and this has been confirmed as the nucleolus organizer locus by in situ hybridization of ^{3}H-labelled *Xenopus* 18 + 28S rRNA to denatured mitotic and lampbrush chromosomes (Nardi et al. 1977). However there are complexities, for subsidiary sites of hybridization exist in *T.v. meridionalis*, as also in *Notophthalmus viridescens* (Hutchison and Pardue 1975). These complications will be considered in Chap. 5.

In *T. alpestris apuanus* two nucleolus organizers have been identified in the haploid lampbrush complement, though attached nucleoli are not always present at these sites. One lies near the middle of chromosome VIII, in the long arm, the other is subterminal in the long arm of chromosome X (Ragghianti et al. 1972). There are secondary constrictions at corresponding sites in mitotic chromosomes VIII and X (Mancino et al. 1972a). In *T. italicus* (Mancino and Barsacchi 1969) and *Salamandra salamandra* (Mancino et al. 1969) attached nucleoli have not been observed at the organizer loci, and this also applies to *T. helveticus helveticus* (Mancino and Barsacchi 1966) for the free nucleoli in this species occur as rings, which would surely have been noticed had attached nucleoli been present.

Lacroix (1968a) has identified two nucleolus organizers at similar sites in the haploid lampbrush complements of *Pleurodeles waltlii* and *P. poireti*, one subterminal in the short arm of chromosome III, the other subterminal in the long arm of chromosome XI (Figs. 3.10 and 3.11). Both sites are in correspondence with secondary constrictions on the mitotic chromosomes, that on chromosome XI being more frequently evident than the constriction on III. Attached nucleoli are not always present on the lampbrush chromosomes, and it may be significant that they are more frequently missing from chromosome III than from XI. When present, the attached nucleoli are morphologically similar to the free nucleoli, though they are generally smaller; usually an organizer carries only a single nucleolus, but occasionally there may be several.

The nucleolus organizers on the lampbrush chromosomes of plethodontid salamanders, except for *Desmognathus fuscus* and *Ensatina escholtzii*, are the most spectacular that have thus far been encountered. Those of *Plethodon c. cinereus* were figured and mentioned by Macgregor in 1965, were further described by Macgregor in 1972, and a definitive account provided by Kezer and

Macgregor in 1973. In this salamander the free nucleoli are solid round objects in oocytes of less than 1 mm diam, but as the oocytes grow they transform, much as do those of *A. mexicanum,* first into rough, compact annular structures and later into beaded rings of contour lengths ranging from 20 to 200 μm. Beaded rings of identical appearance and size range are attached at a nucleolus organizer locus on the short arm of chromosome VII (Fig. 3.4e), but instead of this locus remaining short and compact, as in most other urodeles, and as in young oocytes of this species, where it is under 20 μm long, the organizer becomes greatly extended (to more than 300 μm in some instances), so that the points where nucleolar rings are attached to the chromosome axis are clearly visible. The chromosome axis so exposed may be continuously single or with intermittent double regions, and although it can be broken by DNase it is clearly visible in a phase microscope, and is therefore substantially wider than the interchromomeric strand in more typical regions of lampbrush chromosomes. The beaded nucleolar rings may be attached singly, in pairs or in clusters, without any evident regularity. Kezer and Macgregor (1973) found many examples of double bridges formed by a pair of beaded nucleolar strands within the lengths of the organizer, as also occurs in *A. mexicanum,* and in another plethodontid species, *Eurycea b. bislineata,* they recorded one case where the entire length of the organizer had formed a double bridge of beaded nucleolar material. The most plausible interpretation of the typically chaotic appearance of the nucleolus organizer of *P.c. cinereus* is that throughout its length it is capable of transcribing 18 + 28S rRNA, but intercalated inactive regions of compact rDNP are present and are responsible for the disorder.

Kezer and Macgregor (1973) identified organizers with attached nucleoli, in appearance substantially like those of *P.c. cinereus* and in all cases close to the centromere of one of the middle-sized chromosomes, in eight other species of plethodontid salamanders. In all nine species the organizers' overall lengths on the two chromosomes forming the bivalents were markedly diverse, except in the case of a single individual of *Eurycea b. bislineata.* If it were known that the chromosome with the longer nucleolus organizer was always the same one (i.e. identifiable by some other marker in the vicinity of the organizer), this would be strong presumptive evidence that the overall size of the organizer reflects a difference in the number of ribosomal units on the two chromosomes corresponding to that revealed by in situ hybridization to spermatocytes. No such a chromosomal identification has been made, but a later study (Macgregor et al. 1977) correlating the results of filter and in situ hybridizations of 18 + 28S rRNA and rDNA, has revealed an astonishing variation (7.5-fold) on the amount of rDNA per genome amongst individuals in a single population of *P.c. cinereus,* from 0.012% to 0.090%. They proceed to make an argument for supposing that heterozygosity for the number of ribosomal cistrons in animals which show a marked overall size difference between their lampbrush nucleolus organizers may well be of the order 15-fold, and it would perhaps be surprising if such an order of magnitude difference were not reflected in lampbrush organizer morphology.

From what has been discussed in this section it will be evident that although in some urodele species the nucleoli attached to organizer loci are useful characters for the identification of lampbrush chromosomes, this is certainly not true for all. I will return to consider another aspect of the complex distribution of 18 + 28S rDNA sequences in urodeles in Chap. 5.

Before leaving urodele Amphibia it should be mentioned that the ammoniacal silver-staining techniques developed by Goodpasture and Bloom (1975) and by Howell et al. (1975) for the detection of nucleolus organizers in mammalian mitotic chromosomes, also stain differentially, inter alia, the nucleolus organizers of newt mitotic chromosomes (Ragghianti et al. 1977, Nardi et al. 1978). However when these same techniques were applied to preparations of the lampbrush chromosomes of *T.c. carnifex* (Varley and Morgan 1978), many of the landmark loops, and some others, proved to be "silver-positive", as also the free nucleoli, but not the inactive, chromomeric nucleolus organizers.

Amongst the dozen or so anuran species whose oocytes have been studied in this regard, there is no evidence that nucleoli are attached at the organizer loci on the lampbrush chromosomes of any of them. In oocytes of *Locusta* (Kunz 1967 a) and two grasshoppers, but not such other Orthoptera as *Gryllus* and *Decticus* (Kunz 1967 b), rDNA transcription results in long "pearl-string" threads of nucleoli that remain connected to the lampbrush chromosomes. No working maps of these chromosomes have been constructed. In those animals where little RNA of any kind is synthesized by the oocyte, but where RNA is supplied from another source, the chromosomes may pass through an abbreviated lampbrush stage. However they carry few landmarks, nucleoli or other, that aid in chromosome identification (see Bier et al. 1967, Ribbert and Kunz 1969) and again no working maps have been constructed. It remains to be added that in the one plant where evidence for a lampbrush chromosome organization is indisputable, the vegetative phase of the giant unicellular alga *Acetabularia,* some of the lampbrush chromosomes are physically connected to the large nucleoli present in the single "primary nuclei" of these enormous cells (Spring et al. 1975, 1978), and see Chap. 8.

Lateral Loops as "Landmarks"

In all Amphibia the majority of the lateral loops conform to a "normal" type in which, when viewed in phase contrast, the RNP matrix appears to have a fine fibrous texture. When the nuclear sap has sufficiently dispersed in the saline of a freshly made preparation, the individual components of this matrix, provided they are resolvable, show independent Brownian motion. Each fibre appears to be anchored to loop axis at one (proximal) end but free, and free from other fibres, at the other (distal) end (Callan 1955). Electron microscopy carried out on chromosomes isolated in salines of physiological concentration shows that the fibres in reality consist of 20 to 30 nm particles strung together in chains (Malcolm and Sommerville 1974, Mott and Callan 1975, Sommerville et al. 1978 b) and that in many loops these particles are clustered together in higher order aggregations. The aggregates may be large enough for resolution as granules in the phase microscope, and the granules frequently appear to be spirally disposed around loop axis; this has been corroborated by Angelier et al. (1984) in a study involving scanning electron microscopy. However the extent of particle aggregation varies from one preparation to another, and loops of this character play no reliable part in chromosome recognition.

The distribution of matrix on loops of this "normal" type shows differences in pattern in different loops, patterns that reflect the manner in which RNA transcription, and in some cases RNP processing or shedding, occurs on different loops. Consideration of this topic is deferred until Chaps. 4 and 5; here I will only be concerned with loops of such distinctive morphologies, "landmarks", that they play a major part in chromosome recognition. It is a general feature of landmark loops that their matrix is more refractile than that of normal loops. This greater refractility is a consequence of the accretion of extra material on to the primary RNP transcripts, bonding them together in ways that lead to morphological diversity.

"Working maps", which incorporate the information from landmark loops and other features that enable lampbrush chromosomes to be identified, have been constructed for several Amphibia. Those published up to a few years ago were assembled by Callan and Lloyd (1975); a selection are shown in Figs. 3.5 to 3.11. Two conventions have generally been adopted when constructing these maps. The chromosomes are arranged and numbered in order of length, the

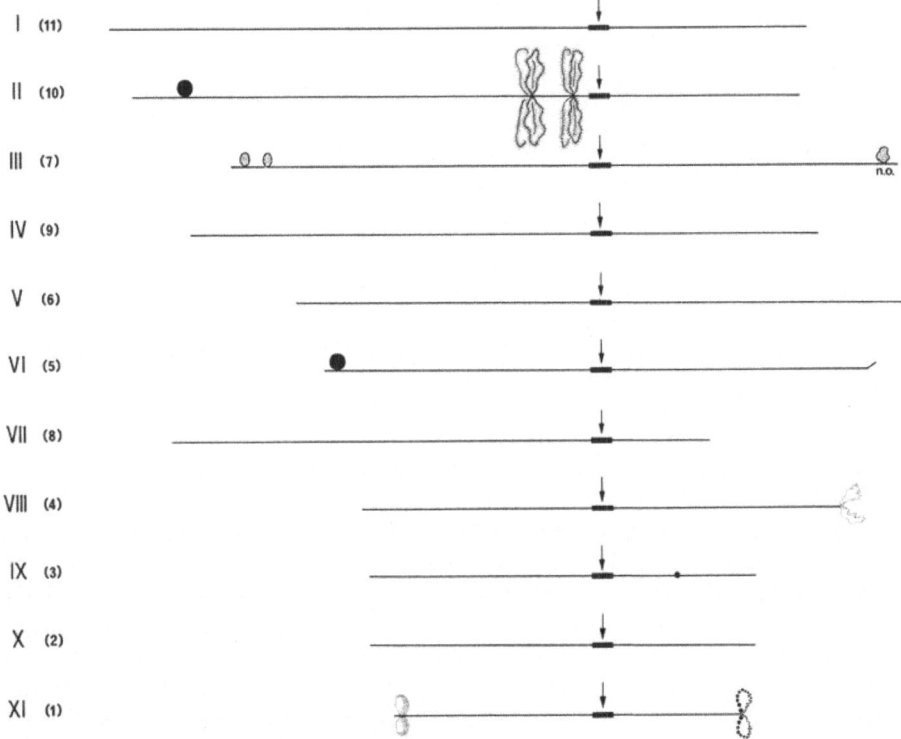

Fig. 3.5. Map of the lampbrush chromosomes of *Notophthalmus (Triturus) viridescens,* from Gall (1954), somewhat modified. Gall's original chromosome numbers are given in parentheses. Positions of the centromeres are indicated by vertically aligned *arrows.* N.O. is the site of the nucleous organizer

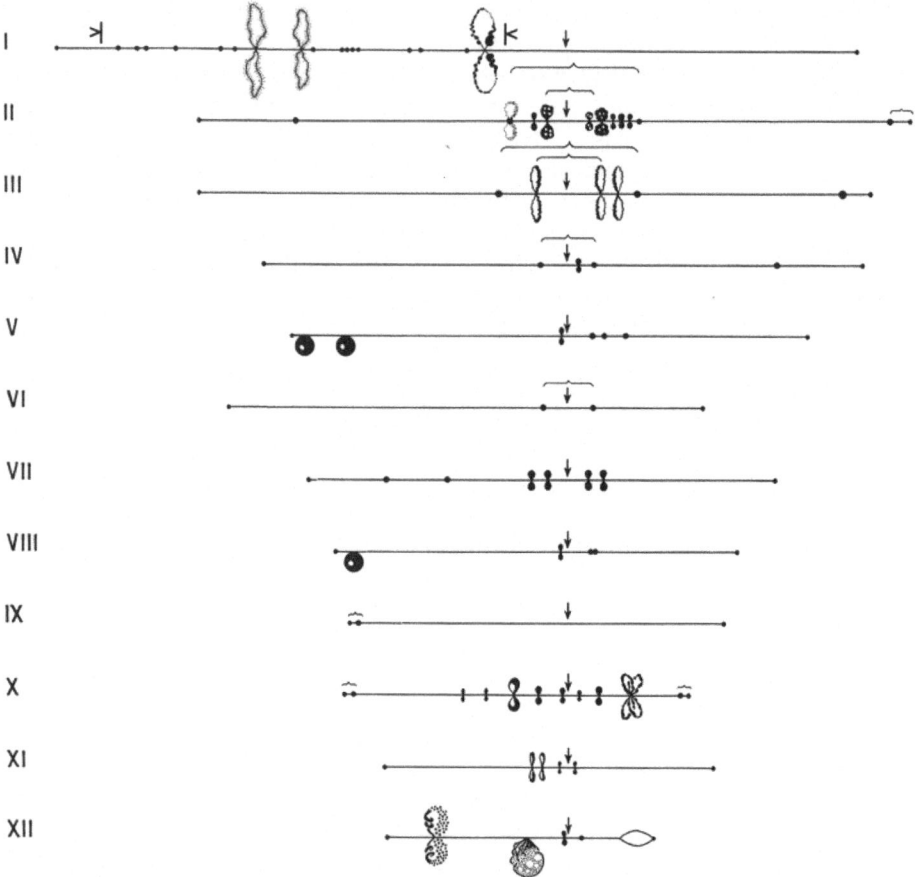

Fig. 3.6. Map of the lampbrush chromosomes of *Triturus cristatus cristatus*. Positions of the centromeres are indicated by vertically aligned *arrows*. Objects that are frequently fused to one another are linked by *brackets*. The limits of the heteromorphic region of chromosome I are marked by *horizontal arrowheads*. (From Callan and Lloyd 1960)

longest being designated chromosome I; and if the centromeres have been identified, the longer arm is designated as its left. At the outset it should be emphasized that of those species so far studied, only a few Amphibia have lampbrush complements that include many landmark loops at regularly identifiable sites. The great majority, including *Notophthalmus, Taricha, Salamandra, Ambystoma,* and plethodontid salamanders generally, have few. Another point concerning some, not all, landmark loops is that they lose their natural morphology when preparations are made in salines of low molarity. The materials that bond together the primary RNP transcripts tend to disperse, and as this occurs such landmark loops may lose, in part or completely, whatever was originally distinctive about them. So if possible, i.e. if sap dispersal permits, for chromosome identification it is preferable to work with preparations made in 100 mM KCl plus NaCl before turning to more dilute salines.

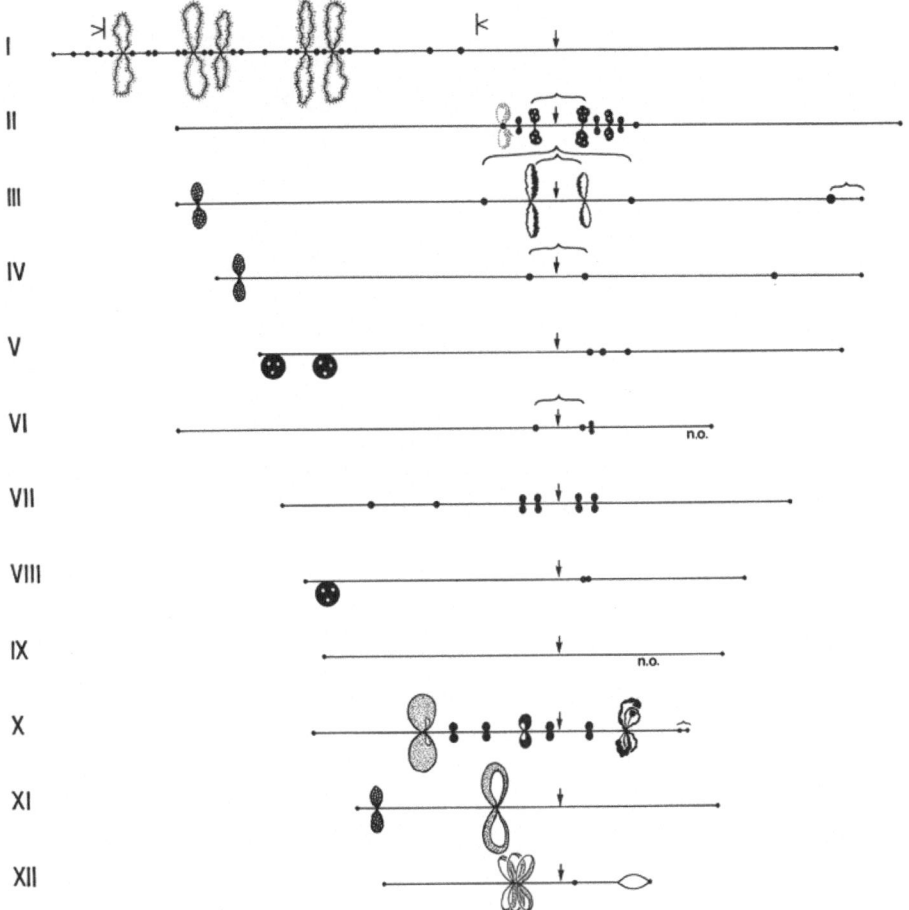

Fig. 3.7. Map of the lampbrush chromosomes of *Triturus cristatus carnifex*. Legend as for Fig. 3.6. *N.O.* are the sites of the nucleolus organizers, but they are not usually occupied by nucleoli. (From Callan and Lloyd 1960)

Amongst landmark loops, some of the most conspicuous are the giant fusing loops (GFLs) of *Triturus cristatus*, *T. marmoratus* and several other Amphibia. A typical example are the GFLs on the left arm of *T.c. cristatus* chromosome XII; they lie close to the middle of this chromosome (Figs. 1.8 b and 3.6). GFLs are invariably present on both homologues forming bivalent XII in *T.c. cristatus*, sister loops are invariably fused together, and homologous GFLs are frequently fused together too. Despite their name these objects generally do not appear in the least like loops. GFLs have sharply defined boundaries, and have compact, though often irregular, rounded outlines. GFLs are heterogeneous in texture, often include vacuoles, are relatively small in young oocytes, but may be bulky objects 30 μm or more in all dimensions in large. The underlying loop form of the GFLs is revealed in lampbrush preparations isolated in saline of low concentra-

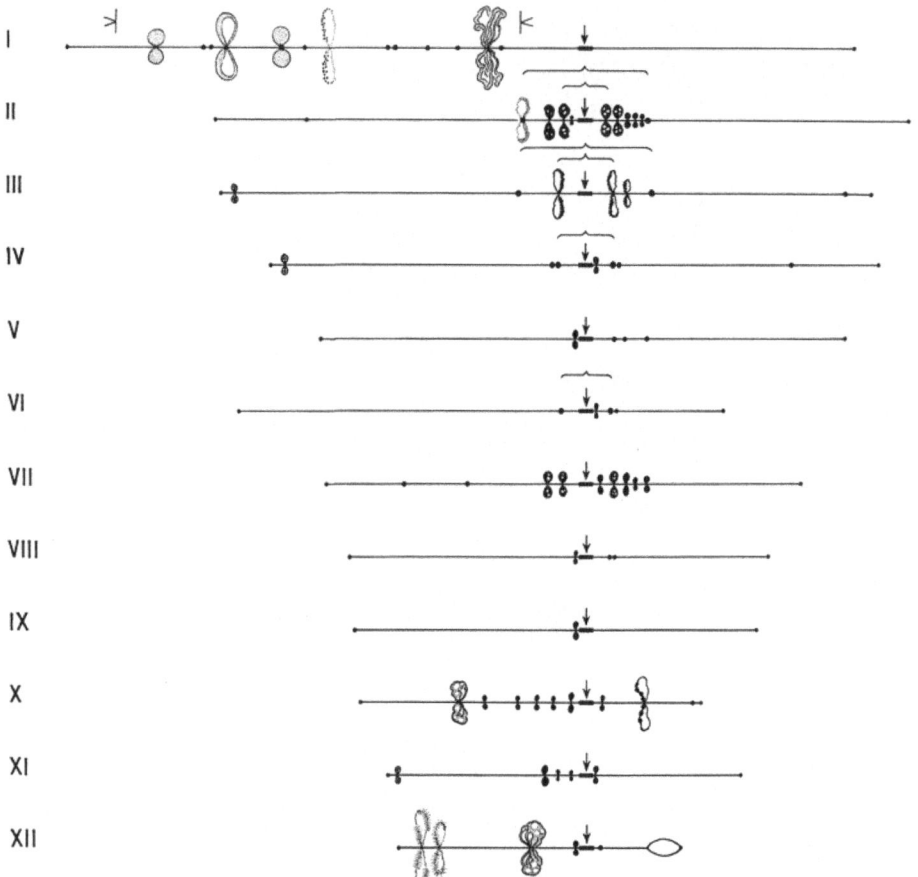

Fig. 3.8. Map of the lampbrush chromosomes of *Triturus cristatus karelinii*. Legend as for Fig. 3.6. (From Callan and Lloyd 1960)

tion. In these conditions the bonding material disintegrates, much of it disperses, and long contorted loops with more or less normal matrix texture and distribution then become visible (Fig. 1.8c; the homologous GFLs of *T. marmoratus* XII).

The bulky portions of the GFLs of *T.c. cristatus* are rigidly attached to the axis of chromosome XII, but there are also tapering conical protuberances from the main mass which are likewise attached to the chromosome axis. When two sister loops are fused together there are two such protuberances; when two homologous loop pairs are involved in the fusion, there are four. The tips of these protuberances are morphologically comparable to the thin end insertions of normal loops into chromosome axis.

As mentioned in the introductory chapter a most unusual feature of the GFLs of *T.c. cristatus* is that they frequently span a discontinuity in the chromosome axis, in other words, the bulky portion of the GFL is attached to part of

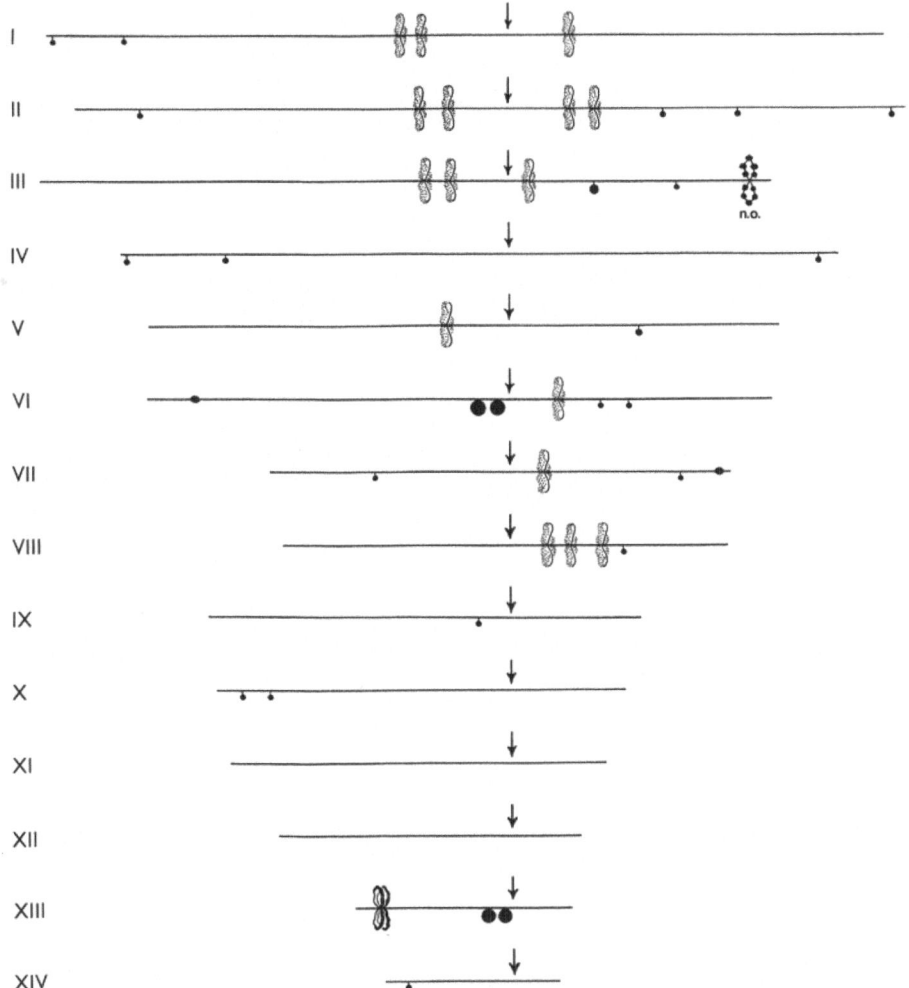

Fig. 3.9. Map of the lampbrush chromosomes of *Ambystoma mexicanum*. Positions of the centromeres are indicated by vertically aligned *arrows*. *N.O.* is the nucleolus organizer locus. (From Callan 1966)

the chromosome axis which, when followed through, includes the centromere and the shorter (right) arm, while the conical protuberances are inserted in the remaining portion of the chromosome axis which continues on to the left telomere. Double-loop bridges, produced mechanically, had previously been seen in other regions of the lampbrush chromosomes of this and other species, but the naturally occurring double bridges formed by the GFLs supplied important information when they were first observed. The mechanical production of double-loop bridges had given grounds for supposing that each lateral loop includes an axis which is part of a continuous strand running throughout a chromosome. That double-loop bridges could exist in natural circumstances in a germ-line chromosome

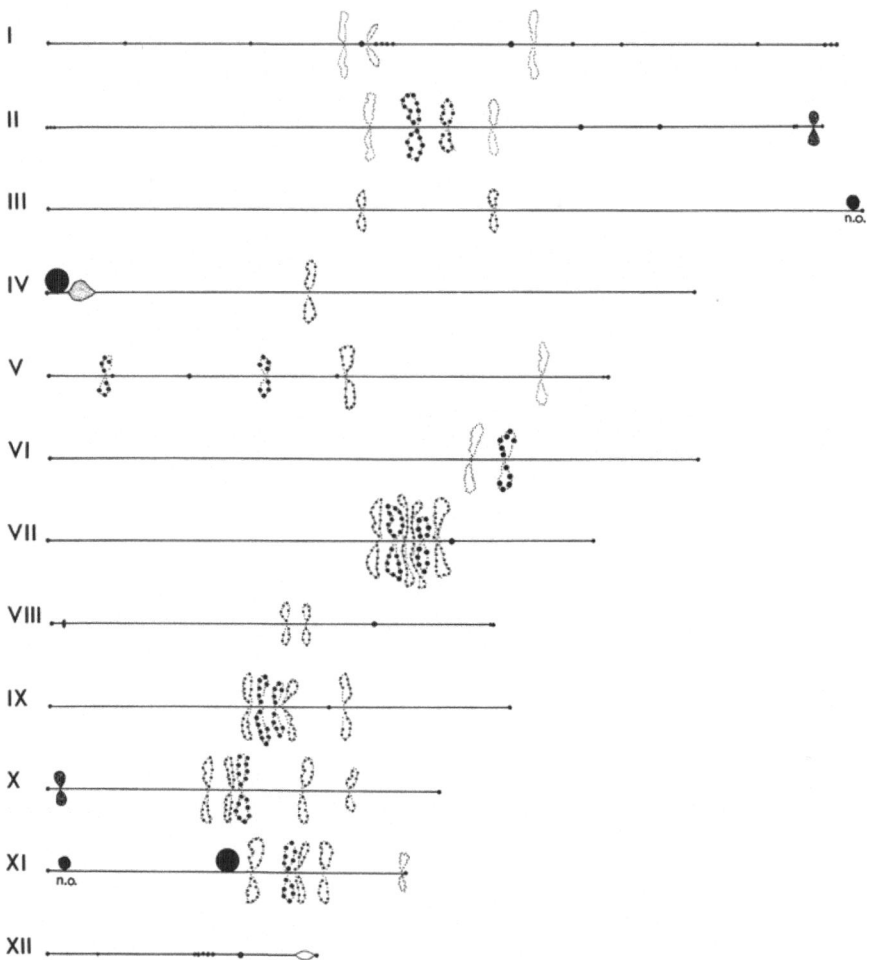

Fig. 3.10. Map of the lampbrush chromosomes of *Pleurodeles waltlii. N.O.* are the nucleolus organizer loci. (From Lacroix 1968a)

without impairing its *genetic* continuity argued against there being some kind of linker material, other than the strand itself running through each loop and compacted in a chromomere at its base, which conferred this continuity. The natural formation of double-loop bridges by the GFLs of *T.c. cristatus,* and by similar GFLs of *T.c. carnifex,* is exceptional. It presumably occurs because of the stiff texture of GFL matrix which, as it accumulates, forces the two loop insertions in the chromosome axis to part company with one another.

There may be GFLs on *T.c. carnifex* bivalent X (Fig. 3.7), but unlike the regular homozygosity for the GFLs on *T.c. cristatus* bivalent XII, individual females of *T.c. carnifex* may have GFLs on both chromosomes X (Fig. 3.12b–d), or on only one (i.e. are heterozygous for this trait) or on neither (i.e. are homozygous

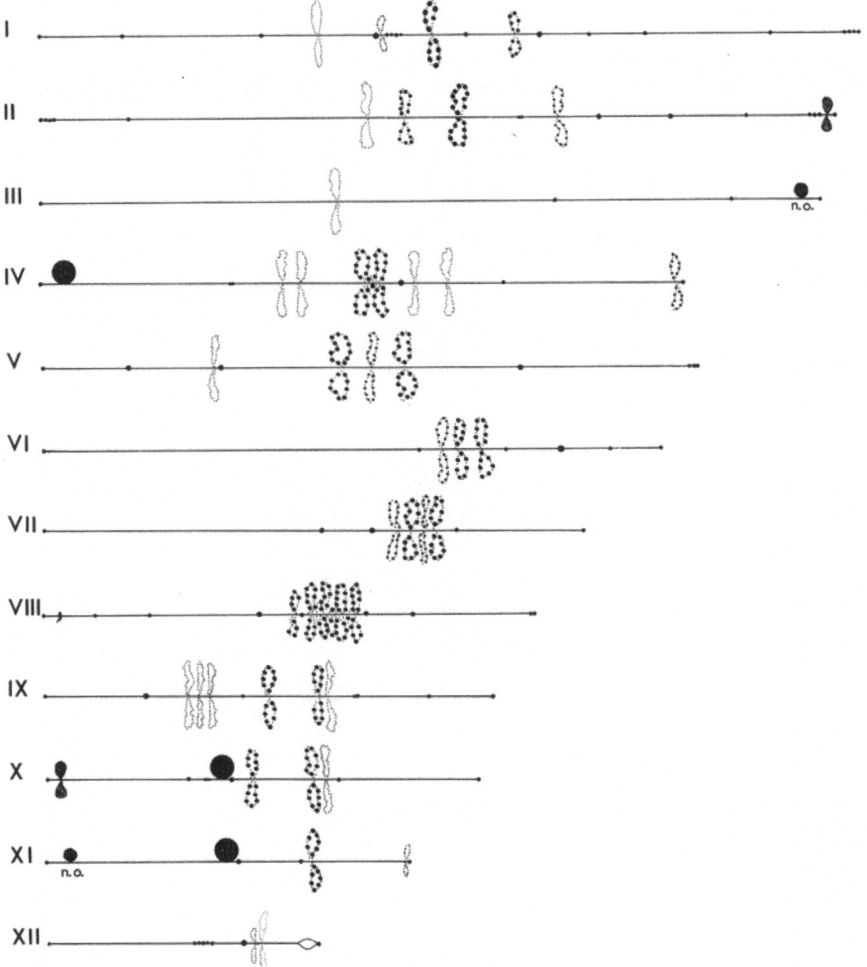

Fig. 3.11. Map of the lampbrush chromosomes of *Pleurodeles poireti*. *N.O.* are the nucleolus organizer loci. (From Lacroix 1968 a)

for its absence). This was recorded by Callan and Lloyd in 1956, and a more extensive analysis was given by Callan and Lloyd in 1960. To describe this situation in terms of presence or absence is not entirely accurate, for though in some individuals the heterozygosity is of an absolute kind, in others it is matter of relative magnitude. Moreover as some individual newts are heterozygous in having small fusing loops on one chromosome X and nothing at all comparable on the partner chromosome, there must be at least three potential alternatives in this site.

The GFLs on *T.c. carnifex* chromosome X sometimes form double bridges; when these occur the thin ends of the loops are connected to that part of the chromosome which includes the centromere and shorter (right) arm, while the bulky ends of the GFLs remain connected to the remainder of the longer (left) arm.

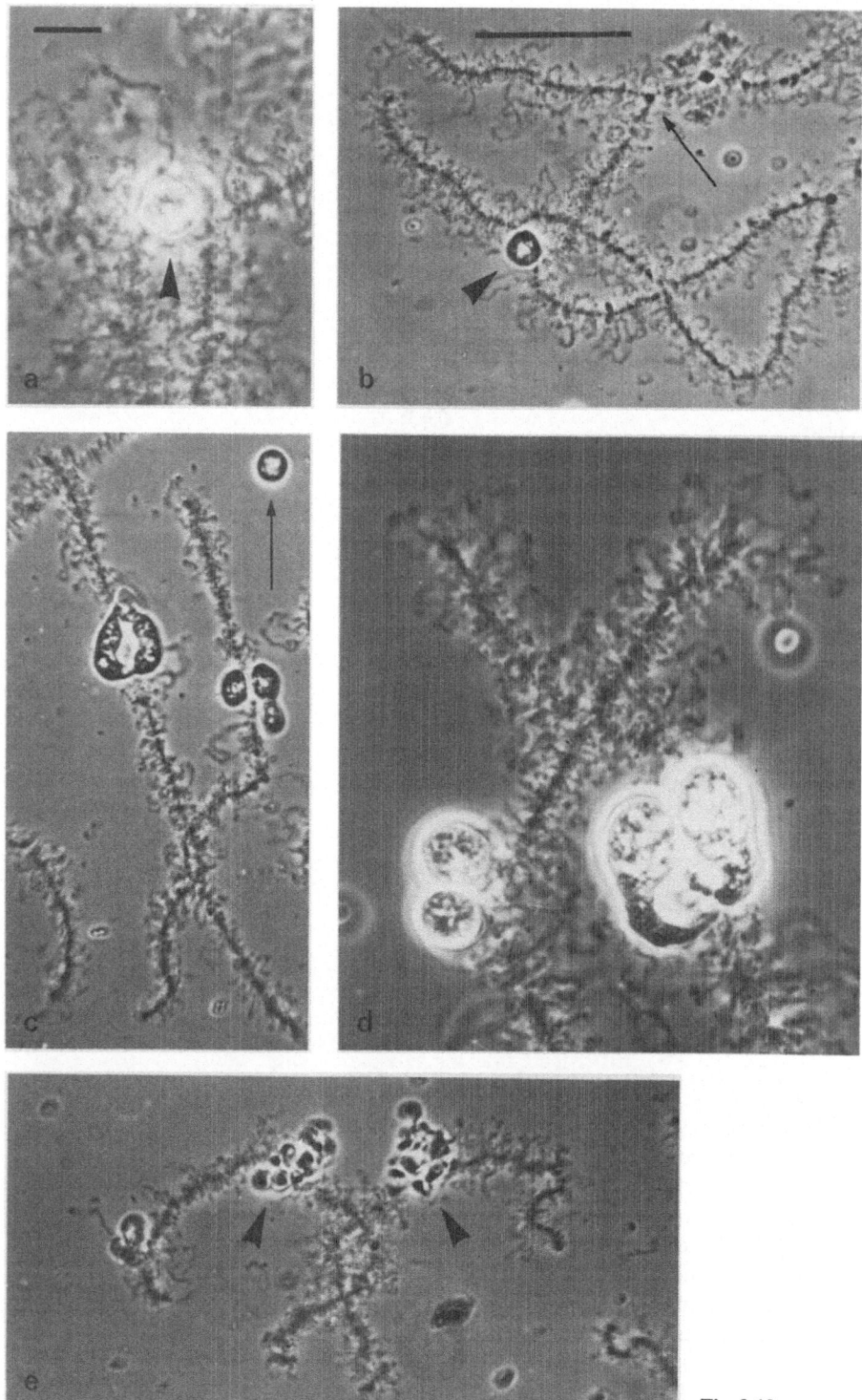

Fig. 3.12 a–e

Their polarity is therefore the reverse of that of the *T.c. cristatus* GFLs on chromosome XII. Like the latter, however, the fusions may involve homologous loops as well as sisters.

There are also GFLs on *T.c. carnifex* chromosome XI, always present on both chromosomes forming a bivalent (Fig. 3.7). In small oocytes (0.8 mm diam and less) these GFLs are almost invariably fused homologue with homologue (Fig. 3.12 a), but in older oocytes the homologues generally become disconnected from one another (Fig. 3.13 a), as also do the sister loops. There may also be GFLs about halfway along chromosome XII of *T.c. carnifex,* at a locus in the longer (left) arm that corresponds to that of the GFLs of *T.c. cristatus* (Fig. 3.7). However whereas females of *T.c. cristatus* are regularly homozygous for GFLs on bivalent XII, individual females of *T.c. carnifex* may have GFLs on both chromosomes XII (Fig. 3.12 e), or on one only (Figs. 3.13 a and 3.14 a), or on neither. Furthermore the GFLs of *T.c. carnifex* chromosomes XII, when present, are regularly multiple structures, i.e. they consist of several contiguous pairs of loops. The extent of matrix fusion within and between these loops is variable, and consequently the complex is a grossly irregular structure. Double bridges can occur at this site, and when present they establish that each component loop of the complex has the same polarity as the GFLs of *T.c. cristatus,* i.e. the thin loop insertions are always directed towards the end of the longer (left) arm (Fig. 3.14 a).

The matrix of the GFLs in all the races of *T. cristatus* is structurally complex, and it consists of two major components. One of these components is that which the GFLs share with the generality of normal loops, namely transcript RNA and the protein directly associated with this RNA, an association that leads to the compaction of RNP as strings of particles each some 20 to 30 nm diam. The other component consists of protein that condenses on the primary transcript RNP, causing the long loop axes and their associated transcripts to aggregate. The two components can be distinguished in thin section through GFLs when these are examined in the electron microscope, the transcript RNP displaying its typical 20

Fig. 3.12 a–e. Phase contrast photographs of unfixed lampbrush chromosomes of *T.c. carnifex* to show giant fusing loops (GFLs). **a** *Arrowhead* points to homologous GFLs of bivalent XI, from an oocyte of 0.6 mm diam, with sister and homologous loops all fused together. **b** Bivalent X entire from an oocyte of 0.9 mm diam; *arrowhead* points to sister and homologous GFLs all fused together; the *arrow* points to a telomere of chromosome X fused to an axial granule in the heteromorphic region of chromosome I; just to the right lie a pair of long, contorted loops with an associated axial granule that is a conspicuous landmark on chromosome I. **c** Bivalent X entire, homozygous for GFLs, from an oocyte of 1.4 mm diam; sister and homologous loops are separate, and an *arrow* points to a round mass of matrix shed from one of these loops. **d** Region at the left end of bivalent X, connected by a chiasma and homozygous for GFLs, from an oocyte of 1.0 mm diam; sister and homologous GFLs are separate; their difference in size is a consequence of recent shedding of matrix from the loops on the left. **e** Bivalent XII entire from an oocyte of 1.1 mm diam; *arrowheads* point to multiple GFLs on both chromosomes; its female parent was heterozygous for these landmarks, and analysis of several families involving its male parent showed that it too was genotypically a heterozygote; the object close to the end of the chromosome arm on the left is probably a mass of matrix detached from one of the multiple GFLs. **a** and **d** are of the same magnification, *bar* = 10 μm. **b**, **c** and **e** are of the same magnification, *bar* = 50 μm. **a, c–e** From Callan (unpublished); **b** from Callan and Lloyd (1960)

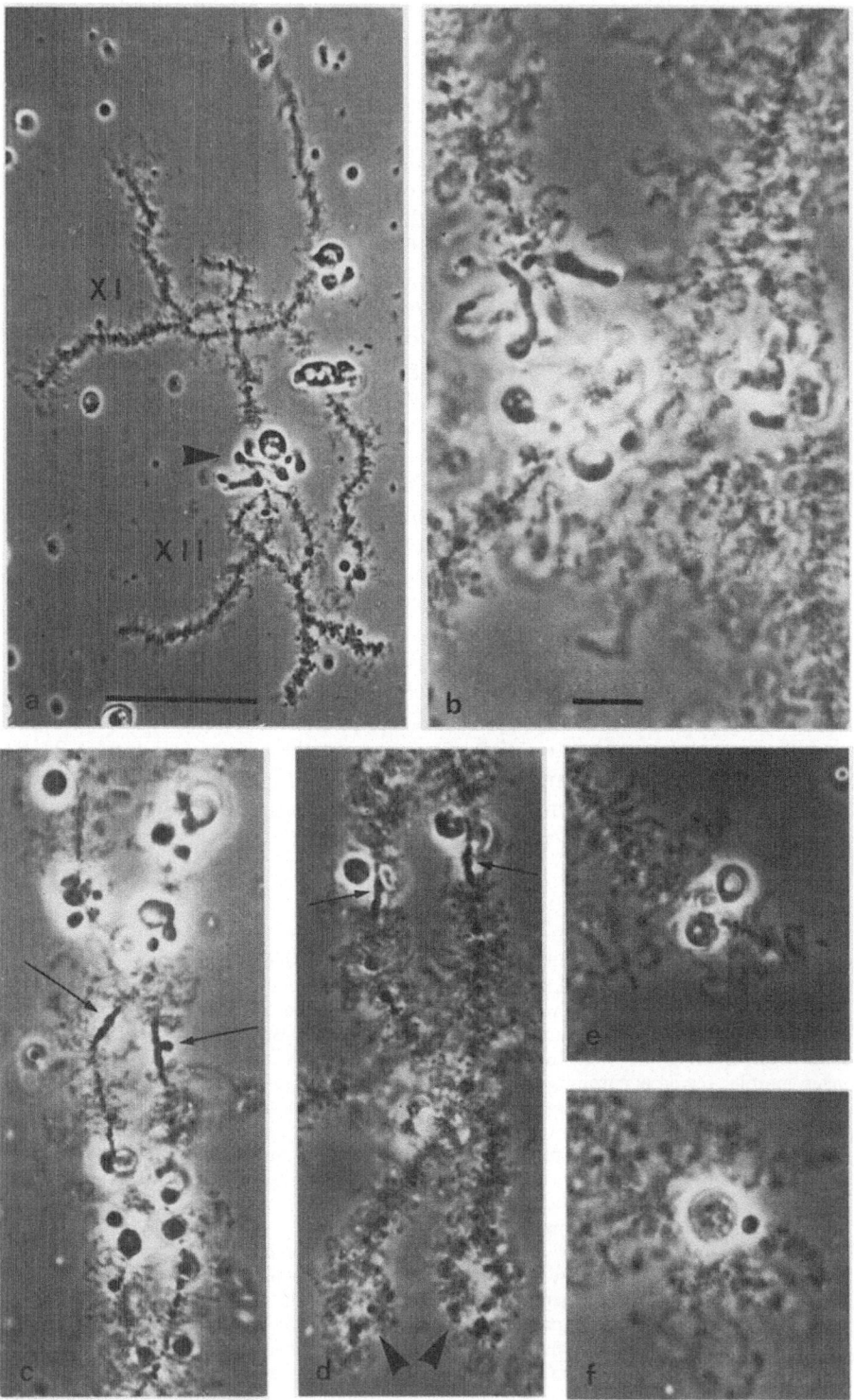

Fig. 3.13 a–f.

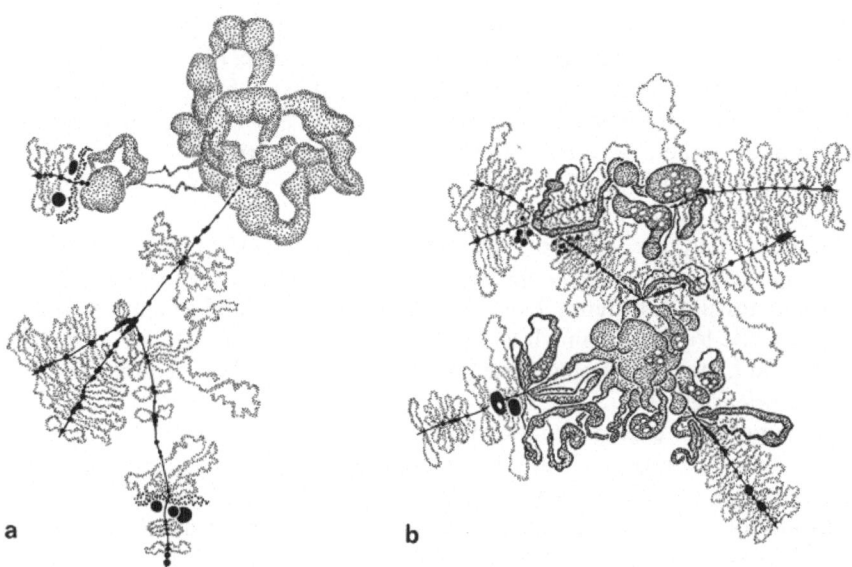

a   b

Fig. 3.14a, b. Camera lucida drawings of regions of unfixed lampbrush chromosomes of *Triturus cristatus carnifex* including giant fusing loops (GFLs). **a** Part of bivalent XII from an oocyte of 0.7 mm diam; in the region above the chiasma there are multiple GFLs, and two sister loops within the complex have formed a double bridge; there are no such loops in the corresponding region of the partner chromosome. **b** Parts of bivalent XI (*above*) and of chromosome XII from an oocyte of 1.0 mm diam; the complex and irregular mass in the lower region is the result of fusion between the GFLs of one chromosome XI and the multiple GFLs of one chromosome XII (cf. Fig. 3.13 b). The drawings were made by Mrs. L. Lloyd (× 960). (From Callan and Lloyd 1960)

◄──

Fig. 3.13a–f. Phase contrast photographs of unfixed lampbrush chromosomes of *Triturus cristatus* to show various fusing loops. **a** Bivalents XI and XII of *T.c. carnifex* from an oocyte of 1.3 mm diam; XI is homozygous for giant fusing loops (GFLs), while XII is heterozygous for multiple GFLs, marked by an *arrowhead*. **b** Parts of bivalents XI and XII of *T.c. carnifex* from an oocyte of 1.0 mm diam; the complex refractile object on the left consists of the GFLs of one chromosome XI fused with the multiple GFLs of one chromosome XII, while the GFLs of the partner chromosome lie on the right. **c** Part of bivalent VII of *T.c. karelinii* from an oocyte of 1.0 mm diam; several dense, lumpy loops (LLs) lie near the centric regions, which are marked by *arrows,* in both arms of the chromosomes. **d** Part of bivalent XII of *T.c. karelinii* from an oocyte of 1.0 mm diam; just above the centric regions, which are marked by *arrows,* are pairs of dense LLs; at the bottom are the double-axis ends of the shorter arms of these chromosomes, marked by *arrowheads,* characteristically smothered in granules. **e** End of the longer arm of chromosome XI of *T.c. carnifex* from an oocyte of 1.2 mm diam, showing a pair of small subterminal fusing loops similar in texture to GFLs. **f,** as **e,** but from an oocyte of 0.8 mm diam, in which the sister loops are fused with one another. *Bar* in **a** = 50 µm; in **b** = 10 µm; and **b–f** are of the same magnification. **a** From Callan (unpublished); **b–f** from Callan and Lloyd (1960)

to 30 nm granules, the other component appearing amorphous (Mott and Callan 1975). It is this amorphous material which causes the matrix of sister and homologous loops to fuse together, and it is significant that such fusions are not restricted to homologous loops; any of the GFLs of *T.c. carnifex* can, on occasion, fuse with one another, i.e. GFLs of chromosome X with those of XI, of X with XII, or of XI with XII (Figs. 3.13 b and 3.14 b). However the fusions are specific; they only involve GFLs with other GFLs.

Although there are GFLs on chromosomes X and XII (though not XI) of *T.c. karelinii* (Fig. 3.8), at sites homologous to those bearing GFLs in *T.c. carnifex,* they are texturally different from *T.c. carnifex* GFLs and the degree to which sister loops fuse together is much less. Fusion between homologous GFLs is infrequent, and I have only seen one non-homologous fusion between the GFLs of chromosomes X and XII.

The lampbrush chromosome complement of *T. marmoratus* also includes GFLs, two on chromosome X and one on chromosome XII (Nardi et al. 1972); heterozygosity at these loci has not been recorded. The two GFLs on *T. marmoratus* chromosome X differ in matrix texture; one shows "sequential" incorporation of ^{3}H-uridine, a circumstance that will be considered in Chap. 4. The other appears to be homologous to the single GFL site on *T.c. carnifex* chromosome X, and similarly the GFL on chromosome XII is almost certainly homologous to the GFL on *T.c. carnifex* chromosome XII; non-homologous fusions between these two GFLs are frequent in *T. marmoratus.*

I have gone into the characteristics of GFLs in some detail, not merely because they are useful landmarks for chromosome identification, but also because of a particular problem to which they draw attention. The GFLs of *T.c. carnifex* and *T.c. cristatus* are sites from which, at least in oocytes of greater than half the mature diameter, large material aggregates are periodically shed (Fig. 3.12 c and d). This shedding process is to a considerable degree responsible for the variation in gross morphology shown by GFLs. The main constituent of this shed material is protein, presumably including the same material whose accretion on these loops brings about fusion. GFLs are rapidly digested, and their fusions disrupted, by proteolytic enzymes; however they are relatively insensitive to ribonuclease digestion in conditions when the RNP transcripts of normal loops are rapidly stripped from loop axes (Macgregor and Callan 1962). GFLs are also morphologically insensitive to the metabolic disturbance produced by actinomycin D. When oocytes of *T.c. cristatus* are incubated for 1 or 2 h in saline containing a concentration of actinomycin-D sufficient to inhibit RNA synthesis on all the lateral loops, GFLs included, the normal loops shed their transcript RNP and their axes retract to the main chromosome axes, but the GFLs on chromosome XII show scarcely any morphological alteration (Snow and Callan 1969). Both these observations suggest that material other than primary transcript RNP determines the gross form of the GFLs, though the site-specific accumulation of this material is presumably determined by the nature of the primary transcript.

Macgregor (1963) has shown that the size of the GFLs of *T.c. carnifex* reflects the physiological state of female newts. Particularly large GFLs and unusually fluid nuclear sap characterize newts with a low gonadotropin level, and when such newts are injected with gonadotropin there is a dramatic reduction in size of the GFLs, coupled with an increased rate of ^{32}P incorporation and stiffening of the nuclear sap. Now the packing density of RNP fibrils on transcriptional units is positively correlated with rate of RNA synthesis both in nucleoli (Scheer et al. 1976b, c) and in lampbrush chromosomes (Franke and Scheer 1978) so if the size of a GFL were primarily determined by the number of RNP fibrils it contained one might anticipate finding larger GFLs in gonadotropin-stimulated newts. In fact Macgregor found precisely the opposite. Thus again we may reason that a secondary accretion of protein on to transcript RNP is the prime determiner of GFL size and form. Evidence from other landmark loops points in a similar direction.

Studies on lampbrush chromosomes by the Miller-spreading technique at very low ionic strength have led to a widely accepted view that the morphologies of "normal" lateral loops primarily reflect the sites of origin and termination of RNA transcription, the rate of transcription, and the processing of transcripts. Up to a point this view is certainly valid, but it is not the whole picture. As mentioned earlier, all the loops in certain regions of the lampbrush chromosomes of *T.c. carnifex* may in some preparations appear to have a "normal", i.e. fine fibrous, matrix when viewed in the phase microscope, but the matrix of these same loops in other preparations can instead be decidedly granular. It may be that this higher order aggregation of the 20 to 30 nm RNP transcript granules into larger objects visible in the light microscope is also mediated by secondary accretion of proteins; I will return to this question in Chap. 9.

There are a variety of smaller landmark loops which share with GFLs the characteristic of compaction by matrix fusion. One class ranges in size from about 10 μm downwards. For want of a better term I will use the inelegant adjective "lumpy" to designate these loops (LLs), because it is a fair description of their morphology. The degree of matrix fusion in LLs is variable, but is often so extreme that all suggestion of loop form is obliterated. When sister loops are not fused together a pair of lumpy objects occupy a given site; when sisters are fused together, only a single object is present.

In all the races of *T. cristatus,* and in *T. marmoratus* as well, the distribution of LLs serves at once to identify bivalents II and VII; they are, to a first approximation, symmetrically disposed about the centromeres of these two bivalents (Figs. 1.11 a, 3.6, 3.7, 3.8, 3.13 c). Interhomologue fusion is frequent between LLs, as also is "reflected" fusion between non-homologous LLs lying on either side of the centromere, which causes the chromosome axis including the centromere to bend back on itself. LLs on chromosomes II and VII occasionally fuse with one another, but just as in the case of GFLs there is a specificity to this non-homologous association; LLs of chromosomes II and VII only fuse with other LLs, not with GFLs or other types of loop, nor for that matter with the LLs on bivalent X of *T. cristatus,* though morphologically these are similar objects.

LLs are also particularly useful for the identification of the lampbrush chromosomes of *Triturus alpestris apuanus* (Mancino and Barsacchi 1965, Mancino et al. 1972a, Ragghianti et al. 1972). In the first of these papers mention is however made of the potentially confusing circumstance that in some preparations a given LL may manifest the dense compact structure typical of this class of object, whereas in other preparations the same loops may be more diffuse, and in the extreme condition may assume the morphology of normal loops.

Much the same variability applies to several of the LLs of *T. cristatus* chromosome X, furthermore in this species it is a "phenotypic" variability that overlays genuine genetic diversity at the LL loci, there being regular heterozygosities for size of homologous LLs in individual animals. Thus although the general distribution of the LLs of *T. cristatus* chromosome X is an immediate recognition character for this chromosome, the distribution in fine detail may not be reliable when it comes to comparing specific regions of this chromosome from one individual to another. The degrees of matrix fusion displayed by the six major LLs of *T. cristatus* chromosome X show site-specific features (Figs. 3.6, 3.7, 3.8). Thus the LLs which lie about halfway along the short arm of this chromosome, although they carry matrix as dense as that of the five other LLs, always retain the general form of loops in which thin and thick axial insertions are recognizable, despite irregularities in the width of matrix between insertions. Moreover although the LLs of *T. cristatus* chromosome X show most of the features typical of loops in which accretion of protein on to the primary RNP transcripts results in total or partial obliteration of loop form, producing structures that have sharply defined boundaries, fusion between non-homologous LLs on this chromosome, or to LLs on other chromosomes, does not occur; this suggests that there is diversity amongst them in regard to the material that aggregates on each.

In *Pleurodeles waltlii* from Spain (Lacroix 1968a) there is a landmark situated subterminally on the left arm of chromosome IV, "structure M", larger in younger than in older oocytes, that resembles in several ways an LL of *T. cristatus* (Fig. 3.10). This landmark, like the GFL of *T.c. carnifex* chromosome X, may in some individuals be present on both partners, and fusion can occur between them, or on one only, or on neither. It is absent from the related species *P. poireti* (Fig. 3.11). This, and another nearby landmark (a sphere) that differentiates Spanish from Moroccan *Pleurodeles,* was used by Lacroix and Capuron (1966) to identify oocytes that developed from primordial germ cells grafted from Spanish to Moroccan *Pleurodeles* embryos at the tail bud stage.

The phenotypic variability of loops with fusing matrix is also well exemplified by the objects located subterminally in the longer (left) arms of chromosomes III, IV and XI of *T.c. carnifex* and *T.c. karelinii* (Figs. 3.7 and 3.8). These objects appear more heterogeneous in texture than the LLs of chromosomes II and VII, indeed their matrix resembles that of GFLs, though interhomologue fusion does not occur. They are regularly present, and are valuable recognition characters for these three chromosomes, in small oocytes of 0.6 to 0.8 mm diam (Fig. 3.13 f), and in larger oocytes of 1.1 mm diam and upwards (Fig. 3.13 e). However they are inconspicuous, or absent altogether, from these three chromosomes in oocytes of a middle-size range, being replaced by loops of normal morphology; this presumably reflects the absence of secondary protein accumulation upon them at this developmental stage. When these subterminal landmark loops are evident, small unattached masses of matrix of similar texture are often present in their vicinity, indicative of a shedding process comparable to that of GFLs. Landmark loops that are similar in appearance and variability of phenotypic expression to those just considered are also present subterminally on two of the lampbrush chromosomes of *Pleurodeles* (Figs. 3.10 and 3.11) (Lacroix 1968 a, «boucles de type D»).

Macgregor (1963) has shown that the subterminal landmark loops of *T.c. carnifex* chromosomes III, IV and XI do not express their characteristic phenotype in gonadotropin-stimulated females, but in hypophysectomized females they are large and dense, even in oocytes of the intermediate size range where they are not so expressed in untreated newts. Thus, just like GFLs, protein accumulation (or retention) on these loops is inversely correlated with hormonal state, and therefore with the overall rate of RNA transcription.

In *Pleurodeles waltlii* and *P. poireti* by far the most striking landmarks are beaded loops, termed «boucles globulaires» by Lacroix (1968 a), of which examples are present at distinctive sites on several of the chromosomes of both species (Figs. 3.10 and 3.11). They differ from the giant granular loops of *T.c. cristatus* (see later) in that the loop matrix is compacted into irregular, or spherical dense masses which intermittently surround the axis of each loop throughout its length. At a few sites these loops occur as single pairs interspersed with normal loops, but more frequently they are massed together in groups where they entirely obscure the chromosome axis.

Beaded loops like those of *Pleurodeles* are present on two of the lampbrush chromosomes of *Taricha granulosa* and *T. torosa* (Dr. J. Kezer, personal communication; Fig. 3.17 d, e). Some individuals of *T. torosa* have bunches of beaded loops of similar bulk on one bivalent, whereas in other individuals this bivalent is regularly heterozygous, one bunch being smaller than its homologue, or may be altogether lacking.

In *P. poireti* from Algeria Lacroix (1970) found that all females were heterozygous for the length of a region near the middle of bivalent IV; in the chromosome where this region is longer there is a group of these beaded loops that are absent from the shorter region of its partner (Fig. 3.15 a). This suggested that in *Pleurodeles* the female sex is heterogametic (ZW), and Lacroix proved this to be so by sex-reversing genetic males with oestradiol treatment of embryos. In these sex-reversed animals (ZZ) bivalent IV was symmetrical, the region carrying beaded loops being absent (Fig. 3.15 b). When these sex-reversed animals were crossed with normal males, all the offspring were male. As will be discussed in the following paragraphs of this chapter, although at first sight females of *T. cristatus* also appeared to be heterogametic, in fact it is the male sex that is heterogametic in this species.

In the lampbrush complements of *T. cristatus* (Callan and Lloyd 1960) and *T. marmoratus* (Nardi et al. 1972) there is a long intercalary region in the longer arm of chromosome I where none of the sites of landmark loops match with those on the partner chromosome; chiasmata are not formed within these "heteromorphic" regions (Figs. 3.6, 3.7, 3.8). In *T. marmoratus* the achiasmate region extends beyond the centromere, into the short arm (Sims et al. 1984). I will digress at this point to consider a general question concerning these most unusual heteromorphic regions, which are not present in other European triturid newts. In 1960 Callan and Lloyd tentatively assumed that they might be sex-determining, and that in *T. cristatus* the female is heterogametic. Interracial male F_1 hybrids within

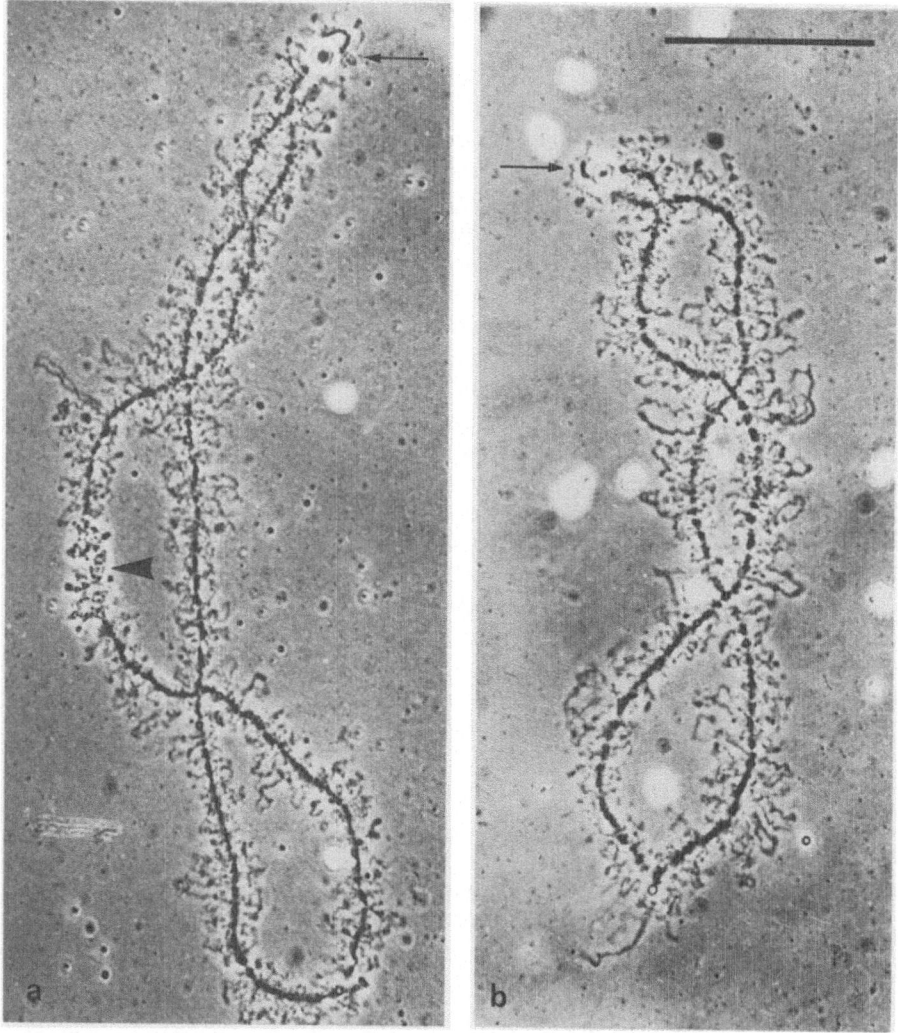

Fig. 3.15 a, b. Lampbrush bivalents IV of *Pleurodeles poireti*. **a** From a normal female, heterozygous for a chromosome region that includes a complex of beaded loops, marked by an *arrowhead*. **b** From a sex-reversed genetic male, in which the beaded loops are absent. *Arrows* point to subterminal, fused spheres, distinctive of chromosome IV. *Bar* = 50 μm. (From Lacroix 1970)

the *T. cristatus* group were known to be much less fertile (in fact nearly sterile) than their sister F_1 hybrids (Callan and Spurway 1951), so if the female sex in *T. cristatus* were indeed heterogametic this would have transgressed "Haldane's (1922) rule". The assumption has proved to be wrong, and Haldane's rule upheld.

Morgan (1978) showed that chiasmata are not formed within the heteromorphic regions in male as well as in female meiosis. Giemsa C-banding of mitotic chromosomes (Schmid et al. 1979) has demonstrated that the male is heterogametic in European triturid newts generally, including *T. cristatus* (Sims et al. 1984). Macgregor and Horner (1980), also studying Giemsa C-banded mitotic chromosomes, have shown that the banding patterns of the pair of chromosomes I of *T.c. carnifex* are regularly diverse in both males and females, i.e. both sexes are heterozygous in this regard, and indeed that they must be so if they are to survive. There is a 50% mortality among embryos, and those that die are homomorphic for one or the other chromosome I. Macgregor et al. (1983) have found that the same is true of *T. marmoratus,* and likewise of F_1 hybrids between *T. cristatus* and *T. marmoratus*. This extraordinary situation, which has been explored in detail by Sims et al. (1984), is thought to have originated something like 20 million years ago, in an ancestor common to *T. cristatus* and *T. marmoratus,* and it is remarkable that despite the genetically-determined wastage of 50% of their fertilized eggs these species are still abundant today!

In *T.c. cristatus* and *T.c. carnifex* most of the landmark loops in the heteromorphic regions of chromosomes I are exceptionally long (Fig. 4.5 a), and some of them are contorted and covered with dense matrix of uniform width (Fig. 3.12 b). The density of this matrix is variable, and appears to depend on a newt's physiological state. At all events an immediately identifiable pair of loops in all preparations from a batch of oocytes from one particular *T.c. carnifex* female were of exceptional density, but in her progeny they were not exceptional, nor were they in preparations from another batch of oocytes taken later from the same female (Callan 1963).

In the heteromorphic region of chromosome I of *T.c. karelinii* the conspicuous landmark loops on both partner chromosomes, although they do not occupy corresponding sites, are uniform as regards matrix texture and different from that of the landmarks on chromosome I of *T.c. cristatus* and *T.c. carnifex*. Their matrix consists of coherent masses of fine granular material, which in outline is irregular, though usually not to the extent that loop form is obliterated (Fig. 3.16).

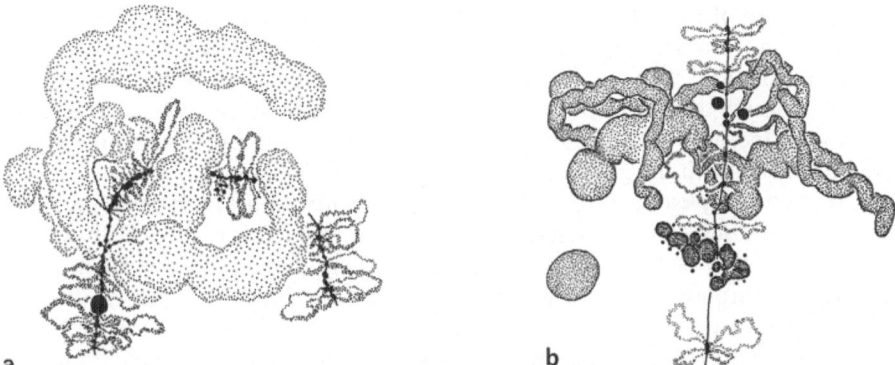

a   b

Fig. 3.16a, b. Camera lucida drawings of parts of the heteromorphic regions of chromosomes I of *Triturus cristatus karelinii* including loops with fine granular matrix that are distinctive of this subspecies. **a** From an oocyte of 1.2 mm diam; masses of matrix, about to be shed, project from such loops. **b** Also from an oocyte of 1.2 mm diam; the outlines of the multiple loops are more easily recognizable than in **a**; the mass of fine granular matrix lying free had recently been shed from these loops, and another such mass was presumably about to be shed. The drawings were made by Mrs. L. Lloyd (× 960). (From Callan and Lloyd 1960)

Lumps of this matrix are shed from these loops, and remain for a time recognizable loose in the nuclear sap. They are another indisputable example, visible by light microscopy, of the shedding of material from identifiable loops.

The uniformity of matrix texture of several loops on one pair of chromosomes (no other chromosomes of *T.c. karelinii* have loops at all similar to these on chromosome I) is striking and remarkable; it invites comparison with the situation in *Pleurodeles* where many landmark loops are of the beaded or «boucle globulaire» type, though here the uniformity of matrix texture extends to loops on several different chromosomes. If similarity in landmark loop matrix texture implies similarity in DNA sequence organization that is transferred to RNA transcripts, which on this account selectively bind specific nuclear protein(s), there should be similar DNA sequences at many chromosomal sites on urodele Amphibia. We will meet with several other lines of evidence (see Chap. 5) which point in this direction. I know of only one other lateral loop whose matrix texture resembles that of the landmark loops of *T.c. karelinii* chromosomes I. This is the sequentially-labelling subterminal landmark loop in the long arm of the smallest chromosome of *N. viridescens* (Fig. 3.5), to which reference will be made in Chap. 4.

As a final example of a conspicuous landmark loop I turn to consider the giant granular loops (GGLs), situated subterminally in the long arm of chromosome XII of *T.c. cristatus* (Figs. 3.6 and 3.18a). They occur in this subspecies only. Each GGL starts as a thin strand projecting from a chromomere, the strand leading into the tip of a dense conical structure. As the cone widens it shows signs of a spiral organization of dense material, which soon gives way to the granular zone, about 10 μm wide, distinctive of the remainder of the loop. Near the dense conical zone of the GGL the granules are larger, about 1 μm diam, but further round, as the loop returns to the chromosome axis, they grade progressively smaller. The GGLs are usually some 60 μm long. They always occur in pairs, the sisters being of similar length; this was one of the first structures to disclose the paired nature of lateral loops generally. Neither sister nor homologous GGLs fuse with one another. The matrix of the GGLs is coherent at the time of chromosome isolation; its constituents do not show independent Brownian motion initially, though they come to do so in the course of time as the texture of the granular zone loosens. While this happens a loop axis becomes plainly visible in the granular zone, with the granules attached on stalks to this axis.

If the freshly isolated contents of a *T.c. cristatus* oocyte nucleus are kept under observation while the nuclear sap disperses, small masses of GGL-type matrix can generally be seen lying free, alongside the granular zone just where it returns to the chromosome axis. While sap dispersal continues these masses dissociate, but their texture and spatial relationship to the GGLs prior to sap dispersal is a clear indication that they were formed on, and have been shed from, the GGLs. I have already stated that certain other landmark loops shed other objects, objects that are amply large enough to be seen in a phase microscope, into the nuclear sap; however because of the characteristic texture of the GGL's matrix this example of shedding is particularly convincing.

Every female of *T.c. cristatus* which I have examined (at least 100), and this includes specimens from France, Germany and Russia as well as from Britain, has always possessed GGLs on both the homologues forming bivalent XII. A single female was exceptional in being heterozygous for a size difference between the GGLs, those on one chromosome XII being of normal size, those on the homologue being only a quarter as long, in all her oocytes (Fig. 3.18 b).

This review of landmark loops is not exhaustive, but it would be tedious to describe other examples. It is difficult to decide what functional significance to attribute to landmark loops on amphibian lampbrush chromosomes. Their value for chromosome identification is undeniable, and not merely amongst urodeles. They have been used to identify the lampbrush chromosomes of *Rana esculenta* by Giorgi and Galleni (1972), and in a study of two closely allied Japanese frogs, *R. nigromaculata* and *R. brevipoda,* Ohtani (1975) was able to identify all 13 chromosomes of both species and to determine the source of origin of homologous chromosomes in the bivalents of F_1 hybrid females. The analysis has been extended to backcross hybrids between these Japanese frogs by Nishioka et al. (1980). The parental species were shown to differ in the electrophoretic patterns of seven polymorphic proteins, and by comparing the patterns of these proteins, extracted from backcross hybrids, with the lampbrush constitutions of the animals from which the proteins had been extracted, they were able to assign the genes encoding these proteins to individual chromosomes. Moreover by recognizing recombinant chromosomes derived from the F_1 hybrid female parent in the backcrosses, to a limited extent the genes involved could be assigned to particular chromosome regions.

However the question posed at the beginning of the previous paragraph remains unanswered. The GGLs are amongst the most conspicuous landmark loops of *T.c. cristatus* lampbrush chromosomes. Evidence that they are heritable will be given later in this chapter. Yet the RNA transcribed on the GGLs, and/or the protein(s) which associate with this RNA, can at most play only a trivial rôle in oogenesis of *T.c. cristatus,* or in morphogenesis following fertilization. GGLs are absent from the other subspecies, but their eggs develop into adults whose phenotypic features differ little from those of *T.c. cristatus,* and it has not been possible to correlate any phenotypic features with the presence or absence of GGLs in backcross subspecies hybrids.

Another paradox is presented by the accumulation of materials, for the most part proteins, on landmark loops such as the GGLs, GFLs and even more strikingly by the structures called "spheres", to be described in the next section, only for these materials to be returned to the nuclear sap from whence they came. This peculiar phenomenon awaits interpretation.

Spheres

It is appropriate to begin a consideration of these particularly useful recognition characters with the original observations of Gall (1952, 1954) on *Notophthalmus* (*Triturus*) *viridescens.* Gall called these objects "knobs", but comparable objects were called "spheres" by Callan and Lloyd (1960); because this latter term has been adopted by most later authors it will be used here. Gall originally numbered the *N. viridescens* chromosome complement from 1, the smallest, to 11, the largest. However because it has become conventional to number the chromosomes of karyotypes in the opposite order, with chromosome I the largest, that convention will be adopted here. With Dr. J.G. Gall's permission a reversed working map of *N. viridescens* lampbrush chromosomes was published by Gould et al. (1976) and this map is shown in Fig. 3.5.

In *N. viridescens* there are two sphere loci, one subterminal in the left arm of chromosome II (Gall's 10), the other closer still to the left end of chromosome VI (Gall's 5). As their name implies, these objects are spherical or nearly so, 7 to 10 µm in diam, and they are attached directly to the chromosome axis. In some oocytes there may be small hemispherical masses of material closely applied to the surfaces of the spheres, but otherwise these structures appear homogeneous in the light microscope. Objects identical in appearance to the spheres attached to chromosomes II and VI are also present free in the nuclear sap.

Fusions between homologous spheres are frequent, the pair of chromosomes involved being held together by their connections to a single "shared" sphere. Gall counted 22 fusions in 29 examples of bivalent II, and 22 fusions in 30 examples of bivalent VI. He found no evidence for correlation between the fused or not fused condition on the two bivalents within single nuclei. Gall found the spheres to be Feulgen-negative, and to stain clear blue (not purple) with toluidine blue at pH 6 whether or not the preparations had been previously digested with RNase, so on this evidence they lack RNA as well as DNA.

In his 1954 paper Gall also described spheres, comparable to those of *N. viridescens*, close to the centromeres of three lampbrush chromosomes of *Ambystoma tigrinum*. In *A. mexicanum* (Callan 1966) there are four sphere loci in the haploid complement, two close together and just to the left of the centromere of chromosome VI, and another two, also close together, just to the left of the centromere of chromosome XIII (Fig. 3.9). In *A. mexicanum* homologous sphere fusions are common (Figs. 3.1 b and 3.3 a); free spheres are generally also present in the nuclear sap, just as they are in *N. viridescens*.

In *A. macrodactylum* (Kezer et al. 1980) there are five sphere loci in the haploid complement. One of these lies in the short arm of chromosome III; there are also two in tandem just to the right of the centromere of chromosome V (Fig. 3.3 b), and a further two in tandem just to the left of the centromere of chromosome XIII. The tandem loci almost precisely equate with comparable tandem loci in *A. mexicanum*. Two other species of *Ambystoma*, *A. gracile* and a triploid member of the *A. jeffersonianum* complex, *A. tremblayi*, have sphere loci distributed in similar fashion on three lampbrush chromosomes. In *A. macrodactylum* free spheres were present in all the preparations examined by Kezer et al. (1980), of about the same size as those attached to the chromosomes, and homologous sphere fusions were frequent. Kezer et al. (1980) examined EM sections through the spheres of *A. macrodactylum* and found them to consist of an inner, spherical, electron-dense core surrounded by a less dense shell or cortex (Fig. 3.17c). Sections through spheres of the newt *Taricha granulosa* proved to be similar.

There are also sphere loci in *Triturus cristatus*, two situated subterminally in the left arm of chromosome V, and one subterminal in the left arm of chromosome VIII (Figs. 3.6 and 3.7). In oocytes of *T.c. carnifex* of less than 0.8 mm diam the spheres only reach a diameter of some 4 µm, but in large oocytes they may reach 15 μm. Small spheres are homogeneous, larger examples generally contain single vacuoles, while extremely large spheres may include a single dense round object within the vacuole. For the most part all the spheres in the chromosome complement of a given preparation look alike, and are of about the same size; this applies to the free spheres also.

Up to this point I have deliberately used the neutral adjective "free" when writing about spheres that lie loose in the nuclear sap, rather than "detached", which implies that they originated on the chromosomes. To my knowledge there is no absolute proof that the spheres are generated on the chromosomes, and that they subsequently detach allowing new spheres to form in their places. I do not think there can be any question about the equivalent composition of attached and free spheres (see Chap. 9), and it is conceptually difficult to imagine them assembling free in the nuclear sap and then migrating to three predestined loci. The strongest evidence against this notion comes from sphere fusions, which can occur at high frequency, and from the rarity of more than one sphere attached to a single locus. However there remains another possibility that cannot be excluded; the material that aggregates to form the spheres may do so not only at the loci on the chromosomes, but also independently within the nuclear sap. One cannot necessarily argue that because there tend to be more free spheres in larger oocytes than in smaller, this results from a continuing process of aggregation at the chromosomal loci, detachment and accumulation of free spheres; it could as well reflect progressive accumulation of sphere-forming material in the oocyte nucleus that aggregates both on the chromosomes and elsewhere.

One observation in support of the view that the free spheres originate by detachment is that although the maximum sizes of the attached spheres can differ widely between different urodele species, and also differ widely between smaller and larger oocytes within one species, the free spheres do not reach sizes larger than the largest of the attached spheres in any given preparation. Another supporting observation is that in occasional oocytes there may be no attached sphere at one of the chromosomal loci, as though detachment had occurred just before the preparation was made; and in *Ambystoma macrodactylum* Kezer et al. (1980) have found on occasion constricted spheres where the part distal to the chromosome axis appears to be in the process of release from the proximal part (Fig. 3.17 b). I assume that free spheres arise by detachment from the chromosomes but this must be recognized as an assumption.

Fusions between homologous spheres are common in *T.c. carnifex,* more so in some animals than in others. Various non-homologous fusions also occur, within bivalent V, and between bivalents V and VIII, but just as with non-homologous fusions between GFLs, such fusions are restricted to spheres only.

All that has been said about spheres in *T.c. carnifex* applies equally to *T.c. cristatus. T.c. danubialis* is different in that fusions are much less frequent, spheres may be missing from one or other locus though not regularly so, and in some newts both free and attached spheres may have one or two concavo-convex lens-shaped "caps" closely applied to their surfaces. This condition recalls Gall's observations on the spheres of *N. viridescens.* In *T.c. karelinii* spheres like those of *T.c. carnifex* were regularly present on chromosomes V and VIII of a single female, but were equally regularly absent from both these chromosomes in nine

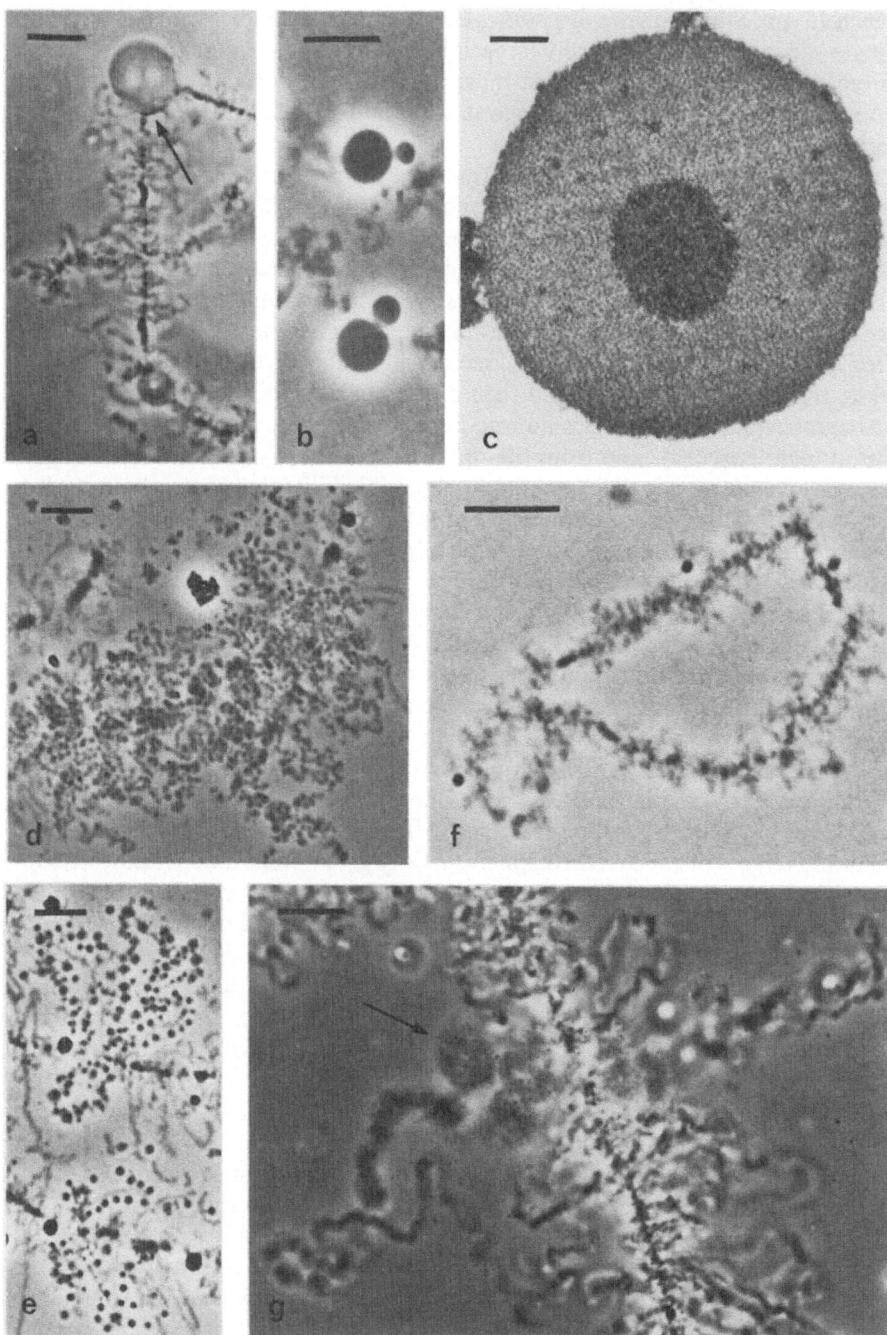

Fig. 3.17 a–g. Except for **c,** all are phase contrast photographs of unfixed lampbrush chromosomes. **a** Part of a bivalent of *Triturus helveticus* with spheres attached subterminally; an *arrow* points to chromomeres lying at the periphery of one sphere. **b** "Budding" spheres in a lampbrush preparation of *Ambystoma macrodactylum*. **c** Electron micrograph of a section through a sphere of *A. macrodactylum*, showing the fine granular texture of the cortex and a core of greater den-

other females. I will return to consider this peculiarity in the section concerning the inheritance of landmark structures.

Lacroix (1968 a) has made a detailed study of spheres in *Pleurodeles*. In these urodeles sphere morphology follows along just the same lines as in *T. cristatus,* and there are detached as well as attached spheres. In *Pleurodeles waltlii* from Morocco there is a sphere locus subterminal in chromosome IV; in *P. waltlii* from the Iberian peninsula there is also a sphere locus about the middle of chromosome XI (Fig. 3.10). In *P. poireti* from Algeria and Tunisia both these sphere loci are present, and also another about the middle of chromosome X (Fig. 3.11); a single *P. poireti* female was heterozygous in this regard, with a sphere locus on one chromosome X but not on its partner. Homologous sphere fusion is almost invariable in bivalent IV, but is less frequent in bivalents X and XI. Non-homologous sphere fusions were not observed in *P. waltlii,* but are frequent in *P. poireti.*

Information that was available a few years ago regarding the number and distribution of spheres in the chromosome complements of urodeles was assembled by Nardi et al. (1972). *Triturus marmoratus* was the primary object of this paper, and on the basis of 13 females studied this species is exceptional in lacking spheres altogether. Within the genus *Triturus,* where the haploid complements of most species include only two or three sphere-organizing loci, *T. alpestris apuanus* is exceptional in having seven loci, on four chromosomes (Mancino and Barsacchi 1965, Ragghianti et al. 1972). In this species, and as previously shown by Gall (1954) for *N. viridescens,* a positional correspondence has been established between the sphere-organizing loci of its lampbrush chromosomes and secondary constrictions in the slender mitotic chromosomes prepared from embryos (Nardi et al. 1972, Mancino et al. 1972 a). These constrictions are not evident in the more compact mitotic chromosomes prepared from adults. Other attempts to relate bulky lampbrush landmark structures (except nucleolus organizers) with mitotic secondary constrictions have been uniformly unrewarding, even when low temperature treatment has been used to induce secondary constrictions that are not otherwise visible (e.g. Callan 1966, on *Ambystoma mexicanum*).

A still larger number of sphere-organizing loci are present in the lampbrush complement of *Salamandra salamandra* (Mancino et al. 1969), 11 or 12 loci on 8 chromosomes within the haploid set of 12. Sphere-organizing loci have recently been demonstrated in yet another urodele genus, *Taricha* (Dr. J. Kezer, personal communication); in *T. granulosa* there are two, both subterminal and on different chromosomes.

Not only are sphere-organizing loci valuable for the identification of amphibian lampbrush chromosomes; they are also exceptional structures in their

---

sity. **d** Large cluster of beaded loops that are a feature of chromosome IX of *Taricha granulosa.* **e** Smaller clusters of beaded loops on bivalent IX of *T. torosa.* **f** Supernumerary bivalent in an oocyte of *Thorius sp.* showing a lampbrush organization just like that of the regular bivalents. **g** A pair of loops with fine granular matrix, characteristic of the heteromorphic regions of chromosomes I of *Triturus cristatus karelinii,* from an F_1 hybrid between *karelinii* ♀ × *cristatus* ♂; an *arrow* points to a mass of matrix presumably about to be shed. *Bars* in **a, b, d** to **g** = 10 μm; in **c** = 1 μm. **a** From Callan (1955); **b** and **c** from Kezer et al. (1980); **d–f** courtesy of Dr. J. Kezer; **g** from Callan (unpublished)

own right, and merit consideration in this regard. Unlike the lateral loops, spheres do not contain RNA; autoradiographic studies have shown that spheres do not incorporate ^{3}H-uridine (Gall and Callan, unpublished, Callan 1963, Mancino et al. 1967). Spheres are also insensitive, unlike normal loops, to digestion with RNase and, again unlike normal loops, they show no immediate morphological response to the inhibition of RNA synthesis by actinomycin-D (Snow and Callan, unpublished, Mancino et al. 1968).

Spheres are digested by proteolytic enzymes. Digestion with pepsin is particularly dramatic if pH is lowered a few minutes *after* pepsin has been applied to a fresh preparation of *T.c. carnifex* lampbrush chromosomes that were originally at near neutral pH, so circumventing the complication of acid-induced denaturation of the chromosomes prior to digestion (Macgregor and Callan 1962); the spheres disintegrate completely within a few seconds of lowering pH, though they withstand a similar pH alteration if pepsin is not present. The digestion of spheres by trypsin at its pH optimum, about 7.8, is much more leisurely and incomplete; sphere fusions survive tryptic digestion, whereas they are disrupted by pepsin. These observations corroborate the evidence from staining that acidic amino acids predominate in sphere protein.

Spheres are attached directly to the chromosome axis, and in favourable examples the axial chromomeric material can be seen to follow around the contour of a sphere (Fig. 3.17a), i.e. the sphere lies lateral to the axis. At first sight this structural relationship might appear unique, but this is not so; in many urodeles there are smaller objects, "axial granules", some of which bear the same relationship to the chromosome axis as do the spheres. I originally thought that spheres are not related to lateral loops, which would indeed make them unique, but I have now come to favour an alternative interpretation. In some preparations of *T.c. carnifex* lampbrush chromosomes there are small lateral loops carrying tiny, refractile globules lying immediately adjacent to the spheres. These loops are not always evident, and they can be visible alongside some spheres but not all. However precisely the same relationship can be observed between such loops and any, often all six, of the spheres both on bivalent V and on bivalent VIII, and they are unique to this location. Indeed on the relatively few occasions in *T.c. carnifex* when a sphere-organizing locus lacks a sphere, the locus can be identified precisely because of the presence of these small loops with particularly refractile globules in attendance (Callan and Lloyd 1960). Similar loops with globules accompany the spheres of *Taricha* (Dr. J. Kezer, personal communication). I suspect that these tiny refractile globules move towards the chromosome axis as transcription proceeds, and when they reach the axis they coalesce with predecessors and give rise to the spheres. This speculation is not as far fetched as may at first sight appear, for certain smaller axial granules, to be described in the section which follows, bear just such a relationship to easily identified but much larger loops.

Subterminally in the left arm of chromosome III (Gall's 7) of *N. viridescens* there are two rounded (though not spherical) objects which, like spheres, are attached directly to the chromosome axis. They are not identical in texture to spheres, and they are less uniform in size, but they show the same capacity for fusion and, like the spheres of *T.c. carnifex,* all of them are accompanied by what Gall (1954) describes as "conspicuous, heavily granular loops". Gall interpreted the various sizes of the axially-attached objects as representing stages in their growth, and it strikes me as probable that here, too, the granules on the adjacent loops contribute to this growth.

As the oocytes of *T.c. carnifex* and other urodeles approach maturity the lateral loops regress, but the spheres remain large and conspicuous, and free spheres become abundant, until just before ovulation. In *N. viridescens* Barsacchi Pilone and Humphries (1975) have observed that unlike the general run of lateral loops in mature oocytes, those immediately adjacent the spheres remain conspicuously extended. This temporal correlation suggests that there is a functional correlation too. However the function of sphere protein is unknown, and likewise its precise chemical nature. Perhaps the transcripts on the globule-bearing loops adjacent to the spheres of *T.c. carnifex* act as specific sites for the sequestration of a protein that is involved in some operation connected with oocyte maturation, or with some subsequent developmental process. However if this is the case the objective can evidently be achieved in some other manner in those few urodele species such as *T. marmoratus* which form no spheres at all.

Sphere-organizing loci are not confined to urodele Amphibia. Four of the 13 chromosomes of the haploid complement of *Rana esculenta* carry spheres of diameters from 5 to 10 μm, homogeneous or vacuolated; homologous sphere fusions occur, and free spheres are present (Giorgi and Galleni 1972). In *R. esculenta* Morescalchi and Filosa (1965) distinguished free spheres from free nucleoli by their different staining reactions to toluidine blue; the former stain orthochromatically, the latter metachromatically, confirming Gall's (1954) observations on *N. viridescens.* I think there can be little doubt that spheres in *Rana* correspond precisely to those of urodeles. Srivastava and Bhatnagar (1962) illustrate what appear to be spheres attached to some of the lampbrush chromosomes of *R. cyanophlyctis,* while Nishioka et al. (1980) show four sphere loci as landmarks in working maps of both *R. nigromaculata* and *R. brevipoda.* According to Müller (1974) there are three sphere loci of landmark value in *Xenopus laevis.*

Whether objects strictly comparable with the spheres of amphibian oocytes are present in the oocytes of other animals is an open question, and will remain so until the chemical nature of sphere protein has been established. Mancino et al. (1969) remark that the „Binnenkorper" present in the oocytes of various insects (Bier et al. 1967) resemble spheres in several respects. These Binnenkorper (or "endobodies") are rounded, largely amorphous (though in some cases vacuolated) concentrations of protein that aggregate by fusion, contain no RNA, and have some as yet ill-defined connection with the oocyte chromosomes, whether or not these are actively transcribing RNA. In the beetle *Carabus* Bier et al. (1969) actually describe these structures as spheres and show four of them attached to a lampbrush bivalent, and even when the oocyte chromosomes of the beetle *Pterostichus* have condensed to form a karyosphere, this maintains attachment to a single enormous (60 μm diam) spherical endobody (Bier et al. 1967).

It can scarcely be accidental (see Chap. 5) that histone genes occupy sites on the chromosomes of three species of newt, *N. viridescens, T. cristatus* and *T. al-*

pestris and also *Pleurodeles waltlii* (mentioned by Lacroix et al. 1985) where spheres are formed on their lampbrush chromosomes; a functional significance for this striking correlation has yet to be found. The sphere loci deserve further study.

Axial Granules

When urodele lampbrush chromosomes are viewed in phase contrast, certain of their chromomeres appear considerably larger than the generality. In fixed preparations of *T. cristatus* stained by the Feulgen technique followed by light green, these enlarged chromomeres, or axial granules, are revealed as compound structures, with a spherical or near-spherical component that stains with light green, confluent with a Feulgen-positive chromomeric component (Macgregor and Callan 1962). In most examples the chromomeric material runs around the edge of the light green staining component, which is nevertheless not displaced laterally from the chromosome axis. In this one respect an axial granule differs from a sphere, but the difference is trivial and probably dependent on the relative sizes of the chromomeric and non-chromomeric components of the two structures.

Like spheres, axial granules can fuse with one another, and the fusions may occur between homologous (Fig. 3.2 f) or non-homologous granules, including also telomeres (Fig. 3.12 b). When non-homologous association occurs within the length of a single lampbrush chromosome, the outcome is a "fold-back" (Gall 1954) or "reflected fusion" (Callan and Lloyd 1960) (Figs. 3.6, 3.7, 3.8). Protein is involved in these fusions, though they may or may not be disrupted by proteolytic digestion. When an homologous axial granule fusion is not disrupted, one might be tempted to speculate that a chiasma is present at the same site. However non-homologous fusions may also withstand proteolytic digestion, this being invariably true of the pericentric reflected fusions that are so often present in chromosomes II, III, IV and VI of *T. cristatus,* so this speculation is not valid.

If fusions between components of lampbrush chromosomes is restricted to like with like by the chemical composition of the material that fuses, and the specificity of fusions between GFLs, and between spheres, in *T. cristatus* suggests that this is so, then in view of the many different non-homologous fusions that have been observed in this species the material involved in axial granule fusions must be common to many chromosomal loci, including telomeres, but different from sphere protein, for axial granules never fuse with spheres.

The relationship between axial granules and lateral loops is problematic. Certain axial granules are invariably associated with particular and easily recognized loops in *T. cristatus,* an example being the granule at the base of the left-most landmark loop on chromosome II (Figs. 3.6, 3.7, 3.8). However other axial granules are equally invariably not associated with any lateral loops at all, an example being the right-most landmark on *T. cristatus* chromosome II. Despite this difference, these particular two axial granules are more frequently reflected fused to one another than separate in *T.c. cristatus* and *T.c. karelinii.*

Although the significance of axial granules is unknown, they are especially numerous in the heteromorphic regions of bivalent I of *T. cristatus,* regions which carry particularly long lateral loops that transcribe a disproportionately large concentration of satellite DNA sequences (Varley et al. 1980a, b). This may not be a fortuitous correlation. Axial granules are more numerous in some urodele species (triturid newts generally, also *Pleurodeles*) than in others (*Notophthalmus, Ambystoma*). At all events axial granules have to be taken into account because those that are invariant in position have proved valuable for chromosome recognition.

Telomeres

The structures situated at the ends of lampbrush chromosomes, in all those organisms where they are large enough to be visible in the light microscope, are telomeric chromomeres. In other words, lampbrush chromosomes do not terminate in lateral loops. Although in unfixed preparations in saline the telomeres usually appear uniformly dense, in fixed preparations stained by the Feulgen technique followed with light green, the terminal Feulgen-positive chromomere is crescent-shaped, closely applied to a spherical object that stains with light green. In this respect telomeres resemble axial granules.

Telomeres differ in size in different urodeles; in most species they are less than 2 µm diam, though some of the telomeres of *T.c. cristatus* are larger, about 4 µm diam, and consistently so. Thus for example the left telomere of *T.c. cristatus* chromosome XII (Fig. 3.18 a) is regularly larger than the right, and larger than the telomeres of any of the chromosomes of other subspecies of *T. cristatus;* this size difference remains recognizable in F_1 hybrids where *T.c. cristatus* was one of the parents (Fig. 3.18 c, f, g). Therefore to a minor extent telomeres can have some value for chromosome identification.

The telomeres of lampbrush chromosomes can fuse with one another, and for the most part these fusions are homologous (Figs. 3.1 a and 3.18 b). They are common in *T.c. cristatus,* 67 examples in 148 bivalents (Callan and Lloyd 1960). They occur, but are less frequent, in the other species of *T. cristatus,* likewise *T. italicus* (Mancino and Barsacchi 1969), *T. helveticus helveticus* (Mancino and Barsacchi 1966), *Pleurodeles waltlii* and *P. poireti* (Lacroix 1968a), *N. viridescens* (Gall 1954), *Taricha granulosa* (Dr. J. Kezer, personal communication), and are decidedly rare in *T. alpestris apuanus* (Mancino and Barsacchi 1965), *T. vulgaris meriodionalis* (Barsacchi et al. 1970), *T. marmoratus* (Batistoni et al. 1974), *Salamandra salamandra* (Mancino et al. 1969), *Ambystoma mexicanum* (Callan 1966), *A. macrodactylum* (Kezer et al. 1980).

There is little information about telomeres in anuran Amphibia, but telomere fusions have been recorded in *Rana esculenta* (Morescalchi and Filosa 1965, Giorgi and Galleni 1972) and in an allotriploid (*R. brevipoda* × *R. nigromaculata*) ♀ × *R. brevipoda* ♂ (Ohtani 1975). I have found no information about telomeres of the lampbrush chromosomes of other organisms, and in papers describing them neither photographs nor drawings show terminal fusion.

The different frequencies of telomere fusions in different species appear to be related to telomere size. This supposition is supported by evidence from *T.c. cristatus;* as mentioned earlier, the left telomere of chromosomes XII is much larger than the right, and in 23 bivalents XII there were 15 fusions involving left (Fig. 3.18 b) but not a single fusion involving right telomeres.

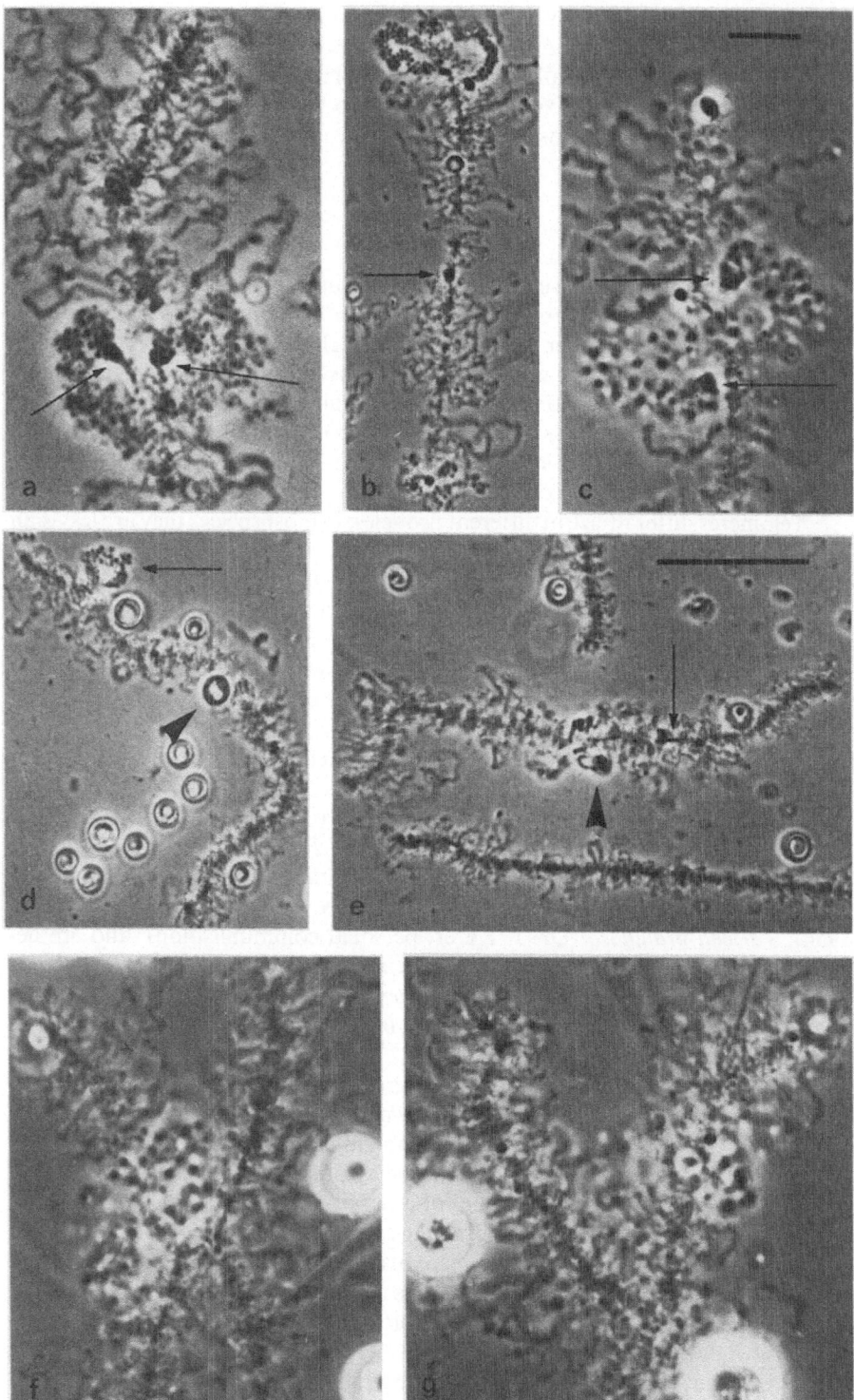

Fig. 3.18 a–g

Digestion of fresh unfixed lampbrush preparations of *T. cristatus* with pepsin or trypsin or pronase or RNase fails to dissociate telomere fusions, though most of the telomeric material, presumably the component which in fixed preparations stains with light green, is put into solution by proteolytic enzymes (Macgregor and Callan 1962). In *T. cristatus* all telomere fusions have dissociated naturally by the time the bivalent chromosomes have moved on to the spindle at first meiotic metaphase (Watson and Callan 1963).

In those species where homologous telomere fusions are frequent, fusions between non-homologous telomeres also occur, and may involve more than two telomeres. In *T.c. cristatus* telomere to axial granule "reflected" fusion, i.e. within one and the same chromosome arm, occurs so frequently in chromosomes II, IX and X as to be of diagnostic value.

A feature of non-homologous telomere fusions is that almost invariably they involve the telomeres of different chromosomes. I have only once observed a fusion between the telomeres of the left and right arms of the same lampbrush chromosome (or that of its homologue). This at first sight seems surprising, for as synapsis begins the telomeres of all the chromosomes of *Triturus* (and those of other urodeles) lie close together and anchored to the nuclear membrane. Homologous telomere fusions are much more frequent than non-homologous, and are presumably a relic of the close apposition of homologous chromosome ends that were previously held together by a synaptonemal complex. Non-homologous telomere fusions must originate in some other manner, and given that in *Triturus* there are 48 telomeres per oocyte, and some more prone to fusion than others, it may simply be a statistical accident that left and right telomeres of the same chromosome very rarely associate.

Fig. 3.18 a–g. Phase contrast photographs of unfixed lampbrush chromosomes of *Triturus cristatus* primarily to show the giant granular loops (GGLs). **a** Terminal region of the longer arm of chromosome XII of *T.c. cristatus* from an oocyte of 0.9 mm diam, with its telomere above; *arrows* point to the dense tips of the GGLs. **b** Terminal regions of the longer arms of bivalent XII of *T.c. cristatus* from an oocyte of 1.0 mm diam; an *arrow* points to fused homologous telomeres; the upper GGLs are of normal size, the lower GGLs are exceptionally small, and the difference in size was a regular feature of all the oocytes of this particular newt. **c** Terminal region of the longer arm of chromosome XII of an F_1 hybrid *karelinii* ♀ × *cristatus* ♂ from an oocyte of 0.9 mm diam; this is part of the *cristatus*-derived chromosome XII, the GGLs are of normal size, and *arrows* point to their dense tips. **d** Univalent *cristatus*-derived chromosome XII of an F_1 hybrid *karelinii* ♀ × *cristatus* ♂ from an oocyte of 1.0 mm diam; an *arrow* points to the GGLs and an *arrowhead* to the GFLs, fused together in a spherical structure; the other round objects are free nucleoli. **e** Univalent *karelinii*-derived chromosome XII from the same oocyte as **d**; an *arrow* points to the centric region, and an *arrowhead* to the GFLs, which are characteristically much more compact in *T.c. karelinii* than are the GFLs of *T.c. cristatus*. **f** Terminal regions of the longer arms of bivalent XII of an F_1 hybrid *carnifex* ♀ × *cristatus* ♂ from an oocyte of 0.9 mm diam; the GGLs on the *cristatus*-derived chromosome are of normal size, and are absent from its *carnifex*-derived partner. **g** Terminal regions of the longer arms of bivalent XII of another F_1 hybrid *carnifex* ♀ × *cristatus* ♂ from an oocyte of 0.8 mm diam; the GGLs on the *cristatus*-derived chromosome are abnormally small and their granules abnormally large. *Bar* in **c** = 10 μm; **a, c, f** and **g** are of the same magnification. *Bar* in **e** = 50 μm; **b, d** and **e** are of the same magnification. **a** From Callan (1982); **b–e** and **g** from Callan (unpublished); **f** from Sommerville et al. (1978 b)

Chromomeres

Chromomeres need to be mentioned in this chapter, but in a negative sense, for in general they are not reliable characters for the recognition of particular lamp-brush chromosomes or for regions within them. Thus even when two chromosomes lie side by side and are anchored together by a chiasma that defines a point of precise homology, the chromomeric patterns in the neighbourhood of the chiasma are rarely identical in the two homologues. There are a few exceptions to this generalization; the axial bars alongside the centromeres of those Amphibia that possess them are reliable diagnostic features, and the limits to the left and right of the heteromorphic regions of bivalent I of *T. cristatus* are clearly defined by the smaller size of the chromomeres in the homomorphic regions that flank them (Sims et al. 1984).

The identification of polytene chromosomes, by contrast, depends primarily on the recognition of their cross-banded chromomeric patterns, and in the larvae of Diptera these patterns are of such constancy that the precise identification of single chromomeres in register with their sisters is standard cytogenetic practice. Not only this; in organisms such as *Chironomus* where "workable" polytene chromosomes are present in several differentiated tissues of the larvae, direct comparison can be made between components of the genome that are active in transcription in different tissues and, in specially favourable cases, between cells within a single organ (Beermann 1952, 1961). The chromomeric patterns of polytene chromosomes appear to be generated during the process of polytenization (Beermann 1962).

The leptotene, zygotene and pachytene chromosomes of many plants and animals also have a chromomeric organization, i.e. dense, compact structures resolvable in the light microscope separated from one another by thinner regions. Belling (1928) proposed that each "ultimate" chromomere resolvable at meiotic prophase represents a single gene. Although the chromomeric organization of these meiotic chromosomes generally appears to be on a grosser scale than that exhibited by polytene chromosomes, i.e. discrete resolvable chromomeres are fewer in number, it was widely assumed for many years that chromomeres in meiotic and polytene chromosomes are essentially comparable structures. This opinion was reinforced when it was found that the numbers of resolvable chromomeres in maximally extended urodele lampbrush chromosomes are of the same order of magnitude (several thousands) as the crossbands of dipteran polytene chromosomes. This in turn led to the speculation that the lateral loops of lamp-brush chromosomes are comparable to the Balbiani rings of *Chironomus* polytene chromosomes (Callan 1955, Gall 1956) and because Balbiani rings expand by the extension of DNA/histone fibres from compact chromomeric crossbands the speculation proceeded on to the assumption that the axes of the lampbrush lateral loops extend at the expense of chromomeric material (Gall 1958, Callan and Lloyd 1960). The speculation was supported by the circumstantial evidence that in urodele oocytes at a stage when the lateral loops are maximally extended the chromomeres are in general small, whereas in older oocytes when the lateral loops are short the chromomeres are larger. As stated by Gall (1958): "Such a reciprocal size relationship suggests that the chromomeres may form at the expense of the loops axis and vice versa". Although this idea appeared plausible there was at the time no reliable information concerning the earliest stages of lateral loop extension, and the status of chromomeres at the transition from pachytene to diplotene. This information could not be obtained by examining the contents of nuclei isolated from oocytes at such an early stage by the methods that work so satisfactorily for larger oocytes; the nuclei are too small for the manual removal of their membranes, and the chromosomes do not disentangle from one another as the nuclear sap disperses in saline. Instead it is necessary to work with conventional sections.

Sections through Sanfelice-fixed ovaries from metamorphosing larvae of *T.c. carnifex* stained with Feulgen and light green show the pachytene chromosomes in nuclei of about 20 to 25 μm diam as whiskery Feulgen-positive strands with irregular and ill-defined axes and no compaction into discrete chromomeres. In these sections light green staining of pachytene and earlier meiotic nuclei is not detectable even when the cells' cytoplasm is well stained. The onset of diplotene in slightly larger nuclei, of about 30 μm diam, is clearly defined, marked by light green staining of material around chromosomes' axes that have suddenly become compact and have acquired smooth outlines. The chromosomes' axes are not of uniform width, but they can be resolved as continuous Feulgen-positive strands over lengths of many microns. If similar sections at this stage are stained with iron haematoxylin and appropriately differentiated, the morphology is identical but much more readily resolved because of the greater intensity of the stain, with chromosome axes black and the lateral material, identifiable as incipient loops, greyish-brown.

Discontinuities in the chromosomes' axes, i.e. places where connections between neighbouring regions can no longer be resolved, first appear in nuclei of some 60 to 70 μm diam, but continuous resolvable axes still range in length up to 15 μm. Discrete chromomeres comparable in size to those visible in isolated preparations of lampbrush chromosomes are not apparent until the oocyte nuclei have reached about 100 μm diam. Even in these relatively large oocytes most of the "chromomeres" are a great deal longer than the longest (1 or 2 μm) chromomeres of the fully-established lampbrush phase, so their total number must increase progressively. This sequence of events implies that chromomeres become separate entities as a consequence of increasing transcriptional activity on the lateral loops. As RNP matrix builds up, the bases of neighbouring loops may be forced apart, and they can only do so by disrupting compact axial DNP. The implication is therefore that far from being units of function in a genetic sense, the chromomeres of lampbrush chromosomes are merely the residual non-transcribed portions of DNP whose size and number is determined not by any intrinsic peculiarity but by the amount and consistency of loop matrix accumulating in the neighbourhood (Callan 1978).

This view of the nature of the chromomeres of lampbrush chromosomes is supported by a study of chromomere number carried out by Vlad and Macgregor (1975) in three closely related species of *Plethodon*. One of them, *P. cinereus,* has roughly half the C-value (20 pg) of the two others, *P. vehiculum* and *P. dunni.* All three species have nearly identical lampbrush and mitotic karyotypes, but overall the chromosomes of *P. cinereus* are considerably smaller than their counterparts in the karyotypes of the other two species. The mean sizes of chromomeres in the lampbrush chromosomes of all three species are much the same, their frequency along the chromosomes' axes likewise, but taken overall *P. vehiculum* and *P. dunni* have some 60% to 70% more chromomeres (about 5500 per haploid set) than *P. cinereus* (about 3500). If lampbrush chromomeres were to represent genetic units, one would expect to find roughly the same number in all three species. This is emphatically not the case. The same consideration, of course, applies to the number of transcribing lateral loops of these salamanders. Many of these are in any case not single units of transcription, but this is another matter, as will be discussed in later chapters.

Double Axis Regions

There are a few exceptional regions of the lampbrush chromosomes of some urodeles where the diplotene nature of these chromosomes is clearly apparent. In these regions the two chromatids lie apart from one another, each consisting of a series of chromomeres connected to one another by interchromomeric strands. Tensions which result in the formation of double-loop bridges in the normal parts of lampbrush chromosomes produce single-loop bridges in these double axis regions. In *T. cristatus* and its close relative *T. marmoratus* there are two double axis regions in chromosome XII (Callan 1955, Callan and Lloyd 1960, Nardi et al. 1972). One of these is regularly evident; it extends over the distal third of the shorter (right) arm (Figs. 3.6, 3.7, 3.8). In *T. cristatus* the lateral loops in this double axis region are unusually short, they are loaded with 0.5 to 2 μm RNP granules (Fig. 3.13 d), and the region is especially fragile in the sense that it often frays out to form a series of loop bridges. The RNP granules on these short loops often conceal the chromomeres from view, but the unique organization of the region is apparent in preparations dispersed in dilute saline (Fig. 1.8 c).

The other region of *T. cristatus* or *T. marmoratus* chromosome XII which may or may not display a double axis lies in its longer (left) arm, near the centromere. When a chiasma happens to have formed in this double axis region, this offers an unusual opportunity for all four chromatids to be clearly seen. As elsewhere, the exchange points are restricted to main axis components of the chromatids; they do not involve the lateral loops.

Although none of the other species of *Triturus* that have been studied display a distal double axis region in the smallest chromosomes of their complements, it is remarkable that chromosome XII of the much more distantly related *Pleurodeles waltlii* and *P. poireti* is like chromosome XII of *T. cristatus* in this regard (Lacroix 1968 a) (Figs. 3.10 and 3.11). The only other recorded example of a double axis that serves as a useful recognition feature is a short interstitial region in chromosome I of *T. vulgaris meridionalis* (Barsacchi et al. 1970).

To conclude this topic, amongst the many investigators who have studied amphibian lampbrush chromosomes by electron microscopy, only one (Ullerich 1970), has succeeded in demonstrating the dual nature of the normal interchromomeric strand, in dispersed and formalin-fixed preparations of *T. alpestris* that were digested with trypsin and RNase after attachment to a filmed grid.

Heritability of Landmark Structures

When I decided to study the lampbrush chromosomes of *Triturus cristatus* rather than some other urodele I did so for two reasons. They carry plenty of landmark structures, which makes chromosome identification simple; and one subspecies, *T.c. carnifex,* is easy to breed in captivity. Adults of this newt do not require training to consume immobile food. The other subspecies, at least when first brought to the laboratory, will only eat moving animals such as earthworms or *Tubifex.*

If *T.c. carnifex* are well fed, not unduly crowded in aquaria, in clean water at about 18 °C, both males and females will come into breeding condition regularly each year. Females of the other subspecies will do likewise, but their males are more difficult to bring into breeding condition, and for unknown reasons this has proved impossible to achieve with male *T.c. cristatus.* The same happens to be true of male *T. vulgaris.* Thus if one wishes to use a male *T.c. cristatus* as a parent in an interracial cross, the only way to do so is to capture one in breeding condition from the wild and introduce him to a receptive female in an aquarium at once. The male will quickly lose breeding condition while in captivity, and will not regain this state thereafter.

The larvae of all races and interracial crosses of *T. cristatus* are robust (though there is a regular 50% mortality to which reference has already been made in this chapter), they can be fed on small Crustacea soon after hatching, thereafter on *Tubifex.* Following metamorphosis they can be fed on small earthworms, and ultimately they can be trained to take immobile food.

A particularly advantageous feature is that after metamorphosis *T. cristatus* does not have an obligatory terrestrial phase, which is an inconvenient aspect of the life history of many urodeles, e.g. *T. vulgaris* and *N. viridescens,* inconvenient because while the newts are terrestrial they are hard to feed, and grow slowly. Given optimum conditions *T.c. carnifex,* and interracial hybrids of *T. cristatus,* mature and reach sexual maturity within a year. The male F_1 interracial hybrids are sterile or at the best much less fertile than non-hybrid males, whereas the female F_1 hybrids, despite lower chiasma frequencies than in the parent races, and the occasional absence of even a single chiasma between homologous chromosomes (Fig. 3.18 d, e), are reasonably fertile and can produce plenty of offspring when backcrossed to non-hybrid males (Callan and Spurway 1951, Callan, unpublished). Regrettably the number of families raised, and the number of female offspring analyzed, was severely curtailed by the outbreak of a catastrophic infection in the newt colony, probably „Molchpest" (Reichenback-Klinke and Elkan 1965), which only after much trial and tribulation proved to be controllable with aureomycin. As a consequence the analysis of landmark inheritance in *T. cristatus* is much less thorough than I hoped it might be, though it has provided sufficient information to be worth recording. Information on inheritance obtained by Lacroix from *Pleurodeles,* another urodele that is easy to breed in captivity, will also be discussed in this section.

That there is a genotypic basis for landmark loop morphology was already apparent before any breeding experiments were undertaken. Callan and Lloyd (1960) classified 22 females of *T.c. carnifex* on a plus or minus basis for the presence or absence of GFLs on their chromosomes X and XII. "Absence" was not

Table 3.1. Twenty-two females of *T.c. carnifex* classified according to the GFL constitutions of bivalents X and XII

Bivalent X	Bivalent XII[a]			
	$+/+$	$+/-$	$-/-$	Totals
$+/+$	1	2	2	5
	(0.15)	(1.01)	(1.74)	(2.90)
$+/-$	0	2	4	6
	(0,53)	(3.59)	(6.07)	(10,19)
$-/-$	0	4	7	11
	(0.46)	(3.12)	(5.32)	(8.91)
Totals	1	8	13	22
	(1.14)	(7.72)	(13.13)	

[a] Hardy-Weinberg expectancy figures are given in parentheses below each entry.

used in an absolute sense, for it included chromosomes with small fusing loops as well as those with none; the distinction between giant and small is easy to make, but because of "phenotypic" variation, especially its developmental component, distinction between small and absent would have been arbitrary and inaccurate.

Using the frequencies of presence or absence of GFLs supplied by this analysis, expectations of the various combinations were calculated according to the Hardy-Weinberg principle (binomial square rule), assuming segregation and assortment at random. Table 3.1 shows the actual and expected combinations, which are in sufficient accord to warrant the assumption that these loop characters behave like typical Mendelian alleles, and that their presence or absence does not have significant influence on viability. In this connection it should be recalled that even those females that are $-/-$ for GFLs on both bivalent X and bivalent XII are $+/+$ for GFLs on bivalent XI, for all *T.c. carnifex* females are so constituted.

The assumption that the GFLs are inherited as Mendelian alleles was supported by later analysis of female offspring in two families of *T.c. carnifex*, the outcome of crossing ♀ J ($-/-$ for GFLs on bivalent X, $+/-$ for GFLs on bivalent XII) and ♀ R ($+/+$ for GFLs on bivalent X, $+/-$ for GFLs on bivalent XII to a common ♂ A, whose "lampbrush constitution" could not be determined in advance. The results of this analysis are given in Table 3.2, and it can be concluded that ♂ A must have had the "potential" constitution $+/-$ for chromosomes X and also $+/-$ for chromosomes XII.

Further evidence that GFLs are inherited in Mendelian fashion came from interracial crosses. *T.c. cristatus* has no GFLs on chromosome XI, whereas the latter are always present in *T.c. carnifex*. Five F_1 female progeny of the cross *cristatus* ♀ H × *carnifex* ♂ E all proved to be $+/-$ for GFLs on bivalent XI. The same was true of five F_2 female progeny of the reciprocal cross *carnifex* ♀ L × *cristatus* ♂♂.

T.c. karelinii, like *T.c. cristatus*, also lacks GFLs on chromosome XI. Four F_1 females of the cross *karelinii* ♀ D × *carnifex* ♂ A were all $+/-$ for GFLs on bivalent XI. One of these, F_1 ♀ A, was backcrossed to *T.c. carnifex* ♂ E. Five

Table 3.2. Inheritance of GFLs on chromosomes X and XII of *T. c. carnifex*[a]

F₁♀	Bivalent X			Bivalent XII		
1. ♀J (Bivalent X −/−, bivalent XII +/−) × ♂A						
57A	+/−			+/+		
57B	+/−				+/−	
57C	+/−			+/+		
57D			−/−			−/−
57E	+/−			+/+		
57F	+/−					−/−
57G			−/−		+/−	
57H			−/−		+/−	
SR 59A	+/−			+/+		
SR 59B			−/−		+/−	
SR 59C	+/−				+/−	
SR 59D	+/−					−/−
SR 59E			−/−		+/−	
SR 59F			−/−	+/+		
SR 59G	+/−				+/−	
Totals 15	0	9	6	5	7	3

F₁♀	Bivalent X			Bivalent XII		
2. ♀R (Bivalent X −/−, bivalent XII +/−) × ♂A						
58A		+/−			+/−	
58B		+/−		+/+		
58C	+/+				+/−	
58D	+/+				+/−	
58E	+/+					−/−
Totals 5	3	2	0	1	3	1

[a] + = present; − = small or absent.

of the backcross female offspring were +/− for GFLs on bivalent XI, and five were +/+; four of the latter were homozygous for *T.c. carnifex* centromeres, but one had axial bars alongside one of its centromeres, inherited from its *T.c. karelinii* grandmother. Thus the chromosome XI inherited from its mother and carrying a fully expressed GFL was the product of a recombination that had occurred somewhere between the GFL and the centromere.

The same picture emerged from the reciprocal cross *carnifex* ♀L × *karelinii* ♂A. Nine F₁ females were all +/− for GFLs on bivalent XI. One of these, F₁♀J, was backcrossed to *T.c. carnifex* ♂C. Two of the backcross female progeny were +/− for GFLs on bivalent XI, and three were +/+. It would be tedious also to describe the inheritance of GFLs on chromosomes X and XII; it suffices to say that their expression, like that of the GFLs on chromosome XI, appeared to be unaffected by hybridity in all the crosses mentioned above.

The most consequential finding that came from the analysis of interracial hybrids, however, was that the morphologies of some landmarks are not determined in a strictly autonomous manner. There are no GGLs on chromosome XII of either *T.c. carnifex* or *T.c. karelinii*, thus all F₁ hybrids between these subspecies and *T.c. cristatus* may be expected to be heterozygous for this character. Six F_1 females originating from the cross *karelinii* ♀ × *cristatus* ♂ showed this heterozygosity, and all their GGLs appeared morphologically normal (Fig. 3.18 c, d). However in five F_1 females originating from the cross *carnifex* ♀ × *cristatus* ♂, and also in five more F_1 females of reverse parentage, the GGLs were either morphologically normal (Fig. 3.18 f), or small (Fig. 3.18 g), or grossly abnormal, reduced to a couple of small lumps and a few granules, or to two lumps only. Out of five females originating from the backcross (*carnifex* ♀ × *cristatus* ♂) F_1 ♀ × *carnifex* ♂, three had inherited a GGL from the maternal grandfather and two had not. The GGLs in those backcross hybrids that possessed them showed much the same range of variation in morphology as was evident in the F_1 generation. It is therefore clear that although the GGLs' morphology is to a degree autonomously determined, the morphology is nevertheless subject to modification by other factor(s) present in a hybrid oocyte environment.

Whereas the left arms of lampbrush chromosomes III, IV and XI generally carry subterminal landmark loops with fusing matrix in *T.c. carnifex* and *T.c. karelinii* (the developmental variability of these loops, their response to hypophysectomy, and to gonadotropin injection, was described earlier in this chapter) in *T.c. cristatus* oocytes from 0.6 mm diam upwards it is most unusual for such loops to be visible. However in oocytes of the appropriate size ranges in two families of F_1 hybrids, one stemming from *carnifex* ♀ × *cristatus* ♂ and the other of reverse parentage, subterminal landmark loops were regularly present on the *T.c. cristatus* as well as on the homologous *T.c. carnifex* chromosomes. In this case, therefore, a hybrid oocyte environment provided the conditions required for the expression of a lateral loop "phenotype" that is not expressed as such in one of the parent subspecies.

The landmark loops that carry a fine granular matrix in the heteromorphic regions of chromosomes I, and which shed lumps of this material into the nuclear sap, are characteristic of *T.c. karelinii* (Fig. 3.16) and are not present either in *T.c. cristatus* or *T.c. carnifex*. They are recognizable in subspecies F_1 hybrids (Fig. 3.17 g). Out of five ♀ F_1 hybrids of parentage *karelinii* ♀ × *cristatus* ♂, and two of parentage *karelinii* ♀ × *carnifex* ♂, three had inherited the long arm of the *karelinii* chromosome I that is drawn on the working map, while four had inherited its partner. Out of eight ♀ F_1 hybrids of parentage *carnifex* ♀ × *karelinii* ♂, four had inherited the arm drawn on the working map, while the other four had inherited its partner. This information refuted the proposal (Callan and Lloyd 1960) that the heteromorphic regions of bivalent I are concerned with sex determination, a conclusion that is now backed up by the other evidence quoted earlier in this chapter. However a specific DNA sequence transcribed at multiple sites on chromosomes I of *T.c. karelinii* cannot be the only factor involved in the accumulation of this most characteristic type of loop matrix, for out of 15 backcross hybrids to *T.c. carnifex* where one or other of the F_1 mothers' parents were *T.c. karelinii*, whereas about one-half of these should have inherited the landmark loops characteristic of *T.c. karelinii* chromosomes I, only three showed the merest hint of loops with such matrix. Perhaps availability of the appropriate protein(s) is at a low level in these animals, for lack of sufficient transcription of their coding sequences elsewhere in their genomes.

A converse interaction was exposed when the inheritance of spheres was followed. In *T.c. karelinii* spheres like those present on chromosomes V and VIII of *T.c. cristatus* and *T.c. carnifex* were regularly present in a single female but were equally regularly absent in nine other females. Three of these latter, lacking spheres, were hybridized, two with *T.c. cristatus* ♂♂ and one with a *T.c. carnifex* ♂. Out of ten F_1 hybrids studied, in nine of them spheres were regularly present on the *karelinii*-derived chromosomes V and VIII, as well as on their partners, though all the spheres were abnormally small in two; in one of the F_1 hybrids spheres were missing altogether. A similar picture was presented by nine F_1 hybrids of constitution *carnifex* ♀ × *karelinii* ♂. Eight of these hybrids had spheres on both chromosomes V and VIII, though in three animals they were abnormally small, and in one F_1 hybrid the spheres were missing from all the loci in all preparations. One can draw two conclusions from these observations. Whatever mechanism that generates the material which forms spheres in *T.c. carnifex* is inoperative, or less effective, in most individuals of *T.c. karelinii;* however the loci on which the prime material of spheres can accumulate are nonetheless present on *T.c. karelinii* chromosomes. This character of *T.c. karelinii* is to a degree carried over into F_1 hybrids, and all sphere loci, whether they be of *T.c. karelinii* or other origin, express this variable phenotype to the same extent.

Lacroix (1968 b) has followed sphere inheritance in F_1 hybrids between *P. waltlii* and *P. poireti*. Chromosomes which regularly lack spheres in one of the parents, i.e. X and XI of Moroccan *P. waltlii*, and chromosome X of Iberian *P. waltlii*, equally lack spheres in F_1 hybrids even though spheres are present on the homologues derived from the other parent. Thus the regular absence of a sphere from a given locus in *P. waltlii* should be regarded as evidence of a genetic deficiency and not, as in *T.c. karelinii*, a locus which for other reasons happens not to express its potential phenotype.

Lacroix and Loones (1971) went on to study X-ray induced mutations in *P. waltlii*. Out of ten females whose male parent had been irradiated, chromosome aberrations were recognized in nine, and in one of these, heterozygous for triple reciprocal translocations, one of the break points happened to have occurred within the "sphere-organizing" region of chromosome IV. It was possible to define the break point so precisely because both translocated portions of chromosome IV formed spheres, one an acentric fragment of IV now connected to a centric part of chromosome XII, the other the centric residue of IV now connected to an acentric fragment of chromosome X. In the translocation heterozygote the sphere on IV/XII often cross-fused with the sphere on IV, while the sphere on IV/X remained separate. And whereas the sphere on IV/XII, when not cross-fused, was evident in young and old oocytes alike (just as in *T. cristatus* the spheres are much smaller in young oocytes of *P. waltlii* than in old) the sphere on IV/X was only developed in older oocytes. From this study Lacroix and Loones were able to conclude that the sphere-organizing region, like the nucleolus-organizing regions of *Chironomus* and on similar evidence (Beermann 1960), consists of repetitive functional units. Despite fragmentation into two by irradiation both fragments retain, though to greater or lesser degree, their original sphere-organizing capacity. There is now other and particularly intriguing evidence for repetitive DNA sequences associated with the sphere loci, histone coding sequences interspersed with a tandemly repeated satellite DNA (Gall et al. 1981), to which I will return in Chap. 5.

I will conclude this chapter by describing the inheritance of the axial bars that lie on either side of the centromeres of the lampbrush chromosomes of *T.c. karelinii* but which are not found in other subspecies of *T. cristatus*. In six F_1 hybrids of parentage *karelinii* ♀ × *cristatus* ♂ all of the bivalents in all of the lampbrush complements examined were heterozygous for centric regions, the centromeres deriving from the female parent being regularly flanked by axial bars while those deriving from the male parent lacked axial bars (Fig. 3.2g). The same condition was found in four F_1 hybrids of parentage *karelinii* ♀ × *carnifex* ♂. In nine F_1 hybrids of parentage *carnifex* ♀ × karelinii ♂ all bivalents were again recognizably heterozygous in their centric regions, though in these hybrids the centromeres flanked by axial bars were derived from the ♂ parent.

Thus the axial bars of *T.c. karelinii* centric regions are inherited and expressed without apparent modification in F_1 hybrids. Furthermore, in bivalents of F_1 hybrids which are favourable for analysis because a chiasma has formed close to the centromeres, the length of chromosome axis, and to a first approximation the number of chromomeres and pairs of lateral loops between the distal end of an axial bar and a neighbouring chiasma on a *T.c. karelinii* chromosome equate with comparable structures lying between the centromere and the chiasma on the *T.c. cristatus* or *T.c. carnifex* chromosome. So the *T.c. karelinii* axial bars represent chromatin that is absent from the other subspecies; they are not formed from condensed and amalgamated chromomeres, present also in *T.c. cristatus* or *T.c. carnifex* chromosomes, but whose lateral loops have regressed precociously.

Two female F_1 hybrid newts, one of parentage *karelinii* ♀ × *carnifex* ♂, the other of reverse parentage, were backcrossed to *T.c. carnifex* ♂♂ and the lampbrush chromosomes of their female progeny examined for centromere inheritance. The outcome of this analysis is shown in Table 3.3.

All bivalents in the backcrosses should have inherited one centromere from the *T.c. carnifex* father, the other from either the *T.c. karelinii* or *T.c. carnifex* grandparent dependent on how the centromeres segregated at the first meiotic anaphase in the oocytes from which they developed. *T.c. karelinii*-type centromeres are still distinguishable from *T.c. carnifex*-type centromeres in the backcross generation, and the data displayed in Table 3.3 show that centromere segregation and assortment in the meioses of the F_1 hybrid parent were at random. This result is in contrast to the observations on "affinity" in mouse crosses made by Michie (1953) and Wallace (1953), and discussed by B. John and Lewis (1965, pp. 252–253), where unlinked markers of similar ancestry tended to assort together in meiosis.

Although, as I have just mentioned, *T.c. karelinii*-type centric regions are always recognizable in these backcross hybrids, in many instances the axial bars in all the backcross hybrids differ in morphology from those present on the lampbrush chromosomes of pure bred *T.c. karelinii* and also F_1 hybrids. The axial bars tend to be less compact, and a pair of lateral loops are frequently present at the sites of the constrictions between the axial bars and centromere granule, either on one side only or both (Fig. 3.2h). This development of lateral loops in places

Table 3.3. Inheritance of *T. c. karelinii* (k) or *T. c. carnifex* (c) centromeres from F_1 hybrid parents

Chromosome		I	II	III	IV	V	VI	VII	VIII	IX	X	XI	XII	Centromere totals	

1. (*T. c. karelinii* ♀D × *T. c. carnifex* ♂A) F_1 ♀A backcrossed to *T. c. carnifex* ♂E

Backcross ♀														k	c
A		k	c	c	k	k	c	k	c	c	k	k	k	7	5
B		c	c	k	c	c	c	c	k	k	k	c	k	5	7
C		k	c	k	k	c	c	k	k	c	k	c	k	7	5
D		k	k	c	k	k	c	c	c	k	c	c	c	5	7
E		k	k	c	k	c	c	k	k	k	k	k	k	9	3
F		k	k	k	k	k	k	k	c	k	c	k	c	9	3
G		c	k	c	k	k	k	k	c	c	c	k	k	7	5
H		k	c	k	c	c	c	k	k	c	k	c	c	5	7
J		c	c	c	k	k	c	k	c	k	k	c	c	5	7
K		c	c	c	k	k	c	k	c	c	k	k	k	6	6
Centromere	k	6	4	4	8	6	2	8	4	5	7	5	6	65	
totals	c	4	6	6	2	4	8	2	6	5	3	5	4		55

Chromosome		I	II	III	IV	V	VI	VII	VIII	IX	X	XI	XII	Centromere totals	

2. (*T. c. carnifex* ♀L × *T. c. karelinii* ♂A) F_1 ♀A backcrossed to *T. c. carnifex* ♂C

Backcross ♀														k	c
A		k	c	k	c	k	c	k	c	k	k	c	k	7	5
B		c	k	c	k	k	c	c	c	c	k	k	c	5	7
C		c	k	c	c	k	k	c	c	c	k	c	k	5	7
D		c	c	k	c	c	k	k	c	c	k	c	k	5	7
H		c	k	k	c	k	k	k	k	c	c	k	c	7	5
Centromere	k	1	3	3	1	4	3	3	1	1	4	2	3	29	
totals	c	4	2	2	4	1	2	2	4	4	1	3	2		31

where they are normally absent is evidently induced by hybridity; it is not an invariable feature, but it was found to be present in more than 90% of the *T. c. karelinii*-type centric regions examined. It is yet another indication of the extent to which lampbrush chromosomes' morphologies depend on the cellular environment in which they engage in transcription.

More refined studies concerning interaction between lateral loops and the genotypic environment in which they develop have been carried out in *Drosophila*. This topic will be discussed in Chap. 7.

CHAPTER 4

Morphological Aspects of RNA Transcription on Lampbrush Chromosomes

It had been apparent to many of the earliest observers of lampbrush chromosomes that their organization, notably the extreme state of dispersion of those nuclear materials that stain with basic dyes in histological preparations, is connected with a preponderance of anabolic activity in oocytes. Many years were to pass before RNA, and the nature of its synthesis, had been discovered, and of its relationship to protein synthesis. Information that had accrued by the late 1950s was reviewed in Chap. 1.

That the lampbrush chromosomes not only contain RNA, but are also directly involved in RNA synthesis, was first demonstrated by Gall (1958) using autoradiography. Gall made lampbrush chromosome preparations from *Notophthalmus viridescens* that had previously been injected with ^{14}C-adenine, and the autoradiographs showed labelling over the chromosomes. Carbon-14 gives only poor resolution in autoradiographs, but with tritium the resolution is a great deal better, so ^{3}H-uridine was used by Gall and Callan (1962) in a similar study carried out on *Triturus c. cristatus*. The autoradiographs over lampbrush chromosomes isolated from oocytes after short-term provision of ^{3}H-uridine showed label confined to the lateral loops, the level of labelling along individual loops appearing uniform (Fig. 4.1 a). However there were noticeable differences in the levels of labelling of different loops (Fig. 4.1 b), and one landmark loop near the end of the left arm of chromosome XII, the GGL (Fig. 3.18 a), was exceptional in being labelled only in the dense tip close to its thin insertion in the chromosome axis (Fig. 4.1 c).

The uniform distribution of labelled RNA on the overwhelming majority of lateral loops, and regardless of the quantitative distribution of RNP matrix on

Fig. 4.1 a–e. Phase contrast photographs of autoradiographs of lampbrush chromosomes of *Triturus c. cristatus* that had incorporated ^{3}H-uridine. The preparations were not centrifuged, so only occasional loops were flattened and well extended. a Ordinary loop from a newt 18 h after injection of ^{3}H-uridine, showing uniform incorporation. b Two neighbouring loops from a newt 26 h after injection, showing different levels of incorporation. c Giant granular loops on chromosome XII from an oocyte after 4 h culture in vitro with ^{3}H-uridine. d Giant granular loops from a newt 6 days after injection of ^{3}H-uridine. *Arrowheads* in c and d point to the unlabelled regions. e Autoradiograph of the giant loops on bivalent II of *Notophthalmus viridescens* showing incorporation of ^{3}H-UTP and ^{3}H-CTP after 20 min incubation of an oocyte nucleus in vitro. Nuclear contents were dispersed in dilute saline before centrifugation, and matrix RNP is well extended from loop axes; silver grains lie over the loop axes. *Bars* = 10 μm. a, c and d From Snow and Callan (1969); b from Gall (1963 b); e from Schultz et al. (1981)

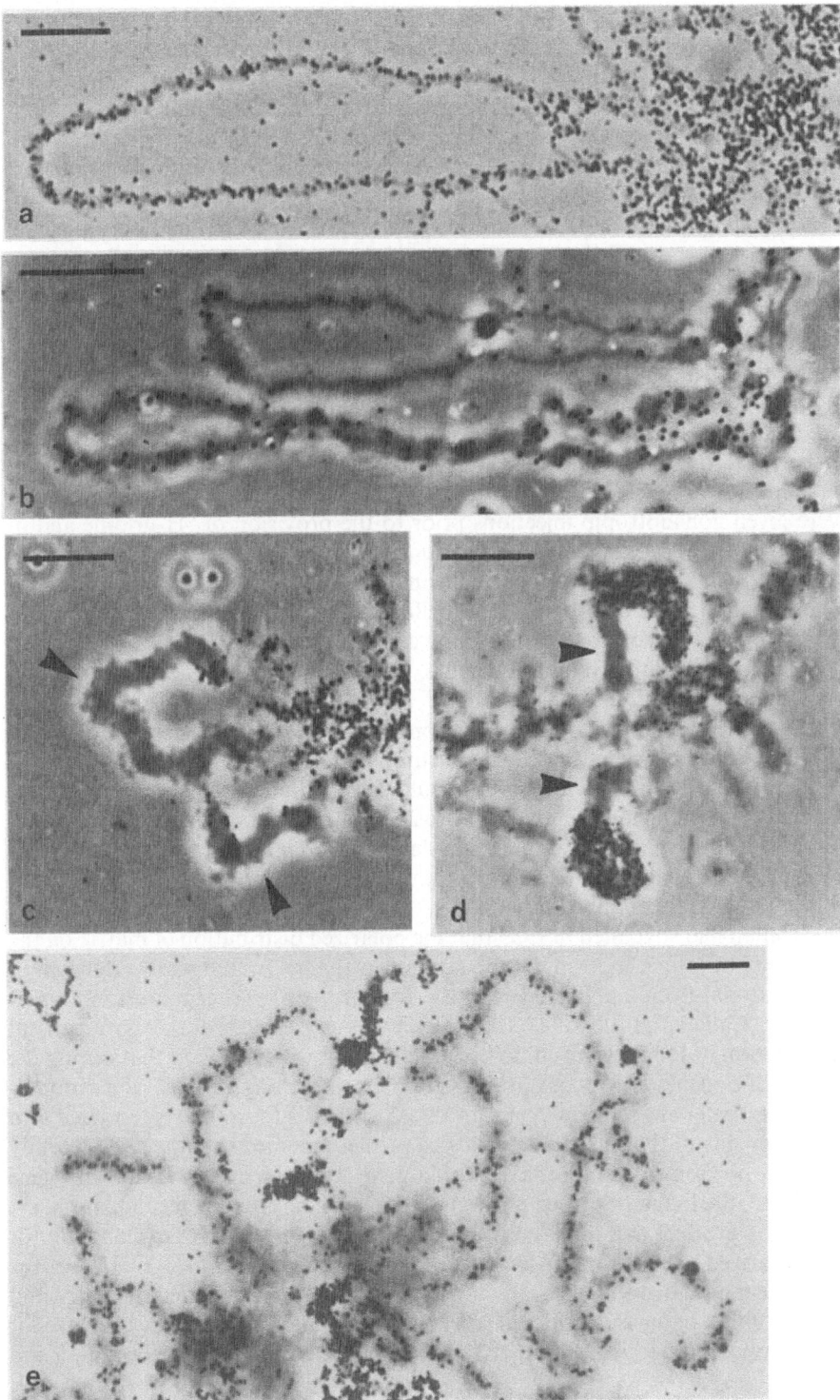

Fig. 4.1 a–e

these same loops, after short-term incubation in vitro of pieces of ovary with tri-tiated ribonucleosides, has been confirmed in several subsequent studies (Gall 1963b, Snow and Callan 1969, Hartley and Callan 1978). It indicates that RNA transcription is simultaneously in progress throughout the length of each lateral loop. However this point was insufficiently emphasized in the Gall and Callan paper of 1962, where instead attention was focussed on the exceptional behaviour of the GGL.

Females of *T.c. cristatus* were injected with ^{3}H-uridine, and chromosome preparations were made from oocytes that had been left in these newts for periods ranging from 12 h to 56 days. Autoradiographs over these preparations showed that on the basis of levels of labelling of the generality of loops, the injected ^{3}H-uridine enters an abundant pool of RNA precursors within each oocyte, and re-mains available for incorporation, though with diminishing specific activity, for more than 28 days after injection. The significant observation concerning the GGL is that it becomes labelled from the thin dense tip for progressively longer distances along the granular zone (Fig. 4.1d), until the loop is labelled throughout its length by about 10 days after injection. The newts were kept at 16 °C; some were given gonadotropin injections prior to the provision of ^{3}H-uridine and in these the rate of progress of labelling along the GGL was accelerated by a factor of about two. Later observations made on Russian specimens of *T.c. cristatus* in-dicate that the rate of progress of labelling along the GGL can be still higher (Makarov and Safronov 1974, Safronov and Makarov 1975).

The observations are not in dispute, indeed Gall (1963b) found that a loop near the left end of chromosome XI of *N. viridescens* labels in similar fashion (consequently acquiring the designation "sequential"), likewise Nardi et al. (1972) found that one of the GFLs on chromosome X of *T. marmoratus* also la-bels sequentially. In parenthesis it should be mentioned that there are no se-quentially-labelling loops in the lampbrush chromosome complement of *T.c. car-nifex*. The interpretation of sequentially-labelling loops considered most plausible by Gall and Callan in 1962 we now know to be wrong, though firm proof that this is so was only obtained in 1977.

Callan had suggested in 1955 that the polarized distribution of matrix on the generality of lateral loops might be accounted for by "...the serial extension of loop material from a polarized source". This proposal was expressed more pre-cisely by Callan and Lloyd (1960), where it was suggested that the asymmetrical distribution of loop matrix could be explained if it were assumed that during the course of oogenesis loop axis progressively extends from one side of a chromo-mere, and after engaging in "gene product" synthesis for a time is retracted into the other side of the chromomere. It was further assumed that the most recently exposed portion of the loop axis would have the least accumulation of "gene product", the longest exposed portion of loop axis the greatest accumulation, i.e. that gene product and loop axis move together. An alternative explanation for the polarized distribution of loop matrix, specifically that of the GGL, i.e. the movement of matrix along a stationary loop axis, was not excluded by Callan and Lloyd (1960), but was considered less likely. Once autoradiographic evidence for the movement of RNA synthesized on the GGL had been established, Gall and

Callan (1962) again considered the two alternative interpretations and preferred the moving axis proposal.

In view of a topic to be considered in Chap. 5, that of the distribution of DNA sequences that are transcribed on lateral loops, mention should be made here that after short-term incubation of oocytes not only is ^{3}H-uridine incorporation restricted to the thin tips of the sequential loops on chromosome XI of *N. viridescens;* so also is incorporation of three other RNA precursors, ^{3}H-adenine, ^{3}H-cytidine and ^{3}H-guanosine (Hartley and Callan, unpublished). Thus the initially unlabelled thicker regions of the sequential loops are genuinely inactive in RNA synthesis.

Another matter raised in the Gall and Callan paper of 1962 was of consequence: the length of DNA in the newt genome. The C-value (Swift 1950) of *T.c. cristatus* was reckoned to be about 30 pg (a more recent determination by Olmo 1983, gives it as 19 pg), equivalent to about 10 m of double helical DNA. The width of the single strand (DNA plus histone) running between the chromomeres of newt lampbrush chromosomes, on later evidence shown to consist of two sister chromatids, had been estimated by Tomlin and Callan (1951) to be about 20 nm, and by Gall (1958) to lie somewhere between 20 and 40 nm. The order of magnitude of these dimensions, coupled with the autoradiographic demonstration by ^{3}H-thymidine incorporation that the DNA of eukaryotic chromosomes replicates by a semi-conservative mechanism, and segregates as though each chromatid contains only one double helix (Taylor et al. 1957, Taylor 1958, 1959) pointed to the remarkable lengths of DNA duplexes in *Triturus* chromosomes, and first raised the problem of the C-value paradox. Although proof of the uninemy of chromatids is not strictly germain to the subject of this chapter, conclusive further evidence for uninemy was soon to be provided by Gall (1963 a, b) with his analysis of the kinetics of breakage of lampbrush chromosome main axes and lateral loops by pancreatic DNase, and by Miller's (1965) determination of the width of deproteinized loop axes, 2 to 3 nm.

Taking studies on RNA synthesis in chronological order, the next important observation was the demonstration by Izawa et al. (1963) that when newt oocytes are incubated in vitro for a few hours in Tris-buffered pH 7 saline containing the antibiotic actinomycin-D at 10 μg ml^{-1}, not only is RNA synthesis inhibited, but the loops are stripped of their already synthesized RNP matrix and loop axes retract to the main chromosome axes; the chromosomes come to resemble those of oocytes that are approaching maturation. Puromycin on the other hand, an inhibitor of protein synthesis, has no such effect on lampbrush structure. These observations led Izawa et al. (1963) to the conclusion that "...the morphology of an active chromosome site is not only closely related to its functional role in RNA synthesis, but is dependent on it". Other observations made in the paper by Izawa et al. (1963) leave room for doubt (see Snow and Callan 1969) but the striking and specific action of actinomycin-D on intact oocytes (though not on isolated lampbrush chromosomes) was soon confirmed by Gall and Callan (unpublished), Ebstein (1967) and Mancino et al. (1968). In passing it should be mentioned that another powerful inhibitor of RNA synthesis, α-amanitin, likewise does not bring about the detachment of loop matrix when chromosomes are isolated directly

into a saline containing this inhibitor (Schultz et al. 1981), though just as with actinomycin it causes matrix to be shed and loop axes to retract if oocytes are incubated for several hours in its presence (Mancino et al. 1971, Bucci et al.
1971).

In an attempt to secure evidence for or against the moving loop axis theory,
Snow and Callan (1969) made further experiments involving actinomycin-D. The
primary aim was to study by autoradiography the recovery of RNA-synthesizing
capacity in ordinary lateral loops of *T.c. cristatus* that had previously experienced
inhibition, though not to the extent that total stripping of loop matrix had occurred. The results were equivocal. However two points emerged from this study, in
experiments where recovery from deliberately severe actinomycin-D inhibition
was followed. Unlike the general run of loops, the GGLs of chromosome XII
(also the GFLs and the smaller dense landmark loops) do not suffer immediate
stripping of RNP matrix. This is presumably because the primary RNP transcripts of these loops are bonded together by accessory protein and, in the case
of the GGLs, transcription, the process that is so sensitive to actinomycin-D and
in normal loops results in matrix stripping, does not occur in the granular zone.
Within a day of recovery a long (ca. 15 µm) strand makes its appearance between
the chromosome axis and the original dense tip of the GGL, after 2 days this
strand becomes contorted and starts to accumulate RNP matrix, the former granular zone meantime progressively shedding matrix from its thicker end, and
after 7 days a new GGL has come to replace the old. This observation was still
no proof that the new GGL had formed on a newly extended loop axis, but it
prompted us to look more closely at the morphology of the GGL. If the lampbrush chromosomes of *T.c. cristatus* are dispersed in dilute saline (40 mM instead
of the usual 100 mM), the granules of the granular zone break up into smaller and
smaller particles so that this region fluffs out, while the dense tip, normally only
4 or 5 µm long, extends to a length of some 100 µm, and displays a matrix texture
not unlike that of the generality of lateral loops. The limits of the two zones remain well defined, and evidently the overall length of loop axis of the GGL that
is engaged in RNA synthesis is much the same as that of normal loops which synthesize RNA throughout their lengths; it is exceptional in being tightly compacted
while transcribing.

The other observation concerns loop retraction. When RNP matrix is stripped from normal loops by any of several agents applied to a preparation freshly
made in pH 7 saline, their axes retract. However when actinomycin-D acts on intact oocytes and a resulting preparation shows retracted loops, the retraction
might have occurred prior to making the preparation, or at the time of making
the preparation when the nuclear sap dispersed, i.e. a conceivable possibility was
that loop axes might have remained extended after matrix stripping had occurred
within an intact nucleus, but only retracted when the chromosomes were isolated.
We were able to rule out the second possibility by comparing sections through
Sanfelice-fixed, haematoxylin-stained oocytes, some of which had been fixed immediately following actinomycin-D inhibition, others allowed 1 day of recovery,
and untreated controls. The oocytes fixed immediately after actinomycin-D treatment showed no loops, while the oocytes fixed after 1 day of recovery showed abnormally short but normally stained loops, not loops of normal length but poorly

stained. So loop axis retraction is a genuine phenomenon that occurs in intact oocytes, and is not a secondary consequence of chromosome isolation. This retraction is now known to be due to the compaction of transcriptionally inactive DNP
into nucleosomes, and structures of yet higher levels of compaction. In the normal
course of oogenesis retraction of lateral loop axes and compaction of chromosomal DNP coincides in time with the shut-down of RNA transcription. This may
be controlled in a positive manner, and not necessarily at the end of oogenesis,
as is exemplified by elasmobranch and birds, where oocyte growth continues long
after the lampbrush phase has terminated. Loones (1979) exposed *Pleurodeles* oocytes to γ-irradiation and found that whereas the lampbrush chromosomes of
mid-vitellogenic oocytes showed no obvious reaction to irradiation, in older oocytes the lateral loops collapse and chromosome axes condense within a few
hours. However, just as with actinomycin-D inhibition, the process is reversible,
and 3 days after irradiation the lampbrush chromosomes have reverted to their
normal state. Loones suggests that this inhibition of transcription is brought
about by a "maturation factor" released from the cytoplasm as a result of irradiation, and that this factor is not present in younger oocytes, hence their failure
to respond.

In 1965 Miller described his first electron microscopic observations on surface spread and microcentrifuged lampbrush chromosomes of *N. viridescens,* and
in this paper he included a remarkable photograph showing several loops intact
from their thin to thick insertions, and with RNP fibrillar matrix of similar texture throughout, but with short fibrils close to the thin insertion, the fibrils becoming progressively longer as they approached the thick insertion. Miller therefore proposed: "... that the synthesis of RNA molecules is initiated repeatedly at
the thin insertion end of typical loops and continues around the entire length of
a stationary loop axis with each molecule continuously being unzipped from the
loop axis as it is synthesized".

I was at the time reluctant to accept Miller's proposal, because although it
sufficiently accounted for matrix distribution in those loops where there is a uniform and relatively steep gradient in width of matrix from the thin to the thick
insertion, such loops, though conspicuous, are few (Fig. 4.5 a). In unfixed preparations that have not been flattened by centrifugation most of the loops of urodele
lampbrush chromosomes appear to carry far less matrix at their supposedly thick
insertions than would be anticipated from their lengths. An example, convincingly demonstrated in the form of a double-loop bridge, is shown in Fig. 1.8 a. In
my opinion this argued for a combination of processes, the lengthening of transcripts as synthesis proceeds, together with loop axis extension from the thin insertion, retraction at the thick insertion, which might be expected to diminish the
slope of the matrix gradient. Such a notion called for not one, but many initiation
sites for RNA synthesis within the DNP axes of individual loops. At the time this
did not present itself as an insuperable problem, at least in theory, because there
was the possibility that coding sequences within the DNP might be serially repeated, according to the Master/Slave organization of genetic units proposed by
Callan (1967) to account for the C-value paradox. Moreover it was a notion reinforced by the assumption that each loop is a unit of transcription. The assumption
was wrong, and the Master/Slave theory is no longer tenable.

The question as to whether or not there is loop axis movement has finally been resolved as a result of the in situ hybridization of labelled probes to RNA transcripts on lampbrush lateral loops (Old et al. 1977, Macgregor and Andrews 1977, and see Chap. 5). Although loop axes must extend as transcription gets under way in young oocytes, once transcription has been fully established they extend no more, and they only retract as transcription diminishes and ceases. The distribution of matrix on most ordinary loops is determined by other factors than I had at the time appreciated, notably multiple transcription units, including switches in the direction of transcription, within single loops. These complications will be considered later in this chapter and in the one following.

In the course of a few years following 1965 the microcentrifugation technique applied to oocyte nuclear contents was refined in various ways, and Miller and Beatty (1969 a) published the first dramatically convincing EM photographs of RNA transcription in progress along tandemly arranged segments of rDNA, with intervening untranscribed "spacer" regions, originating from the free nucleoli of *N. viridescens.*

The first, and equally dramatic, EM photographs of RNA transcripts on the lateral loops of *N. viridescens* lampbrush chromosomes, were published by Miller and Beatty (1969 b) and were more fully documented by Miller and Hamkalo (1972). In conformity with the impressions formed from light microscopy, the loop axis near the thin insertion end of a loop is densely clothed with RNP fibrils that steadily increase in length the further they lie from the thin insertion (Fig. 4.2). Where each matrix fibril is attached to the loop axis there is a granule some 12.5 nm in diam, presumed to be a molecule of RNA polymerase, and these molecules lie very close to one another, with a centre-to-centre spacing of approximately 35 nm. Comparison between RNP fibril lengths and the loop axis lengths from which they had been transcribed indicate that the RNA molecules do not remain fully extended as nucleotides are added; they are foreshortened, presumably because of their immediate association with protein as synthesis proceeds. Miller and Hamkalo did not put a figure on the degree of foreshortening of the RNP fibrils that they observed, nor were they able to assess the lengths of RNP fibrils at the thick insertion ends of loops, but they did observe RNP fibrils up to 25 μm long at intermediate positions on loops.

"Miller-spreads" of the contents of oocyte nuclei have proved of great value for the study of various aspects of transcription, and many papers where this material and this technique have been used have been published in recent years. Some of these papers, on organisms as diverse as the unicellular alga *Acetabularia* (in this case the contents of the so-called primary-nucleus), the cricket *Acheta,* the beetle *Dytiscus* and various amphibians, have been primarily concerned with the transcription of 18+28S rRNA. They demonstrate the ubiquity of the system, al-

---

Fig. 4.2. Part of a transcription unit (TU) in a Miller-spread of the lampbrush chromosomes of *Pleurodeles,* including its origin. The larger dense objects on the axis of the TU are RNA polymerase II particles. Beyond the origin (*arrow*) the non-transcribed DNP axis is beaded, and is compacted into nucleosomes. The several non-transcribed long beaded strands running down the micrograph are also nucleosomally-compacted DNP, presumably of chromomeric derivation. *Bar* = 1 μm. (From Scheer and Dabauvalle 1985)

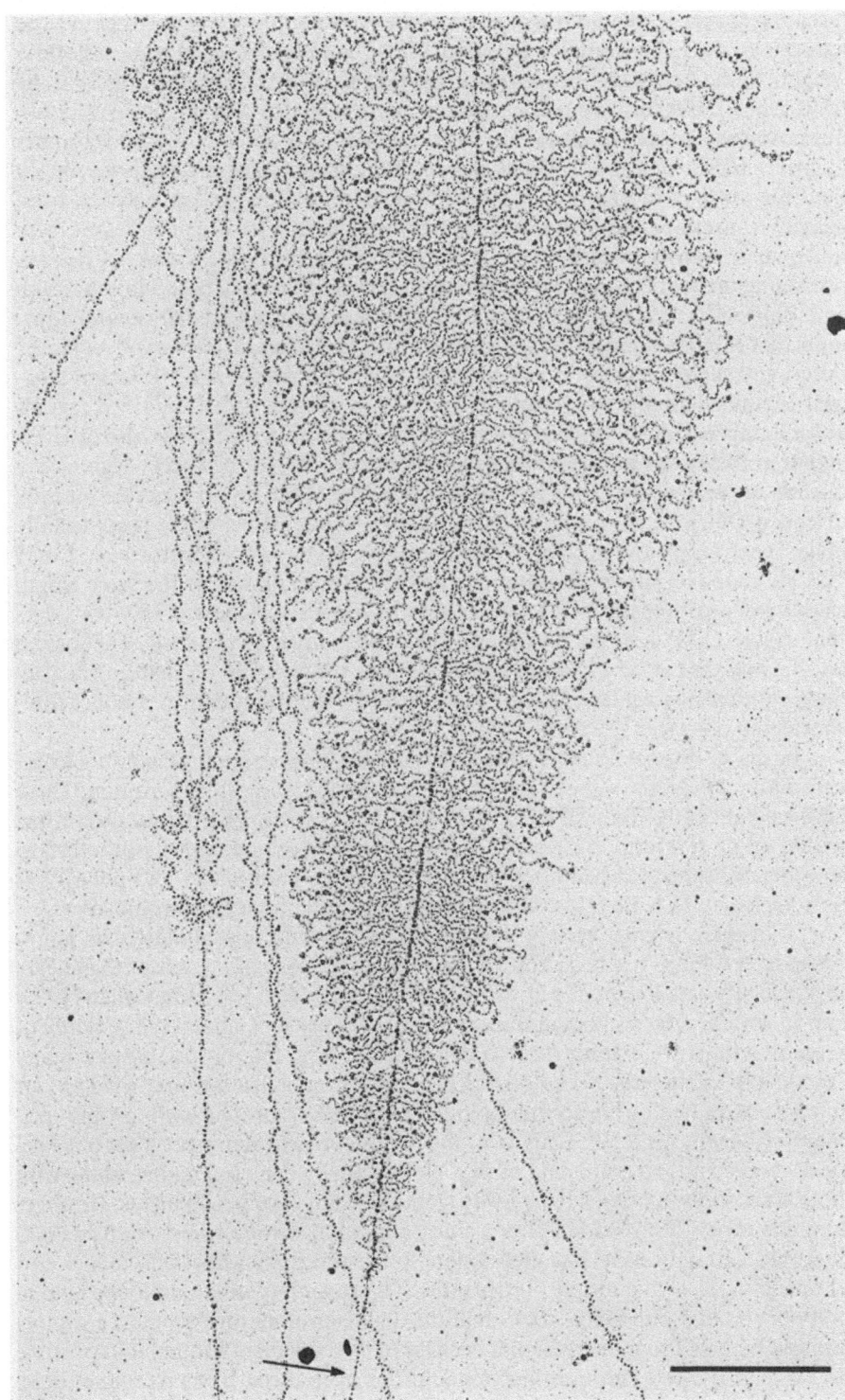

Fig. 4.2

ternating spacer DNP and transcribed segments of lengths that are either in close agreement with, or even longer than, those anticipated from the molecular weights of the first stable product of rRNA transcription. This indicates that the transcribed DNA in these preparations is in its extended β-form, not coiled into nucleosomes (Franke et al. 1976). These authors also observed no nucleosomes in spacer DNP. Scheer (1978) has shown that during transcription in the amplified free nucleoli of amphibian oocytes nucleosomes are not only absent from heavily transcribed units and their adjoining spacers; they are also absent from units where transcripts are sparse, as in previtellogenic oocytes and oocytes approaching maturity, likewise in vitellogenic oocytes that have been inhibited with actinomycin-D. It is only when transcription has ceased altogether on individual units that rDNP compacts to form nucleosomes and dense structures of yet higher order, and such inactive compacted units may lie on the same DNP strand adjacent to units with few transcripts in which both spacers and transcribed regions are still fully extended. A source of confusion that Scheer (1980) has also clarified is that in Miller-spreads of transcriptionally active rDNP the spacer regions frequently appear beaded. However these regions are of the length expected for fully extended DNA, so the beads cannot be nucleosomes. Miller-spreads are usually made at extremely low electrolyte concentration. At higher concentrations, 1 mM NaCl and above, spacer chromatin is uniformly smooth but of the same length as in standard preparations made at very low salinity. Scheer (1980) considers that spacer DNP is normally covered by an even layer of proteins of unknown nature, and that at very low salt concentrations the particles on spacers are artefacts, residual aggregates of these proteins; at all events they do not produce foreshortening of spacer DNP.

In his 1978 paper Scheer also examined the organization of non-nucleolar chromatin in regions engaged in less than maximum rates of transcription. These differ significantly from rDNP in that between sparsely distributed transcripts within single transcription units (TUs) nucleosomes are present. I will return to consider Scheer's evidence after first discussing the multiplicity and polarity of non-nucleolar TUs, i.e. TUs on the lateral loops of lampbrush chromosomes.

That there is great variety in the form and distribution of matrix on lampbrush lateral loops has been known for many years. An early attempt to indicate this diversity was shown in a drawing, reproduced in Fig. 1.8, from Callan (1955) and in several schematic diagrams in a review paper by Callan (1963). Some of these diagrams were concerned to demonstrate how the morphology of many "landmark" structures is based on a lateral loop organization that is wholly or in part obliterated by the accumulation of adventitious protein amongst the "primary" filamentous RNP. Only one, then considered to represent a reversed repeat, showed a loop consisting of two symmetrically disposed matrix units, with both transcription origins lying away from the main chromosome axis. However had this review been written after a study of axolotl chromosomes (Callan 1966) it would certainly have included amongst the schematic diagrams others also showing loops with multiple matrix units. The generality of lateral loops in axolotl oocytes of 1 mm diam, about half the diameter of mature oocytes, are enormously long and so numerous that because of entanglement I found it impossible to make analyzable chromosome preparations from them. It is a feature of these very long loops that matrix does not show marked asymmetry from one axis insertion to the other (Fig. 3.1 b), and within the lengths of the majority there are several thin to thicker matrix units, i.e. multiple TUs.

Angelier and Lacroix (1975) used Miller-spreads to study the nuclear contents of the oocytes of the urodele *Pleurodeles,* and included several EM photographs of intact lampbrush TUs. One of these, reproduced here as Fig. 4.3 a, shows a TU about 20 μm long, with evident polarity, transcripts densely packed, the longest appearing shorter than the total transcribed length of DNP, though in view of their contorted nature RNP fibril lengths cannot be accurately assessed. Another illustration (Fig. 4.3 b) shows two neighbouring TUs of similar polarity but dissimilar in their lengths, with an untranscribed gap between of about 4 μm, and the authors make the point that this discredits an earlier view where each lateral loop was envisaged as a unit of function. Another photograph shows parts of two TUs of opposite polarity, with only 2 μm separating the two sites of initiation, both likewise taken to be components of a single loop.

This matter of the number and variety of TUs within single lateral loops as seen in Miller-spreads by electron microscopy was further investigated by Scheer et al. (1976 a), who exposed lampbrush chromosomes of urodele amphibians and *Acetabularia* for varying periods of time to low salt concentration before centrifugation, and compared the resulting morphologies. They show photographs and present diagrams (Fig. 4.4) of lateral loops of several different categories. The "classic" kind is defined as possessing matrix throughout the length of the loop, clearly recognizable thin and thick insertions without intervening "interruptions", the matrix gradually increasing in width from the thin insertion end, but becoming of maximum width, and thereafter remaining uniform, before reaching the thick insertion. A derivative of the "classic" kind is represented by loops which comprise a single TU of clearly recognizable polarity, but preceded and/or followed by an untranscribed region or regions. Then there are loops including two or more TUs of different lengths but the same polarity, also loops which include two or more TUs of opposite polarity, arranged either head-to-head or tail-to-tail, where by "head" is intended the initiation, and by "tail" the termination of transcription. On evidence from light microscopy, similar categories of matrix distribution on the lateral loops of *T. cristatus* have been described by Makarov and Safronov (1976).

Wherever two adjacent TUs are of opposite polarity, there must either be intra-axial switches in DNA polarity (Wolff et al. 1976) or transcription is from one strand of the DNA in one unit and from the other in its neighbour; Scheer et al. (1976 a) prefer the second of these two alternatives, and evidence that their preference is correct has recently been established by in situ hybridization (Diaz et al. 1981, and see Chap. 5).

Scheer et al. (1979 a), in a general review of studies of transcription on the loops of lampbrush chromosomes, have in particular focussed attention on how observations of Miller-spreads in the electron microscope relate to observations with the light microscope. They show several light micrographs (Fig. 4.5 b–d) of loops that include multiple TUs from oocytes of *Triturus alpestris,* both with and without reversal of polarity, which tie in precisely with what can be seen in Miller-spreads at low ionic strength. They discuss whether a length of nucleosomal DNP

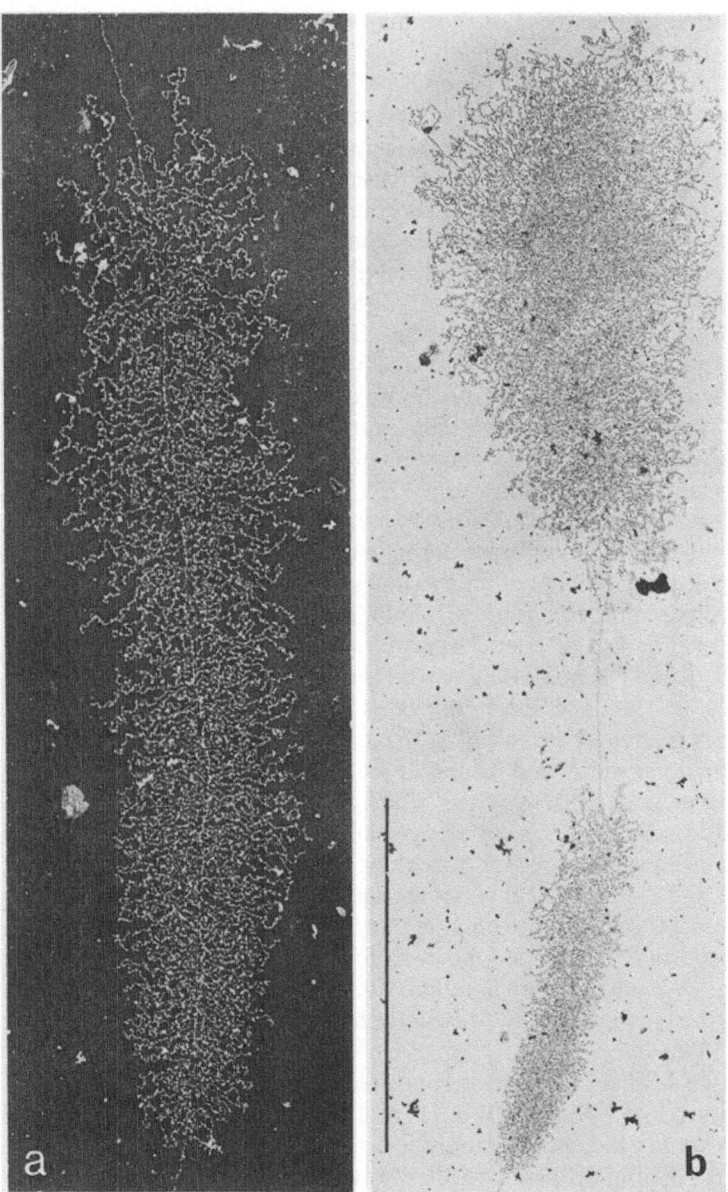

Fig. 4.3 a, b. Miller-spreads of transcription units (TUs) in lampbrush chromosomes of *Pleurodeles*. **a** Single TU printed as a negative. **b** Two TUs, of similar polarity, in tandem. *Bar* = 10 μm. (From Angelier and Lacroix 1975)

connecting neighbouring TUs in a Miller-spread originated by dispersion of a chromomere lying at the base of two or more loops, each of which in its native state consisted of a single TU, or represents an interruption in a loop that included more than one TU. They conclude that when the transcript-free regions are short (1 μm or less) they are most unlikely to have originated from chromo-

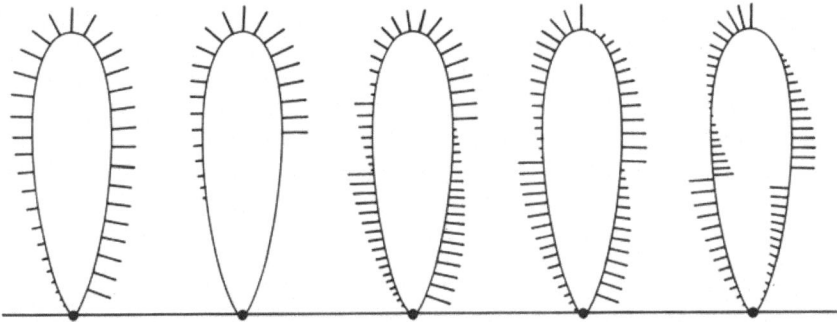

Fig. 4.4. Various arrangements of transcription units within the loops of lampbrush chromosomes. (From Scheer et al. 1976 a)

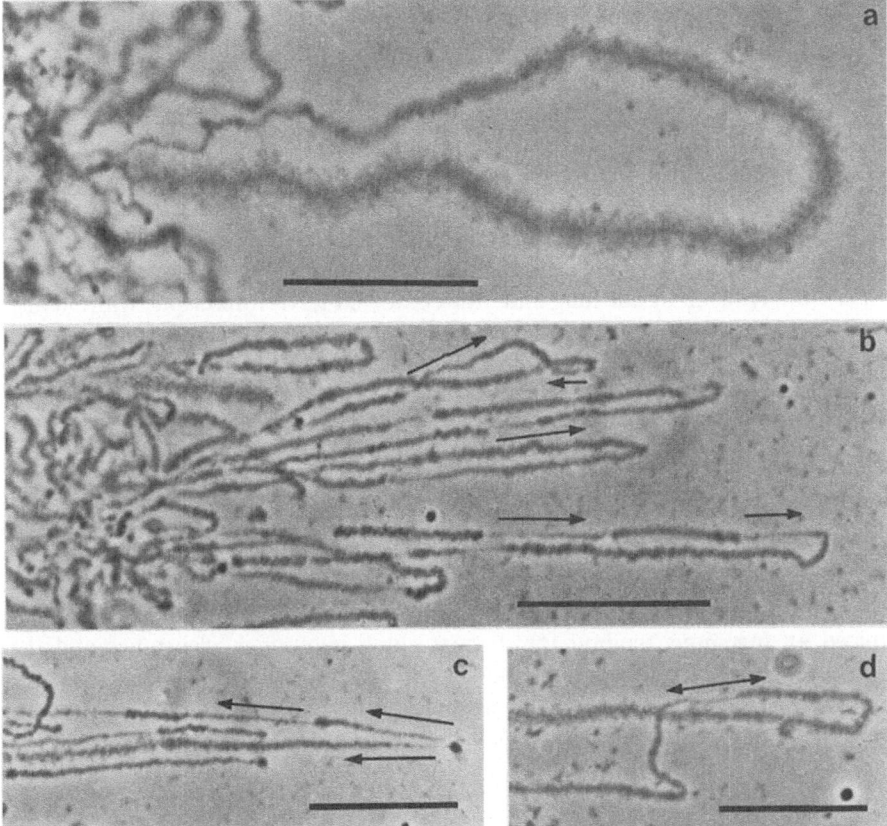

Fig. 4.5 a–d. Phase contrast photographs of lateral loops of urodele lampbrush chromosomes showing various arrangements of transcription units (TUs). **a** Loop consisting of a single, long TU in *Triturus c. cristatus*. **b, c** and **d** Loops including multiple TUs in *T. alpestris;* where clearly recognizable, the polarities of transcription are indicated by *arrows.* Some adjacent TUs are of similar, while others are of opposed polarities. *Bars* = 20 μm. **a** From Snow and Callan (1969); **b, c** and **d** from Scheer et al. (1979 a)

meric DNP. The distinction should not be stressed unduly, for it is certainly not absolute; the giant loops on chromosome II of *N. viridescens,* in different oocytes from a single female, can in some preparations all appear as single TUs, but in others as exceptionally long loops each comprising multiple TUs (Hartley and Callan unpublished). In this example the stretch of DNP between adjacent TUs sometimes is, and sometimes is not, included in the axial chromomere, and presumably depends on variability as to where transcription terminates. Precisely the same applies to the multiple GFLs of chromosome XII of *T.c. carnifex* (Callan and Lloyd 1960). The reversed repeat loops drawn in Callan's 1963 review and mentioned earlier in this chapter illustrate a further complication; they are evident landmarks on chromosomes V and VIII of *T.c. carnifex* but only when the dense tip, from which transcription in both TUs is directed back towards the chromosome axis, has been "expelled" from a chromomere. More frequently the transcription origins remain lodged within this chromomere, and the distinctive morphology then passes unnoticed. A clear example, from a Miller-spread, of two adjacent TUs of reverse polarity that in their native state must have existed within a single loop is shown in Fig. 4.6 for the transcript-free gap in the DNP is only 0.1 μm long, and without nucleosomes.

Scheer et al. (1976a) found many TUs where RNA polymerase density on loop axis was fully as high as that of the polymerases present on maximally active rDNA in free nucleoli, but they also illustrate one example of a loop axis bearing long (5 μm or so), but sparsely and irregularly distributed RNP transcripts, reminiscent of rDNA with reduced transcriptional activity (Scheer et al. 1976b). This particular photograph shows many examples of putative RNP circles, diam 0.05 to 0.1 μm, lying alongside long and little contorted RNP transcripts, without any visible connection to the latter (Fig. 4.7b). "Annelets" of similar dimensions, but connected in some manner to RNP transcripts, have also been noted by Angelier and Lacroix (1975). Whether these structures have any connection with RNP "processing" is not known. They are suggestive of intra-transcript RNA/RNA annealing, discussed in Chap. 5.

For several years there had been controversy as to whether the chromatin within TUs is packed into nucleosomes, or is fully extended, and as already mentioned, in the case of rDNA transcription this question was substantially settled by Franke et al. (1976); it is fully extended. Whether the same applies to transcription on lampbrush lateral loops remained uncertain until Scheer (1978) introduced an ingenious method for distinguishing between nucleosomes and the slightly larger RNA polymerases. Inactive chromatin, and this includes the chromomeric DNP of lampbrush chromosomes, appears beaded in Miller-spreads, and the beads are nucleosomes. If Miller-spreads are made in the presence of the anionic detergent Sarkosyl et 0.5%, most chromosomal proteins, including histones, are dissociated, and inactive chromatin fibrils assume a smooth morphol-

---

Fig. 4.6. Parts of two closely adjacent transcription units of opposite polarities in a Miller-spread of the lampbrush chromosomes of *Pleurodeles*. The larger granules lying on the DNP axis, each associated with an RNP transcript, are RNA polymerase II particles. There are fewer, relatively smaller granules between some polymerases, without associated transcripts; these are nucleosomes. *Bar* = 1 μm. (From Scheer and Dabauvalle 1985)

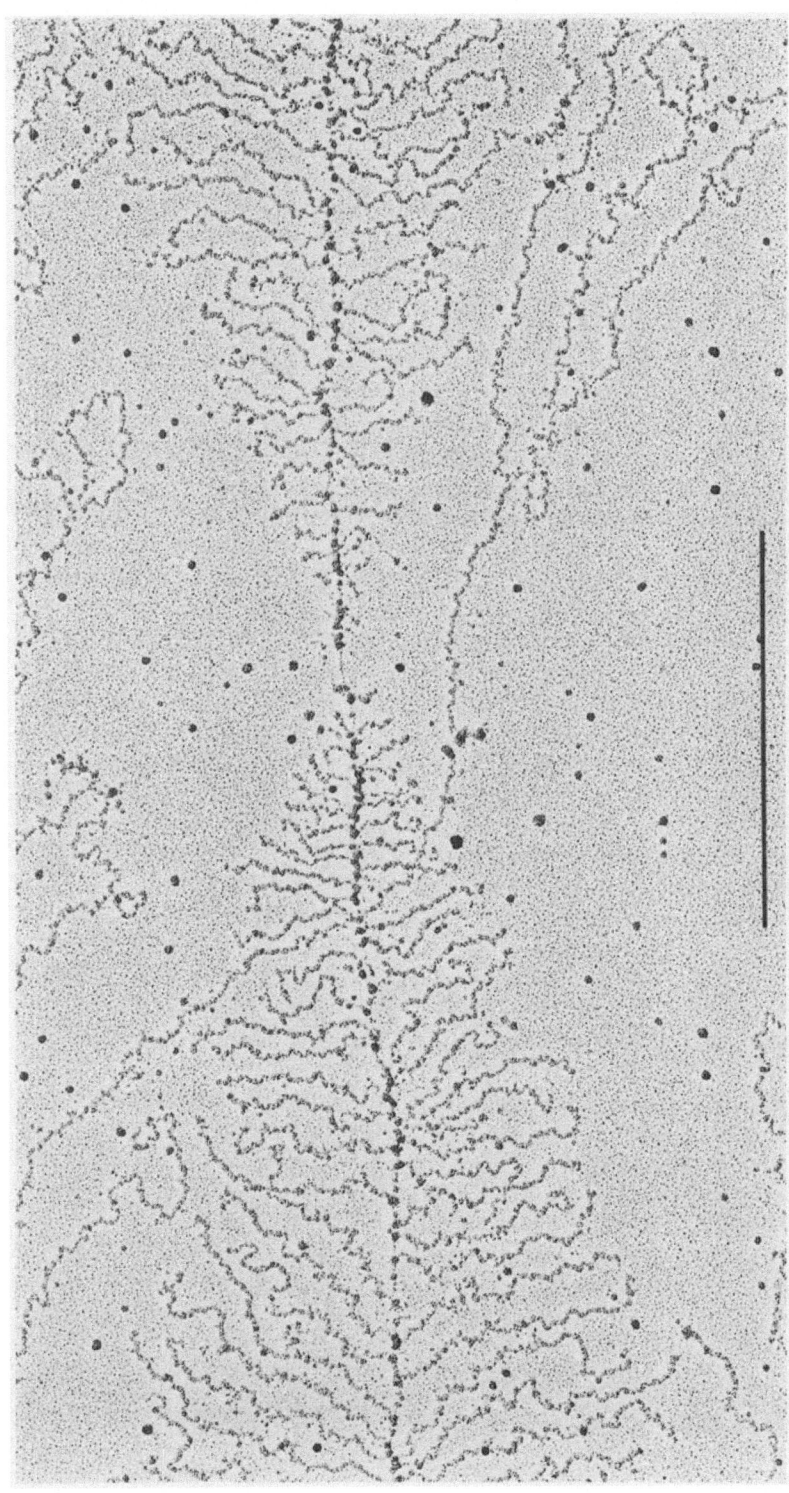

Fig. 4.6

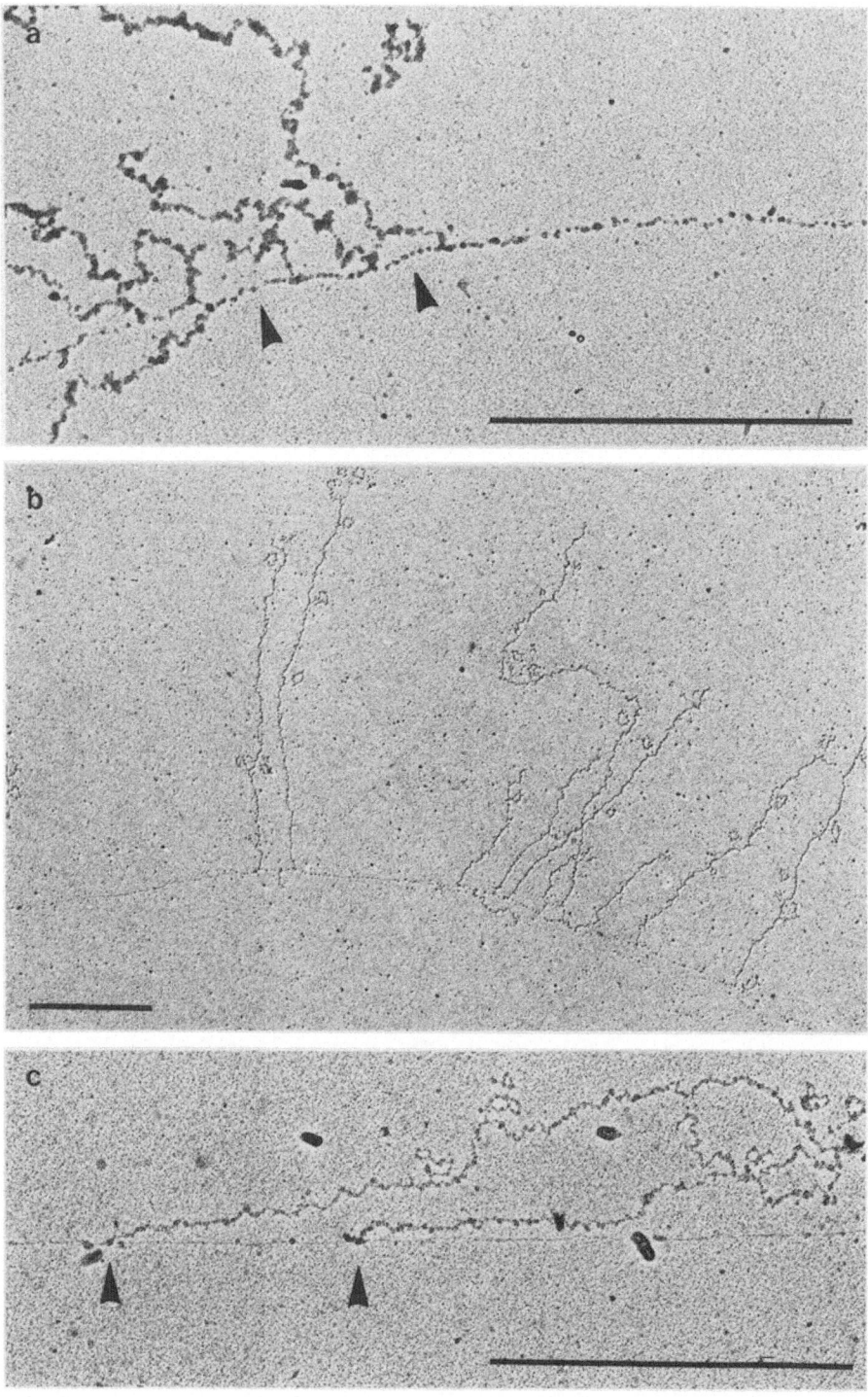

Fig. 4.7 a–c

ogy. However RNA polymerases and their associated transcripts are not destroyed by Sarkosyl, and they remain attached to the DNP axis that was being transcribed. In spreads made without Sarkosyl the axial intercepts of TUs between sparsely distributed transcripts are beaded (Fig. 4.7a), whereas in similar TUs spread in the presence of Sarkosyl the axial intercepts are uniformly thin and not beaded (Fig. 4.7c). When transcripts are at the limit of close-packing on a lampbrush TU, the axial DNP is fully extended and not nucleosomal. However when transcript density is low, intercepts of axial DNP between transcripts are compacted to form nucleosomes. In other words nucleosome packing is a reversible process that can occur repeatedly between acts of transcription, provided one RNA polymerase that has just transcribed a certain length of DNA, perhaps as short as 200 base pairs (bp), is not immediately followed by another (Fig. 4.6). This conclusion is supported by observations of Osheim et al. (1978) on *Xenopus*. Evidently it is this process, and higher levels of DNP compaction, that is responsible for general loop retraction towards the end of the lampbrush stage, and also for the stage-specific contraction of individual and particular loops, several examples of which are known in *T. cristatus*.

Some loops of urodele lampbrush chromosomes are remarkably thin, close to the resolving limit of the light microscope, and they show no sign of matrix polarity. They are noticeable because of the exceptionally violent Brownian motion that they show when dispersed in saline and lying free. Interpretation of their organization remained obscure until Scheer (1981), when studying Miller-spreads of *Pleurodeles* lampbrush chromosomes, observed tandemly repeated, short TUs occurring in clusters of more than 100 units (Fig. 4.8a). Some 12 polymerase molecules are attached to each unit, whose length includes about 940 bp (about 300 nm) of fully extended DNA. In the non-transcribed spacers between units, each about 400 nm long, the DNP is compacted as nucleosomes. The dimensions suggest that these tandemly repeated short TUs correspond with the exceptionally thin loops that show no matrix polarity when seen in the light microscope.

Scheer (1982) has recently discovered an even shorter tandemly repeated TU in *Pleurodeles*. In Miller-spreads these appear as an alternating pattern of thick and thin regions of chromatin, without apparent laterally associated transcripts, in clusters of some 10,000 units (Fig. 4.8b). Each thick region, 45 nm long, contains two juxtaposed particles of RNA polymerase size, while each intervening thin region, 80 nm long, includes on average two granules of nucleosome size. Assuming that in the thick regions two adjacent RNA polymerases cover fully ex-

---

Fig. 4.7 a–c. Miller-spreads showing parts of transcription units from *Triturus* oocytes in which transcripts are sparsely distributed. **a** From *T. alpestris; arrowheads* point to regions on the loop axis, between transcripts, where nucleosomes are evident. **b** From *T.c. cristatus* in a spread made in the presence of 0.1% Sarkosyl. At this detergent concentration most of the chromosomal proteins are removed. The RNP transcript granules have dispersed to produce long fibrils, but these remain attached to loop axis, which has retained its nucleosomal organization. Notice the ring structures adjacent to the fibrillar transcripts. **c** From *T. alpestris* in a spread made in the presence of 0.5% Sarkosyl. At this detergent concentration nucleosomes have disintegrated, leaving the loop axis as a thin, non-beaded strand. The transcripts remain attached to this strand by polymerase granules, indicated by *arrowheads*. *Bars* = 1 μm. **a** and **c** From Scheer (1978); **b** From Scheer et al. (1976a)

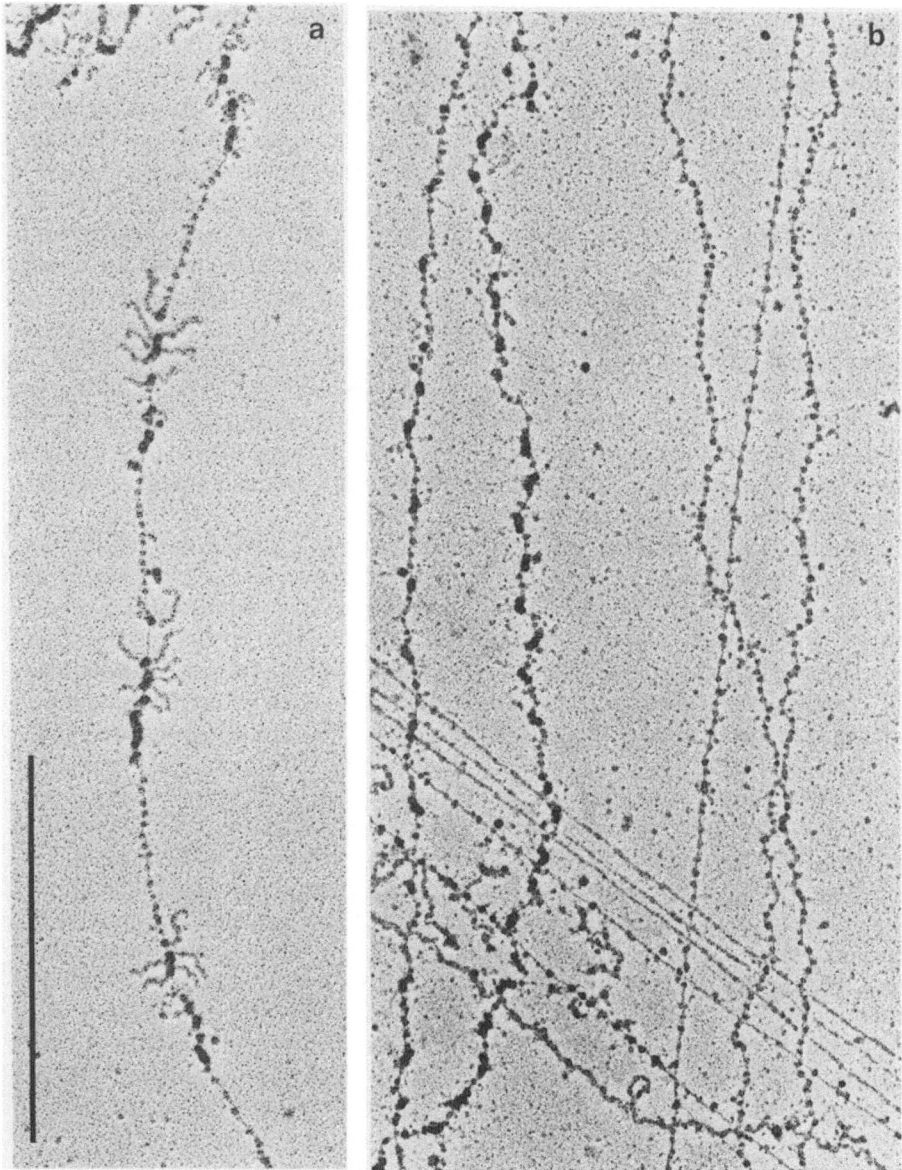

Fig. 4.8 a, b. Tandemly repeated short transcription units (TUs) on the lampbrush chromosomes of *Pleurodeles*. **a** Each TU is covered by about 12 RNA polymerases. **b** The two strands on the *left* contain extremely short TUs, each covered by two RNA polymerases; the three strands on the *right* consist of non-transcribed chromatin, presumably originating from a chromomere. *Bar* = 1 μm. **a** From Scheer (1981); **b** from Scheer (1982)

tended DNP, each TU comprises about 130 bp of DNA, while the non-transcribed spacer regions consist of some 240 to 400 bp of DNA, the number being dependent on the degree to which the DNP is nucleosomally compacted. Such short TUs would not be expected to show lateral RNP transcripts, for none are visible on the first two RNA polymerases close-packed on longer TUs (Franke et al. 1976, Laird et al. 1976, Scheer 1981). In spreads from single oocyte nuclei only one or two clusters of these highly repeated, exceptionally short TUs are present. In view of their size and degree of repetition Scheer considers that they may represent 5S rRNA genes (encoded by 120 bp of DNA) and/or tRNA genes (encoded by some 90 to 100 bp of DNA). If they do represent the transcription of 5S rRNA then they are unlikely to be extended as long lateral loops in vivo, unless *Pleurodeles* differs from *Notophthalmus* in this respect, for Schultz et al. (1981) have shown by autoradiography that 5S rRNA is transcribed alongside compact DNP in isolated oocyte nuclei of *N. viridescens,* and that this is probably also true of tRNA transcription. I will return to this topic in Chap. 5.

In their review Scheer et al. (1979a) include two photographs from Miller-spreads of *Xenopus* that provide evidence for the processing of nascent RNP transcripts on loops. The evidence comes from TUs showing a progressive gradient in transcript length, then an abrupt reduction in transcript length, though not to zero, followed by a gradient with the same polarity as before. If such processing or cleavage events were to occur repeatedly within the lengths of single TUs this could explain how certain loops, when viewed in phase contrast, have a clearly defined thin insertion and initial matrix gradient, but uniform width of matrix thereafter. Several contorted loops of exceptional length but uniform width of matrix are distinctive landmarks in the heteromorphic regions of chromosome I of *Triturus c. carnifex.* Scheer et al. (1979a) again consider packing densities of RNA polymerases along TUs of lateral loops; in amphibian oocytes and *Acetabularia* they commonly found even higher densities (15 nm centre-to-centre distances) than those recorded by Miller and Hamkalo (1972). Assuming a chain elongation rate of about 30 nucleotides per second established by Scheer (1973) for actively transcribed $18 + 28S$ rRNA in *Xenopus,* they calculate that a newly synthesized RNP fibril is released from loop axis once every 2 s and, making various other assumptions which refer specifically to *Xenopus,* that the maximum rate of hnRNA synthesis reached by *Xenopus* is some 22×10^6 transcripts per hour per nucleus.

An RNA polymerase packing density of 20 nm centre-to-centre is equivalent to 50 RNP transcripts per micron of chromatin, and per micron of DNA if fully extended. This figure is somewhat higher than transcript density on two very actively transcribed structural genes for which comparable information is available. The silk fibroin gene of *Bombyx mori* (McKnight et al. 1976) carries on average 23 transcripts per micron of chromatin, in which the DNA packing ratio is calculated to be 1.1. The two major Balbiani rings on chromosome IV of *Chironomus tentans* transcribing mRNAs for salivary polypeptides (similarly components of silk) carry on average 16 transcripts per micron of chromatin, in which the DNA packing ratio is calculated to be 1.6 (Lamb and Daneholt 1979). The rate of RNA chain elongation in Balbiani rings has been worked out at 31 nucleotides per second, a figure in excellent agreement with Scheer's (1973) estimate for *Xenopus* rRNA quoted earlier.

It should be borne in mind that the figures for rates of RNA synthesis in *Xenopus* oocytes are necessarily averages, and that they conceal systematic differences between rates of RNA synthesis in different components of the genome, some of these being stage-specific. Gall and Callan (1962) found significant differences in the labelling of different "normal" loops of *T.c. cristatus* in autoradiographs over single preparations of lampbrush chromosomes after short-term incorporation of ^{3}H-uridine (Fig. 4.1 b). Some of these differences must be attributable to different proportions of uridylic acid in the RNA transcripts synthesized on particular loops, and this should occasion no surprise when we consider what is known about the peculiarities of some of the DNA sequences being transcribed. However others reflect differences in rates of RNA synthesis.

Hartley and Callan (1978) made grain counts in autoradiographs of *N. viridescens* lampbrush chromosomes, from oocytes that had previously been incubated for 4 h in the presence of ^{3}H-adenine, ^{3}H-cytidine, ^{3}H-guanosine or ^{3}H-uridine, and compared the incorporation rates per unit loop length of these four RNA precursors into the RNP matrix of the giant loops on chromosome II with averaged incorporation rates on "normal" loops. Not only is ^{3}H-uridine incorporation lower than average, ^{3}H-guanosine incorporation is even lower, and taken overall the rate of RNA transcription on these giant loops is, at the most, only half the averaged rate. Yet the giant loops have a particularly liberal clothing of matrix, and they look to be fully extended in oocytes of 0.6 to 0.8 mm diam as judged by light microscopy. It seems plausible to assume that the lower rate of RNA synthesis on the giant loops of *N. viridescens* is a consequence of their exceptional length, for the transport of their very long RNP transcripts may present particular physical problems. It is generally assumed than an RNA polymerase and its associated transcript moves without rotation along a TU, and that this is accomplished by temporary local separation of the transcribed from the complementary strand of a DNA duplex as the polymerase passes by, with compensatory swivelling of the DNA duplex immediately in front of, and behind, the moving polymerase. Evidence from prokaryotes supporting this assumption has been briefly reviewed by Fisher (1982). However although it is well-nigh inconceivable that transcripts on the lampbrush loops of Amphibia rotate around the DNA double helix in loop axis, from phase contrast observations on unfixed lampbrush chromosomes' resolvable granular matrix aggregates appear to be helically distributed on certain loops, as was mentioned in passing in Chap. 3. Is this distribution in some way imposed by the helical organization of the loop axis that is being transcribed, and if so, at what level of organization? The question is not trivial, for Angelier et al. (1984) have confirmed by scanning electron microscopy that matrix aggregates are helically wound around loop axes in *Pleurodeles* lampbrush chromosomes, and moreover have shown this to be characteristic of *all* loops, not merely those with matrix aggregates large enough to be resolved by light microscopy. Angelier et al.'s (1984) observations will be considered later in this chapter.

The differentiated state of eukaryotic cells is certainly accompanied, and probably accomplished, by activating the transcription of different components of their genomes, and controlling their rates of transcription. From a cytologist's view point this is clearly shown by the diverse puffing patterns of the polytene chromosomes of different tissues of larval Diptera, as was first demonstrated by Beermann (1952) in *Chironomus*. It is now widely accepted that differentiation for the most part depends on control exercised over the transcription of mRNAs, rather than over their translation (reviewed by Davidson 1976), though the presence of masked mRNAs in oocytes implies that the expression of some coding sequences is under translational control. Oocytes are themselves differentiated cells and control over their transcriptional activities is to be anticipated. At one level this control presumably operates by initiating the transcription of some DNA sequences, not all. At a higher level the control presumably operates by determining appropriate rates of transcription of different DNA sequences. The relative lengths of the TUs of lampbrush chromosomes may well be one, but is unlikely to be the only factor that determines rates of RNA synthesis on different loops. Maximum packing densities of RNA polymerases on different TUs implies that as soon as the promoter region of a TU has been vacated by a polymerase which has moved on to start transcription, another polymerase takes its place. However this does not mean that rates of RNA synthesis on different TUs are necessarily similar, nor necessarily at rates as high as may be achieved in other cell nuclei, for rates of polymerase movement on different TUs will determine rates of RNP release at their termini. Control over the rates of polymerase movement may depend on diverse states of aggregation of primary RNP transcripts brought about by their interactions with one another or with accessory loop-specific proteins, as exemplified by some of the marker loops discussed in Chap. 3. It is an aspect of the organization of lampbrush chromosomes that has yet to receive the attention it deserves.

Until recently the oocytes of urodele Amphibia have for the most part been deliberately preferred for studying transcription on lampbrush chromosomes by the Miller-spreading technique. However the anuran *Xenopus* offers the advantage that its lampbrush TUs are generally much shorter than those of urodeles and is thus potentially capable of displaying more TUs in their entirety, and of the relationships between neighbouring TUs. Another good reason for paying greater attention to *Xenopus* lampbrush chromosomes is that the biochemistry of *Xenopus* oocytes has been much more deeply explored than that of any urodele.

R. S. Hill (1979) has made a quantitative study of *Xenopus* lampbrush chromatin from vitellogenic oocytes of about 0.5 mm diam, in which the rate of transcription is normally at its peak. He finds dispersed chromomeric DNP that is of great length, organized as nucleosomes, and without transcripts, i.e. just as in urodeles, and what he terms "transcriptive" arrays, 30 to 50 μm long, where two to seven TUs are interspersed with non-transcribed "spacer" DNP. This spacer DNP is also nucleosomal, and is only distinguishable from chromomeric DNP in a descriptive sense by its short though variable (1 to 22 μm) length; it can account for 30 to 80% of the total length of an array. R. S. Hill (1979) gives the lengths of TUs as mostly within the range 2 to 10 μm, the longest being between 30 and 40 μm. However what these figures represent in terms of fully extended DNP is debatable, for Hill finds that polymerase packing densities range from 40 to less than 10 μm^{-1} of chromatin. The 40 μm^{-1} packing density presumably occurs on almost fully extended DNP, but at lower packing densities Hill describes the DNP as being nucleosomal between neighbouring transcripts, so various degrees of DNP compaction must be present. Neglecting what is debatable, R. S. Hill's significant observation is that the TUs of *Xenopus* lampbrush chromosomes tend to be clustered, with long stretches of chromomeric DNP between clusters. It is open to question whether each cluster, i.e. transcriptive array, lies within a single loop, or whether the spacer DNP within an array normally resides in a chromomere. R. S. Hill favours the latter proposal; certainly if most transcriptive arrays were extended as single loops the general run of loops would considerably exceed in length those observed by Müller (1974) and by Jamrich et al. (1983).

R. S. Hill and Macgregor (1980) have used Miller-spreads of *Xenopus* to study the onset of transcription in young oocytes. The difficulties attending the study of small oocytes of known size and developmental stage were overcome by digesting small pieces of ovary from recently metamorphosed tadpoles with collagenase dissolved in calcium-free saline. This frees the oocytes from their investing follicle cells, and allows oocytes of specific sizes to be selected.

In oocytes of 10 to 25 µm diam, which includes all meiotic stages prior to diplotene, there are no signs of transcriptional activity and all the chromatin appears to be nucleosomal. Transcription begins in oocytes of about 50 µm diam, which are at early diplotene. At this stage short RNP transcripts are present, in some regions sparse and irregularly distributed, in others sufficiently abundant for the limits of TUs to be recognizable. In larger though still previtellogenic oocytes of about 100 µm diam (vitellogenesis in *Xenopus* begins when the oocytes have reached about 300 µm diam) transcripts are much more closely packed and regularly spaced, and the TUs generally resemble the "established" TUs of early vitellogenic oocytes except that their transcripts are more compact than those of fully established TUs. Hill and Macgregor claim from their observations that once transcription starts in *Xenopus* oocytes the lengths of the TUs are of the same range as those present when transcription rate is at its peak. This may well be so, but accurate comparison is precluded by the different degrees of nucleosomal compaction in TUs that have different polymerase packing densities. However an uncontroversial observation made by R. S. Hill and Macgregor (1980) has an important bearing on biochemical work on *Xenopus* oocytes. Several studies have demonstrated the synthesis of polyadenylated RNAs, transfer RNAs and 5S rRNA by previtellogenic oocytes of *Xenopus*, but Dumont (1972) stated that lampbrush chromosomes do not start to develop until the beginning of vitellogenesis; this has led to the claim that lampbrush chromosomes are not responsible for their synthesis (e.g. Rosbash and Ford 1974, and Golden et al. 1980, concerning polyadenylated RNAs). That claim is clearly invalid, for lampbrush chromosomes are well-developed in *Xenopus* oocytes much earlier than Dumont had supposed. In this connection Scheer and Dabauvalle (1985) have recently proved that transcription on lampbrush chromosomes does lead to the production of functional mRNAs. They transferred manually isolated nuclei of early vitellogenic *Pleurodeles* oocytes into full grown *Xenopus* oocytes, incubated the host oocytes for a few days in culture medium and then labelled their newly translated proteins by incubation for a further day in medium containing L-(^{35}S) methionine. Two-dimensional gel electrophoresis of proteins from the *Xenopus oocytes* that had acted as hosts displayed some labelled *Pleurodeles*-specific proteins as well as those characteristic of *Xenopus*. Unless they were carried over as contaminants, the mRNAs encoding these *Pleurodeles* proteins must have been synthesized during the period following nuclear implantation, and thus by implication were derived from lampbrush loop transcripts.

Within the Amphibia, species with low C-values have, in general, shorter lampbrush loops and shorter TUs on these loops than species with high C-values when transcription rate is at its maximum (cf. Fig. 3.1 a with b). From observations on Miller-spreads of the lampbrush chromosomes of Amphibia with large C-values and long TUs, it was commonly assumed that many loops 200 μm or longer and with uniformly increasing quantities of RNP matrix throughout their lengths might release transcripts that when fully extended, would be of comparable lengths. Thus Miller and Hamkalo (1972) stated: "If no nicks are made in nascent RNA chains during synthesis the RNA in such fibrils would range up to 125 μm in length". In view of more recent work, the qualification about possible nicking, foreshadowed by Miller and Hamkalo (1972), has proved to be justified.

Sommerville and Scheer (1981, 1982) and Scheer and Sommerville (1982) have to a first approximation quantified the correlation between C-value and loop length for *Xenopus laevis* (C-value 3.1 pg), *Triturus cristatus carnifex* (23 pg), *Amphiuma means* (65 pg) and *Necturus maculosus* (78 pg). They state that in *Xenopus* most of the loops fall within the size range 5–10 μm, in *Triturus* 30–50 μm, while in *Amphiuma* and *Necturus* the mean is >100 μm. However when the heterogeneous nuclear (hn)RNA is extracted from previtellogenic oocytes by a procedure that minimizes molecular degradation, hnRNA molecules of similar length ranges (0.1 to 10 μm) and average lengths (0.6 to 1.7 μm), are found in all four species. This hnRNA includes molecules derived from free RNP, but also a substantial proportion (Sommerville 1973, estimates as much as 50% in *Triturus*) derived from nascent RNP. Degradation during preparation is not responsible for these unexpectedly short RNA lengths, for when similar methods were used to prepare RNA from vitellogenic oocytes, in which there is extensive transcription from 18 + 28S rDNA, rRNA precursor molecules, identifiable by their secondary structure, are well represented and are of the expected length (2.6 to 2.8 μm). Scheer and Sommerville (1982) conclude that cleavage of hnRNA occurs on nascent transcripts, and that the integrity of long nascent RNP fibrils as seen in Miller-spreads must be maintained by their protein components.

Although Miller-spreads have provided impressive views of RNA transcription in progress on the chromosomes of oocytes and several other kinds of cell, in one respect that view is seriously distorted. In preparations made at low ionic strength and high pH (0.1 mM borate buffer, pH 9) RNP transcripts generally appear as linear structures of nearly uniform width, about 5 nm or somewhat thicker. But when lampbrush chromosomes are isolated in isotonic saline in a life-like state, as judged by their appearance in phase contrast while nuclear sap disperses, and then fixed for electron microscopy, sections show the loop matrix RNP to consist of linear arrays of particles of remarkably uniform size, each about 20 to 30 nm wide (Malcolm and Sommerville 1974, 1977, Mott and Callan 1975). The size uniformity of the particles applies both within the length of single

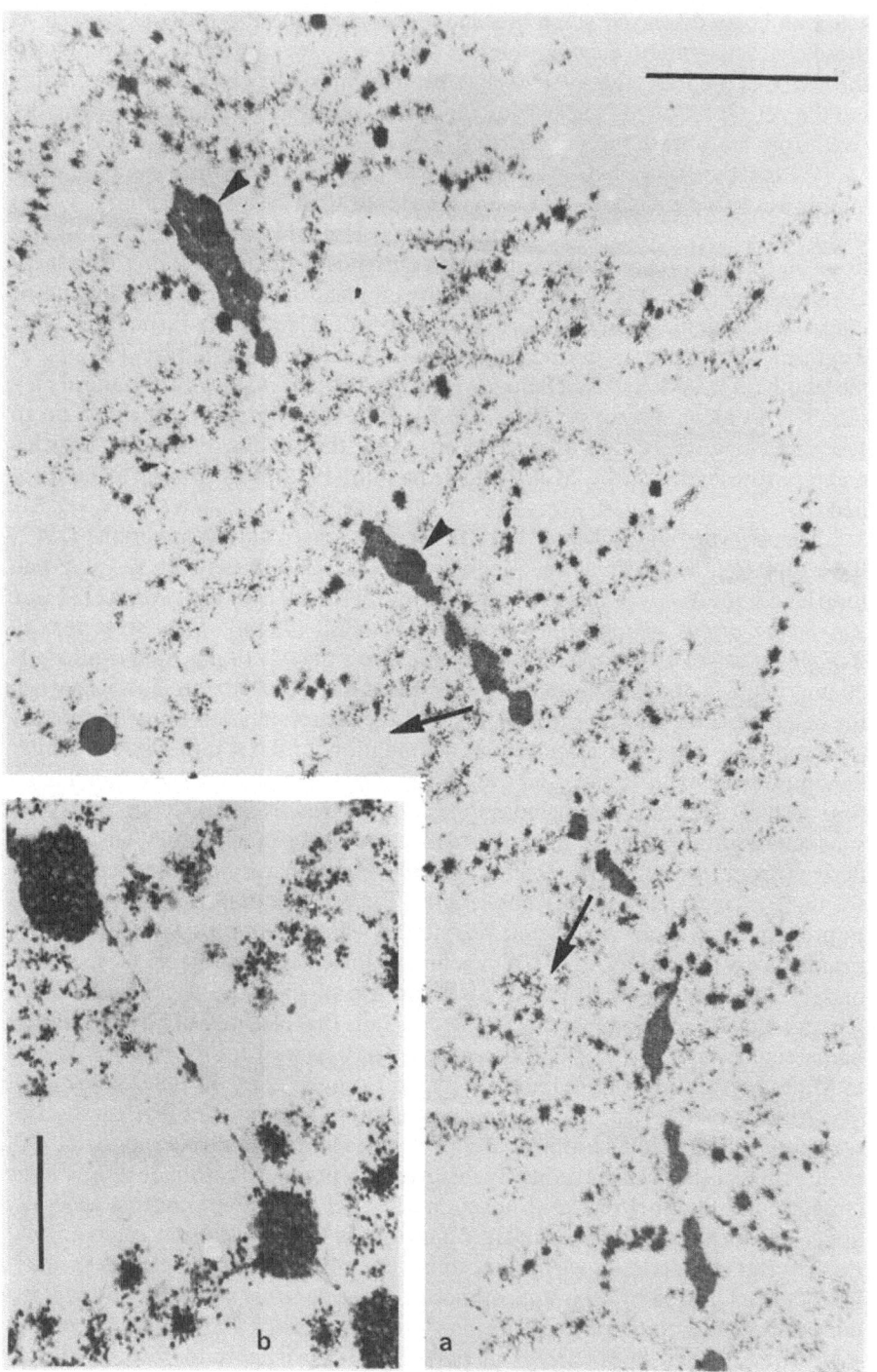

Fig. 4.9

transcripts, between different transcripts on individual loops and between different loops. These RNP particles may be aggregated into tight clusters that are generally less than 1 μm wide; however the width cannot be measured accurately because of the loosely distributed particles at the periphery of each aggregate (Fig. 4.9). Mott and Callan (1975) assumed that the 20–30 nm particles are generated by compaction of a 5 nm fibril as transcription proceeds, and when the particle has reached the standard size another starts to compact alongside. In this way linear arrays of particles are formed, connected to one another by short lengths of 5 nm fibril. If the 5 nm fibrils are tightly packed in 30-nm-wide RNP particles, each particle would contain about 700 nm of primary fibril, implying an RNP compaction ratio of about 23:1, though lower if the fibril packing is less tight.

Malcolm and Sommerville (1974) considered alternative interpretations for the origin of the primary RNP particles, one being that they are artefacts resulting from RNP condensation after the chromosomes have been isolated in isotonic saline, i.e. while the nuclear sap disperses. This is most unlikely, because it would require that loop matrix progressively compacts while nuclear sap disperses, which it manifestly does not if a preparation is kept under observation while dispersal of the sap takes place. Moreover the larger range of particle aggregates can already be seen by phase contrast inside oocyte nuclei immediately after their isolation in isotonic saline, before significant dispersal of nuclear sap has occurred. Furthermore RNP granules of a similar size are evident in sections through urodele oocyte nuclei that were fixed intact, without nuclear sap dispersal (Fig. 4.10 b), likewise in sections through spermatocyte nuclei of *Drosophila* (see Chap. 7).

The relationship between primary RNP fibrils and RNP particles has been further investigated by Malcolm and Sommerville (1977). Tangled aggregates of particulate nuclear RNA can be extracted from homogenates of oocytes by differential centrifugation, and the particles in the aggregates are of uniform (20 nm) diameter. When dispersed in 85% formamide the aggregates extend as linear beaded structures, the beads being disposed at regular intervals. A very thin fibre, to judge from an illustration certainly less than 5 nm wide, links the particles together, and this fibre is cleaved by mild digestion with RNase A. The immediate product of RNase digestion is a population of single particles, but as digestion proceeds the particles themselves disintegrate. From these observations Malcolm and Sommerville (1977) conclude that the particles cannot be independently formed protein complexes of uniform size with RNA wound around the surface (i.e. analogous to the relationship between DNA and histone in nucleosomes), but

---

Fig. 4.9. a Thin section through part of a chromosome from a dispersed, centrifuged and end-embedded lampbrush preparation of *Triturus c. carnifex*. The chromomeres show the partial unravelling that occurs while they lie in 100 mM KCl/NaCl before fixation, though some compact regions, the natural state and indicated by *arrowheads,* are still evident. Notice the clusters of RNP particles on many loops. *Bar* = 5 μm. **b** The region delimited by two *arrows* in **a,** at higher magnification, showing the interchromomeric fibril and the point of entry of the thicker end of one lateral loop into a chromomere. *Bar* = 1 μm. (From Mott and Callan 1975)

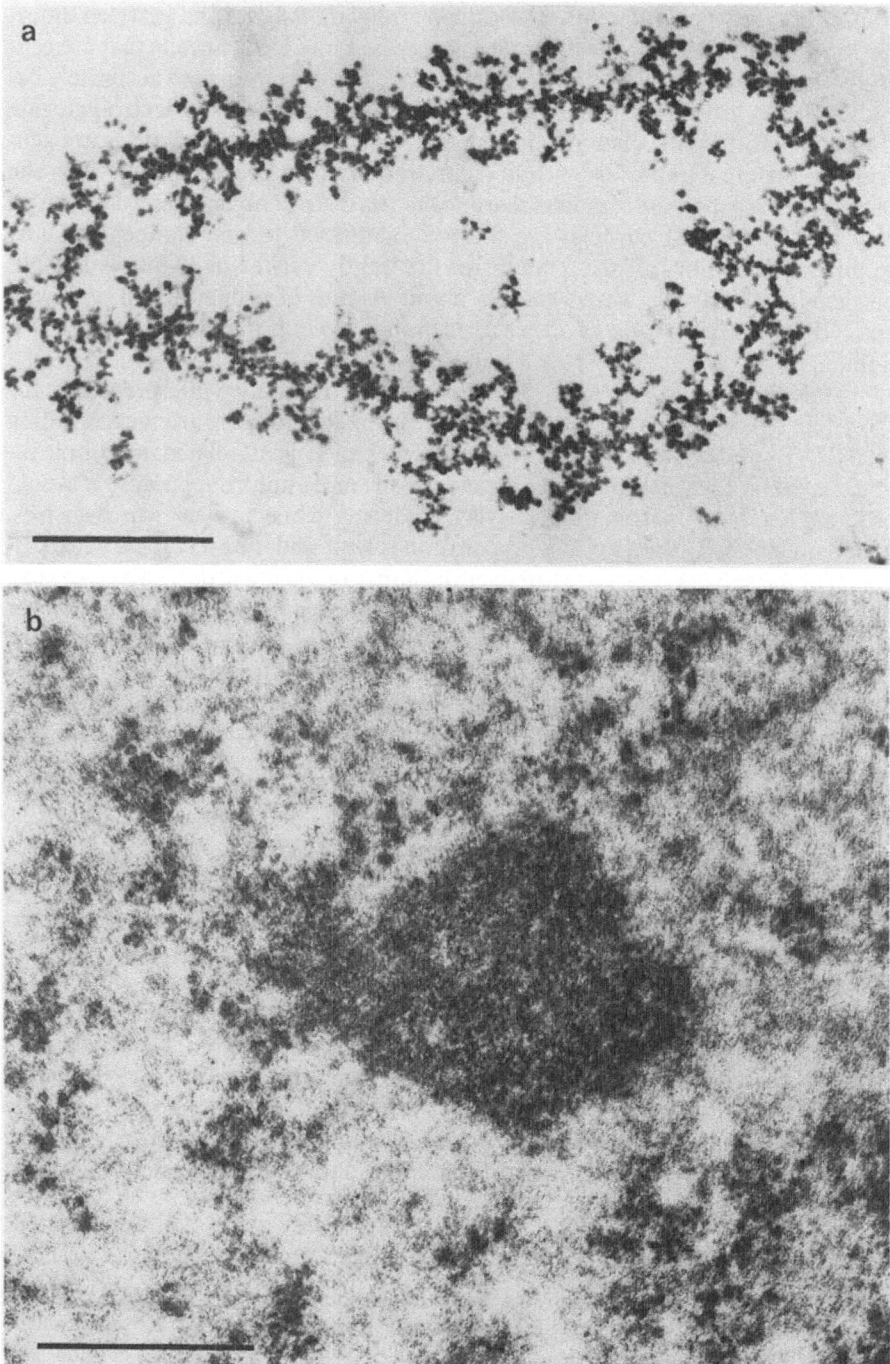

Fig. 4.10. a Thin section including part of a lateral loop from a dispersed, centrifuged, fixed and end-embedded lampbrush preparation of *Triturus c. carnifex;* the transcripts consist of interconnected granules of uniform size. *Bar* = 1 μm. **b** Thin section through part of an isolated oocyte

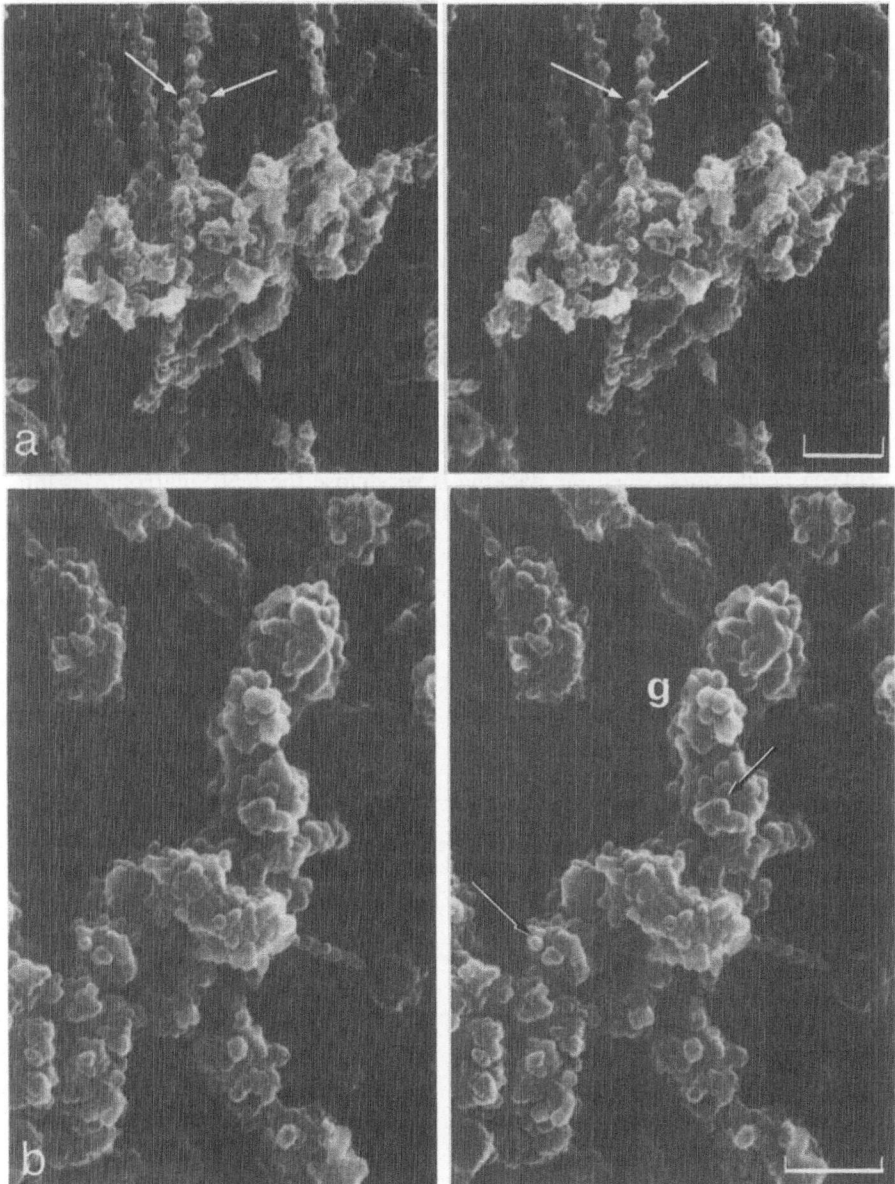

Fig. 4.11 a, b. Photographs of parts of the lampbrush chromosomes of *Pleurodeles* viewed by scanning electron microscopy. **a** Stereopair of a chromomere surrounded by loops with normal matrix; the matrix consists of 200 nm bodies, indicated by *arrows,* helically disposed around loop axis. **b** Stereopair of part of a "globular" or "beaded" loop; the globules (*g*) are aggregates of smaller bodies, indicated by *arrows,* similar in size to those present in **a.** *Bars* = 1 μm. (From Angelier et al. 1984)

---

nucleus of *T.c. carnifex* that was fixed entire and before sap dispersal had occurred. The dense central structure is a chromomere; the granular objects elsewhere are components of RNP transcripts. *Bar* = 0.5 μm. (From Mott and Callan 1975)

that the particles are produced by periodic condensation of linear RNP, much as had been proposed by Mott and Callan (1975).

The aggregation of primary 20–30 nm particles into larger structures on the lateral loops of lampbrush chromosomes has been confirmed by observations of Angelier et al. (1984) using scanning electron microscopy. Lampbrush preparations of *Pleurodeles* were made in 100 mM saline buffered to pH 7.2, centrifuged just as for end-embedding, fixed in glutaraldehyde, dehydrated, critical-point dried from liquid CO_2, sputter-coated and gold-shadowed. Excellent preservation of morphology resulted from this procedure. In "normal" loops the RNP matrix consists of 100–200 nm "bodies" disposed helically around loop axis (Fig. 4.11 a), corresponding to the zig-zag arrangement of structures of this size seen in transmission electron microscopy of thin sections, and Angelier et al. (1984) consider that these are aggregates of the primary RNP particles. Higher order aggregation leads to the formation of granules or globules within the resolving power of the light microscope (Fig. 4.11 b), with variable degrees of packing between neighbours; when the packing is very close the RNP matrix fuses into "... a spiral sleeve encircling the loop axis".

CHAPTER 5

Distribution of Nucleic Acid Sequences in Lampbrush Chromosomes

Although for the most part this chapter will be concerned with sequences that are transcribed in amphibian oocytes, studies in this field have progressed in such a way that transcribed sequences cannot be considered in isolation from sequences that are not transcribed. Neither can lampbrush chromosomes be considered in isolation, for work on other chromosomes has frequently supplied evidence that is essential for the proper appraisal of the primary topic of this chapter. Much of the information at present available has come from the use of in situ hybridization, the singularly valuable autoradiographic technique devised by Gall and Pardue (1969) and by H. A. John et al. (1969). Mention has been made of one outcome of this technique in the section of Chap. 3 dealing with nucleolus organizers, but here I will discuss wider aspects.

The first successful application of in situ hybridization to lampbrush chromosomes was made by Barsacchi and Gall (1972). They hybridized ^{3}H-labelled RNA copied in vitro, after the removal of rDNA, from total *N. viridescens* DNA (cRNA) to denatured interphase nuclei, mitotic and lampbrush chromosomes of this newt. They obtained heavy labelling of heterochromatic lumps (chromocentres) in interphase nuclei, of regions adjacent to the centromeres of mitotic chromosomes, similarly of the axial regions without loops that include the centromeres of the lampbrush chromosomes, the regions of chromosome axes to which the spheres are attached, and some telomeric regions. They also noticed, in preparations from nearly mature oocytes that had been stimulated by gonadotropin, that the loops which in these circumstances become prominent at the sphere loci are also labelled. Because the RNA probe used in these first experiments was heterogenous, the most heavily labelled regions of the chromosomes would necessarily be those containing the most highly repeated sequences complementary to the most highly repeated sequences present in the probe.

In situ hybridization with better characterized probes was soon under way, attention being first directed at target sequences in urodele genomes that are known, on biochemical evidence, to be highly reiterated. One such family of sequences is 5 S rDNA. Barsacchi Pilone et al. (1974) hybridized ^{3}H-labelled 5 S rRNA prepared from *Xenopus* tissue culture cells to denatured interphase nuclei, mitotic and lampbrush chromosomes of *T. marmoratus*. In mitotic chromosomes a single locus is labelled, intercalary in the long arm of chromosome X, and likewise a single locus in the main axis of lampbrush chromosome X, close to a landmark loop with dense matrix. In this paper Barsacchi Pilone et al. (1974) state that the 5 S genes are similarly confined to a single locus on the axis of chromosome

X of *T.c. carnifex*. The 5 S genes of *T. vulgaris meridionalis* are likewise concentrated at a single locus in the long arm of chromosome XI, and in situ hybridization to the lampbrush chromosomes of this species confirmed that they lie in one or a few chromomeres near the middle of this chromosome (Barsacchi-Pilone et al. 1977). Using iodinated 5 S rRNA on denatured mitotic and male meiotic chromosomes León (1976) found a single locus for 5 S rDNA sequences in *Taricha granulosa*, and two loci in *Batrachoseps wrighti*, in both species the loci lying close to centromeres.

Most of the 5 S genes in the karyotype of *N. viridescens* also lie close to centromeres; Hutchison and Pardue (1975) hybridized ³H-cRNA transcribed in vitro from 5 S rDNA of *Xenopus* to denatured mitotic chromosomes of *N. viridescens*. They found sites of hybridization adjacent to the centromeres of four pairs of subtelocentric chromosomes, and one other site in the middle of the long arm of a pair of one of the two smallest chromosomes. Pukkila (1975) confirmed these findings, and extended the investigation to the lampbrush chromosomes. Using ¹²⁵I-labelled 5 S rRNA isolated from *N. viridescens* as a probe, and hybridizing to denatured preparations, she found labelling confined to the centromeric regions of four lampbrush chromosomes (I, II, VI, and VII) and, though less consistently, to an intercalary site on the long arm of chromosome X. The seven other centromeric regions were unlabelled.

Pukkila (1975) went a step further by hybridizing ¹²⁵I-labelled 5 S rDNA prepared from *Xenopus* erythrocytes to non-denatured lampbrush preparations of *N. viridescens*. The intention here was to hybridize the DNA probe to nascent RNA, and loops near the centromeres of chromosomes I, II, VI, and VII were labelled. Control preparations digested with RNase before hybridization were unlabelled, whereas in preparations digested with DNase before hybridization, labelled loops were still observed. This seemingly convincing demonstration that 5 S RNA is transcribed on loops large enough to be seen by light microscopy raised problems that were recognized by Pukkila. The labelled loops ranged in length from 15 to 200 μm. However the 5 S rDNA coding sequences are so short (120 bp) that they can accommodate only one or at the most two polymerases at any instant, and the 5 S genes are separated from each other by spacer DNA (Kay et al. 1981, Kay and Gall 1981). So unless transcription were to proceed uninterrupted through many 5 S plus spacer sequences, the product would not be expected to form a lateral loop with enough matrix to be visible by light microscopy. Working with Miller-spreads Scheer (1982) has found clusters of highly repetitive, very short TUs, of the order 10,000 per cluster, arranged in tandem in *P. waltlii* (see Chap. 4). Each unit is covered by two polymerases only, the spacer intervals are likewise short, and Scheer considers it likely that these TUs represent 5 S rDNA. Though uninterrupted transcription through 5 S plus spacer sequence might account for Pukkila's labelled loops (see later), her findings were not confirmed in a study by Kay and Gall (unpublished).

Schultz et al. (1981) tackled this problem from a different angle. They developed an incubation medium (see Chap. 2) which supports RNA synthesis, extending over several hours, by isolated oocyte nuclei of *N. viridescens*. ³H-labelled ribonucleoside triphosphates were included in the incubation media, and in some experiments α-amanitin was also included, at 0.5 μg ml⁻¹ or 200 μg ml⁻¹. *Xeno-*

pus oocyte nuclei contain, in great abundance, the three kinds of RNA polymerases that occur in other cells of vertebrates (Roeder 1974, 1976). Polymerase I transcribes the large $18+28$ S rRNA precursor and is not inhibited by high concentrations of α-amanitin. Polymerase II, which is extremely sensitive to α-amanitin, transcribes heterogeneous nuclear RNAs. Polymerase III, which is sensitive to α-amanitin, but less so than polymerase II, transcribes several low molecular weight RNAs, including 5 S rRNA and tRNAs.

After incubation, isolated nuclei or "nuclear gels" (in the media used by Schultz et al. (1981) the nuclear contents remain as a gelatinous ball after removal of the nuclear membrane) were transferred to a more dilute medium promoting sap dispersal, lampbrush chromosome preparations were made by centrifugation, and autoradiographed. Preparations from nuclei incubated in the absence of α-amanitin showed labelled loops, some labelled and some unlabelled regions including centromeres (Fig. 4.1 a, b), and labelled nucleoli, in other words, just as if intact oocytes had been incubated with ³H-labelled ribonucleosides. In preparations where the incubation medium contained α-amanitin at 0.5 μg ml^{-1}, the lateral loops were unlabelled, but the regions including the centromeres of 3 or 4 bivalents, also some 15 to 20 sites in the chromosome axes of several bivalents, and the nucleoli, were labelled (Fig. 5.1 c). In preparations where the incubation medium contained α-amanitin at 200 μg ml^{-1}, only the nucleoli were labelled. On the basis of these observations Schultz et al. (1981) concluded that polymerase II is responsible for transcription on the lateral loops, that polymerase III is responsible for the transcription of 5 S, transfer, and possibly some other small RNAs, but that the products of this transcription are not visible in the light microscope as matrix on lateral loops, and that polymerase I is responsible for transcription in the nucleoli. In the preparations from nuclei incubated in the presence of α-amanitin at 0.5 μg ml^{-1} the labelled regions including the centromeres are taken to be sites of synthesis of 5 S rRNA, because this agrees with the results of in situ hybridization where radioactive 5 S rRNA or rDNA had been used as probes, and the additional labelled sites on the chromosome axes are thought likely to be places where tRNAs are synthesized.

In the dilute dispersing medium used by Schultz et al. (1981) loop matrix is more diffusely extended from loop axis than when 100 mM saline media are used for sap dispersal. In autoradiographs of preparations made after only 20 min in vitro incubation, silver grains were concentrated over loop axis (Fig. 4.1 e), consistent with the assumption that RNA synthesis is restricted to the DNA in the axis.

Until recently it had been generally assumed that there is a strict differentiation within lampbrush chromosomes between lateral loops with extended DNP axes, where transcription occurs, and compact chromomeric DNP that is not transcribed. In the light of what is now known about the transcription of low molecular weight RNAs, in particular 5 S rRNA, this view is no longer tenable. Moreover if Scheer's (1982) clusters of very short repetitive sequences, all transcriptionally active, do indeed represent 5 S rDNA, then one must conclude from their high repetition frequency that transcription is not restricted to units that lie on the surface of compact DNP, but also includes those that lie within the chromomeres; thus these chromomeres are not so densely compacted (Mott and Cal-

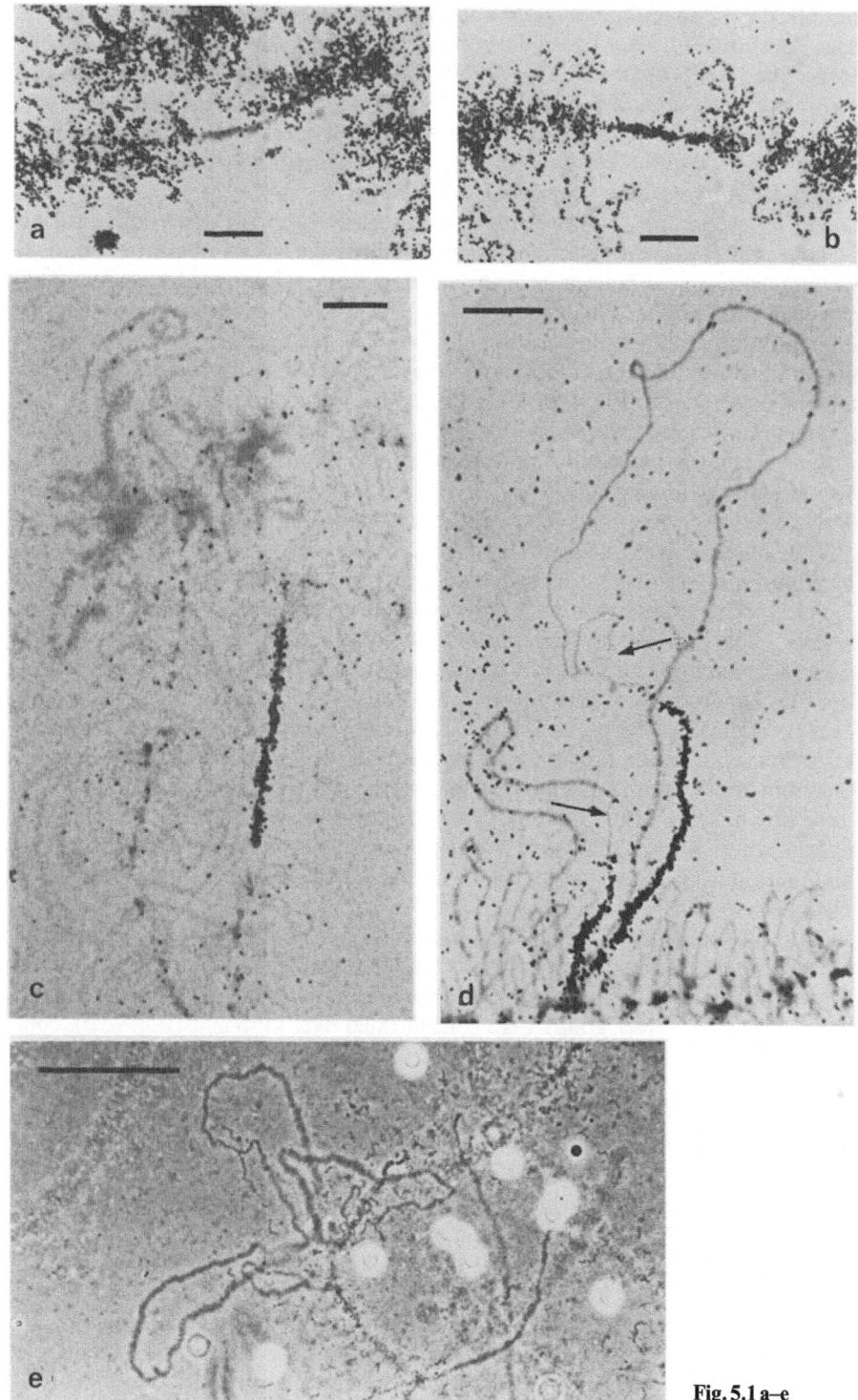

Fig. 5.1 a–e

lan 1975) as to prevent the entry of polymerases and the exit of low molecular weight RNP transcripts.

Kay and Gall (1981) have isolated, cloned and analyzed 5 S rDNA of *N. viridescens*. They have detected two types of repeating units. The major type, 231 base pairs (bp) long, consists of a 120 bp coding region with a 111 bp spacer, while the minor type, 269 bp long, consists of a similar coding region following by a 36 bp segment, corresponding to the terminal third of the coding region (a "pseudogene") and a 113 bp spacer. From the detailed analysis of different plasmid inserts two arrangements of 5 S rDNA tandem repeats are inferred, one composed entirely of 231 bp units, the other with 269 bp units interspersed among 231 bp units; these two arrangements of repeats may be at different loci in the *N. viridescens* genome, for it includes four (possibly five) different 5 S gene loci. A notable feature of the 36 bp pseudogene is that its sequence is identical to that of its counterpart in the terminal third of the 5 S coding sequence. Kay and Gall (1981) state: "The conservation of the pseudogene sequence may be a consequence of the mechanisms which maintain sequence and length homogeneity in repeated gene families." What such mechanisms may be are unknown, but the statement has a familiar ring for I invoked something similar when proposing the Master/Slave theory as a general explanation for the C-value paradox (Callan 1967).

The single 5 S rDNA site at a similar chromosomal locus within the karyotypes of several different species of *Triturus* is suggestive of conservative evolution, and it would be of interest to know whether there is sequence homogeneity amongst the tandem repeats at this single site. The number of 5 S rDNA sites in the karyotypes of different Amphibia is not related to C-value, for *Xenopus laevis*, with a much lower C-value than any of the urodeles, has 5 S rDNA loci on most if not all of its chromosomes (Pardue et al. 1973). Another conclusion is that 5 S rDNA sites may lie at a variety of chromosomal loci, at the telomeres in *X. laevis*, mostly alongside the centromeres in *N. viridescens*, and at intercalary sites in other urodeles.

When dealing with the identification of nucleolus organizer regions in lampbrush chromosomes in Chap. 3 I mentioned some of the information concerning

---

Fig. 5.1. a–e Autoradiographs of lampbrush chromosomes of *Notophthalmus viridescens* showing incorporation of ^{3}H-UTP and ^{3}H-CTP after 20 min incubations of oocyte nuclei in vitro. **a** Labelled loops, but unlabelled centric region of one chromosome. **b** Labelled loops and also labelled centric region of another chromosome. **c** Part of chromosome II from an oocyte nucleus where the incubation took place in the presence of 0.5 µg ml^{-1} of α-amanitin; the giant loops (*upper left*) and all other loops are unlabelled, whereas the centric region is conspicuously labelled. **d** Autoradiograph showing a pair of sister loops in the heteromorphic region of chromosome I of *Triturus c. carnifex* after DNA/RNA transcript hybridization in situ with nick-translated *Xenopus* rDNA; *arrows* point to interstitial regions with least matrix in both loops, presumably the origins of transcription units that have opposite (head-to-head) polarities; in both loops only one transcription unit is labelled. The sister loops are unusual in that they differ markedly in length, though their labelled to unlabelled length ratios are similar, suggesting that they differ in overall transcript density. **e** The contents of an oocyte nucleus of *N. viridescens* dispersing in a medium containing restriction endonuclease *Hae* III; whereas other loops and chromosome axes are digested by this enzyme, the giant loops on chromosome II remain intact. *Bars* in **a–d** = 10 µm; in **e** = 50 µm. **a–c** From Schultz et al. (1981); **d** from Morgan et al. (1980); **e** from Gould et al. (1976)

the distribution of 18 + 28 S rDNA sequences that has been obtained from in situ hybridization. I want now to cover some other aspects of this topic.

Hutchison and Pardue (1975) in situ hybridized ³H-labelled 18 + 28 S rRNA from *Xenopus* to mitotic chromosomes of *N. viridescens,* and found single labelled sites on three different pairs of chromosomes. None of the animals studied showed all these sites labelled; the pattern of labelling was constant within an individual, but different from one individual to another. However the site that corresponds with the nucleolus organizer region identified by Gall (1954) in lampbrush chromosomes on morphological criteria was regularly labelled in both homologues. This unanticipated complexity was also found in *T. vulgaris meridionalis* by Nardi et al. (1977), using similar methods. Of 20 animals 13 had individual-specific 18 + 28 S rDNA sites on several different chromosomes, over and above the recognized and regularly labelled nucleolus organizer locus subterminal in the long arm of chromosome XI. Batistoni et al. (1978) followed the inheritance of these extra 18 + 28 S rDNA sites and showed that they segregate and reassort according to regular Mendelian rules.

Extra 18 + 28 S rDNA sites are also present in the karyotype of *T.c. carnifex,* but their chromosomal distribution is unlike that in both *N. viridescens* and *T.v. meridionalis* (Morgan et al. 1980). 18 + 28 S rDNA from the ovaries of recently metamorphosed female *Xenopus* was purified by buoyant density centrifugation and used as a template for nick-translation (Macgregor and Mizuno 1976) to give an ³H-labelled rDNA probe. Four other nick-translated probes were prepared, two from cloned *Xenopus* rDNAs and two from purified restriction fragments derived from one of these clones. The probe rDNAs were denatured and applied to preparations of lampbrush chromosomes, the aim being to hybridize to nascent RNA transcripts on the lateral loops. Adequate controls were run; all five probes gave essentially similar results.

In the autoradiographs some 10 to 15 pairs of loops in the heteromorphic arms (HTAs) of bivalent I were labelled, differing in location between the two arms, reasonably consistent in position in different preparations from a single female, but with some identifiable differences between females. Apart from these loops on bivalent I, some of which were only labelled over parts of their lengths (Fig. 5.1 d), a few other regularly labelled loops were present on bivalents X and XI. Loops close to the known nucleolus organizers were unlabelled, whereas the extrachromosomal nucleoli were heavily labelled. Despite the identification of rRNA transcribed at several discrete sites in the HTAs of bivalent I, neither rDNA nor rRNA probes when hybridized with denatured preparations of mitotic chromosomes produced significant labelling of these regions, implying that not many rDNA sequences are present within them.

In a further series of experiments, isolated germinal vesicles were incubated in media containing ³H-labelled nucleoside triphosphates, with or without α-amanitin (for details, see Schultz et al. 1981). Autoradiographs of control preparations were normal, with lateral loops and nucleoli labelled, but preparations from germinal vesicles incubated in the presence of α-amanitin at 1 or 10 µg ml⁻¹ showed labelled extrachromosomal nucleoli, but no incorporation on the lateral loops. From these observations Morgan et al. (1980) drew two conclusions. Although whole or partial 18 + 28 S rDNA coding sequences must be present at several loci over and above those at the nucleolus organizers, they are not present in concentrated clusters. Moreover as their transcription is sensitive to α-amanitin at low concentration, they are being transcribed by RNA polymerase II, not polymerase I.

These observations raise the question as to what signals determine that in *T.c. carnifex,* and those other Amphibia which generally lack nucleoli attached to the lampbrush chromosomes, RNA polymerase I recognizes and transcribes 18 + 28 S rDNA sequences in the extrachromosomal nucleoli but rarely those chromosomal sequences from which they were derived, whereas polymerase II transcribes similar sequences that lie at chromosomal sites other than the nucleolus organizers. Whether those sequences that are transcribed by polymerase II contribute to ribosomes is unknown; their RNP matrix is like that of the generality of loops and shows no textural resemblance to free nucleoli.

J. G. Gall (personal communication, 1985) has recently confirmed the original observation of Pukkila (1975), previously in dispute, that occasional *loops* of *N. viridescens* become labelled after in situ hybridization with 5 S rDNA (Jamrich and Gall, unpublished). They ascribe this to the presence of some 5 S rDNA sequences that happen to have lodged downstream of active RNA polymerase II promoters, therefore comparable in status to sequences on the loops of *T.c. carnifex* that hybridize with 18 + 28 S rDNA probes.

When transcription rate is at its highest level, the lengths of single TUs in urodele lampbrush chromosomes are enormous. The units transcribed by polymerase II range in length from a few microns to over 100 μm. Urodeles have high C-values, and they have high percentages of repetitive DNA sequences, so high that single copy sequences are difficult or impossible to detect in studies of reassociation kinetics (reviewed by Sommerville 1977). Taken together, these two features of urodele genomes suggest that most if not indeed all of the lateral loops must be transcribing repetitive, perhaps even simple satellite repeat sequences, though these latter, from evidence in other sytems, had been thought not to be transcribed at all.

Suggestive early evidence for the transcription of highly repetitive DNA came from Gould et al. (1976). They digested the lampbrush chromosomes of *N. viridescens* with four restriction and two non-sequence-specific nucleases. One of the restriction enzymes, *Hae* III, which cleaves double-stranded DNA within the sequence G-G-C-C.C-C-G-G, attacked all the loops and interchromomeric strands except for giant loops clustered in two regions near the centromere in the long arm of chromosome II (Fig. 5.1 e). The loops resistant to *Hae* III, though not to the other enzymes, are each some 200 μm long, and they may be transcribed as single units, or as two or three units. In a DNA molecule of random base sequence a sequence sensitive to *Hae* III would be expected to occur once in every 256, i.e. $(\frac{1}{4})^4$ bp, therefore just less than 0.1 μm long. If this sensitive sequence is missing from 200 μm of DNA, a probable explanation for its absence would be that the giant loops' axes consist of a short sequence lacking the sensitive site, repeated on a huge scale, yet it is certainly transcribed (Hartley and Callan 1978).

In situ hybridization of repetitious DNA probe sequences to RNA transcripts on lampbrush chromosomes has produced further revealing results. Macgregor and Andrews (1977) labelled middle repetitive DNA (C_0t 0.2–50) of *T.c. carnifex* by nick-translation and hybridized this heterogeneous probe to non-denatured lampbrush preparations. After short exposure times of the autoradiographs the loops that were heavily labelled by this probe (some 20 to 40) were concentrated to a remarkable degree in the HTAs of bivalent I. The distribution of labelled loops was different in the two arms of the bivalent, and differences in distribution were also observed between preparations from smaller and larger oocytes. Several loops were only labelled through part of their lengths. After longer exposure times many more loops were demonstrably labelled, particularly in regions around the centromeres, though with few exceptions none to compare in degree of labelling with those on the HTAs of bivalent I.

Macgregor (1979) labelled even more highly repetitive DNA (C_0t 0–0.2) of *T.c. carnifex* and hybridized this probe to denatured mitotic chromosomes. The labelling pattern matched well with the chromosomal distribution found in the DNA/RNA transcript hybridizations, i.e. intensely labelled HTAs of chromosomes I, and heavy labelling around the centromeres of, in particular, chromosomes X, XI and XII.

The analysis was taken further by Varley et al. (1980a, b). As well as a *T.c. carnifex* DNA probe similar to that used by Macgregor (1979), i.e. purified by renaturation in the C_0t 0–0.2 range, probes were prepared from two heavy satellite DNAs (called H1 and H2) purified by buoyant density centrifugation, and further probes (TcS1 and TcS2) were prepared from cloned satellite DNAs isolated after restriction endonuclease digestion. In situ hybridization of these probes to lampbrush chromosomes gave autoradiographs, all after short exposure, consonant with those described previously. The C_0t 0–0.2 probe, and one of the heavy satellite probes, H1, labelled more than 60 loop pairs on the HTAs of bivalent I, and a dozen or so loop pairs elsewhere (Fig. 5.2a). The other heavy satellite probe H2 labelled more than 30 loop pairs on the HTAs of bivalent I, and some 10 loop pairs elsewhere.

When these DNA probes, including TcS1 and TcS2, were in situ hybridized to denatured mitotic chromosomes, all produced heavy labelling of the HTAs of chromosomes I, with some minor differences in the relatively few labelled sites on other chromosomes. When they were in situ hybridized to denatured lampbrush chromosomes, many chromomeres on the HTAs of bivalent I were labelled, likewise a few chromomeres at other sites (including some centromeres), all in good correspondence with the patterns of hybridization to mitotic chromosomes (Fig. 5.2b). These experiments were well controlled in all respects, one valuable internal control being hybridization to lampbrush chromosomes of a probe nick-

---

→

Fig. 5.2a, b. Autoradiographs of lampbrush bivalents I of *Triturus c. carnifex,* after hybridization in situ with a denatured ^{3}H-labelled probe nick-translated from a 0–0.2 C_0t fraction of *T.c. carnifex* DNA, as seen in dark field. Most of the *white dots* are silver grains. **a** DNA/RNA transcript hybrid; the heavily labelled loops are virtually confined to the heteromorphic regions of the longer arms. **b** DNA/DNA hybrid; the preparation was digested with RNase and denatured before hybridization; the chromomeric axes of the heteromorphic regions are conspicuously labelled. In both **a** and **b** the regions of the longer arms proximal to the centromeres are unlabelled, as also are the shorter arms, which are indicated by *arrowheads*. *Bar* = 200 μm. (From Macgregor et al. 1981)

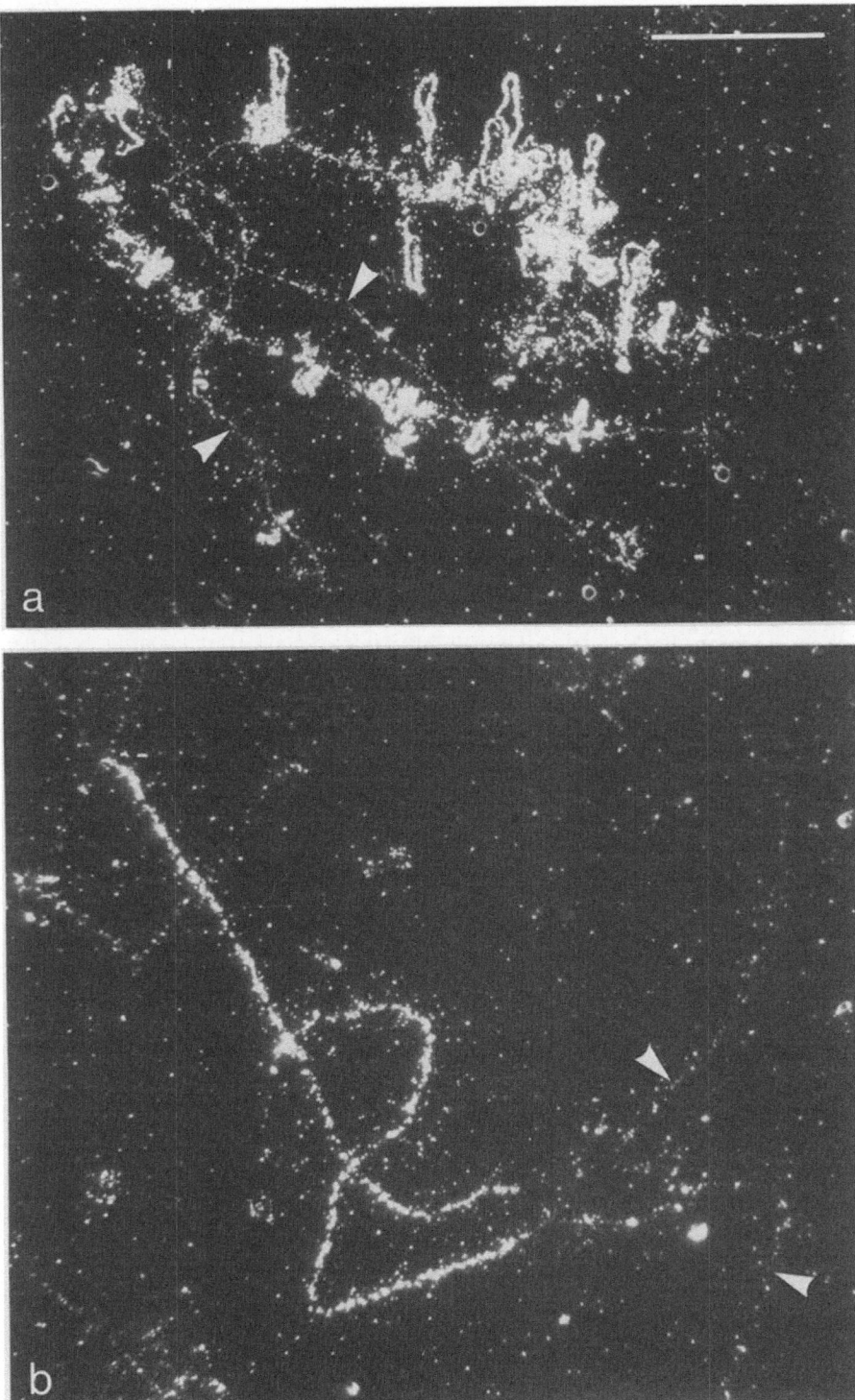

Fig. 5.2 a, b

translated from "heavy main band" *T.c. carnifex* DNA from which the two heavy satellite sequences had been excluded by buoyant density centrifugation. The resultant autoradiographs required much longer exposure times to demonstrate labelled loop transcripts, and loops on the HTAs of bivalent were not exceptional in the density of their labelling.

Several deductions have been drawn from these observations. One is that the HTAs of chromosomes I are exceptional in carrying many loops on which identical or at least similar satellite sequences are transcribed. The concentrations of these sequences differ from one loop to another, as judged from relative labelling intensity, and in some loops the transcribed satellite sequences are confined to parts of the loops only. Similar sequences are also present in the chromomeres on the HTAs of chromosomes I, though whether there is precise correspondence between labelled loops and their associated chromomeres remains to be demonstrated.

The general run of the lateral loops in the heteromorphic regions of chromosomes I of *T. cristatus* are distinctly longer than loops elsewhere on the lampbrush chromosomes of this newt; indeed if they were typical of the whole complement one would expect the C-value of *T. cristatus* to be considerably higher than it is in reality. The HTAs of chromosomes I of *T. cristatus* are also exceptional in that chiasmata never occur within the heteromorphic regions either in females (Callan and Lloyd 1960) or in males (Morgan 1978), though in lampbrush bivalent I there are short regions both distal and proximal to the heteromorphic regions in which chiasmata can occur. The transcribed and untranscribed satellite sequences are mostly confined to the heteromorphic regions and this correlation with the absence of meiotic recombination can scarcely be accidental. It is significant that Varley et al. (1980 b) were able to demonstrate that these satellite DNA probes from *T.c. carnifex* also hybridize with exceptional intensity to RNA transcripts concentrated in comparable regions of bivalent I of *T. marmoratus,* whereas there are no such regions in *T. alpestris* and *T. vulgaris* nor any such concentration of intensely labelling loops either. I will return to consider the interpretation to be put on loops that are only labelled in parts of their lengths later in this chapter.

Satellite TcS2 consists of 330 bp, not 275 as originally estimated, and it or closely related sequences are present in several other species of European newts. It is furthermore identical in length and remarkably similar in sequence to satellite 2 (Nv 2) of *Notophthalmus viridescens* (Epstein, Mahon and Gall, personal communication, cited by Macgregor and Sessions 1985). Both are transcribed on many lampbrush loops in oocytes, and monomers and multimers of their transcripts are detectable in oocyte cytoplasm. Epstein et al. (1985) have also found monomers and multimers of satellite 2 transcripts in somatic tissues of *N. viridescens* (liver and intestine). Moreover they have shown that transcripts copied from the complementary strands of satellite 2 are both present on lampbrush loops; they propose that this is a consequence of read-through transcription from upstream structural gene promoters, a topic to which I will return later in this chapter. It is not yet known whether transcripts of satellite TcS1, which consists of 380 bp, not 330 as originally estimated, are present in the cytoplasm of *T.c. carnifex* oocytes, but like satellite TcS2 it or closely related sequences are present in several other species of European and American newts (Macgregor and Sessions 1985).

Two much shorter satellite sequences have been identified in and cloned from the DNA of *T.c. karelinii* (Baldwin and Macgregor 1985). The smaller, TkS1, is a sequence of 33 bp and represents 5% to 10% of the genome. The larger, TkS2, consists of 68 bp and represents 0.5% to 1% of the genome. Both are located in the axial bars alongside all the centromeres of *T.c. karelinii* lampbrush chromosomes. They are both present beside the centromeres of *T.c. cristatus* and *T.c. carnifex*, though much less abundant. These satellite sequences are sometimes transcribed in oocytes, but on a trivial scale.

I turn now to in situ hybridization where labelled DNAs containing histone-coding sequences were used as probes. The sequences coding for the five histones are tandemly repeated and clustered in *Drosophila*, and in several species of sea urchin. This topic has been reviewed by Hentschel and Birnstiel (1981). Old et al. (1977) hybridized ^{3}H-labelled lambda clones containing inserted histone sequences, originating from three species of echinoderms, with the lampbrush chromosomes of *T.c. carnifex*. Four loop pairs in one of the HTAs of bivalent I were labelled, similarly four loop pairs, at different loci, in the other HTA (Fig. 5.3 a), also two loop pairs in the long arm of chromosome VI, one loop pair in the short arm of chromosome X, and one loop pair subterminal in the short arm of chromosome XI (in the paper this locus was wrongly assigned to the long arm of XI). The obvious controls were run; preparations digested with DNase 1 were labelled and normal, and preparations hybridized with ^{3}H-labelled lambda DNA, but lacking histone gene inserts were unlabelled.

Three different histone probes all gave identical sites of hybridization in preparations from a single female, but individual-specific differences were found in the four females from which lampbrush preparations had been made. The most striking individual-specific features were heterozygosities for labelled loops in bivalents VI and X.

These experiments were undertaken on two assumptions. One was that there would be sufficient sequence homology between echinoderm and *Triturus* histone-coding sequences for hybridization to occur. The other was that the spacer sequences in between coding sequences in the probes would not be directly involved in any such hybridization. While the first assumption may have been valid, the second assumption was certainly not.

Because relatively few loops were labelled in each preparation, it was possible to identify each labelled loop without ambiguity. A notable feature of some of them was that label was not present throughout the loop's length. Thus a particularly long loop on chromosome I (I_{24}) was labelled for about one-half to two-thirds of its length with a steep rise in labelling intensity from the thin axial insertion followed by a long plateau, in all four newts, likewise a shorter loop nearby (I_{29}) in preparations from three of these newts, but with label confined to a very short interstitial region in the corresponding loop of the fourth (Fig. 5.4 a, b). We interpreted these partially labelled loops on the basis that histone-coding sequences are confined to a relatively short region of loop axis; as these are transcribed the transcripts move on round the loop, at first remaining attached to

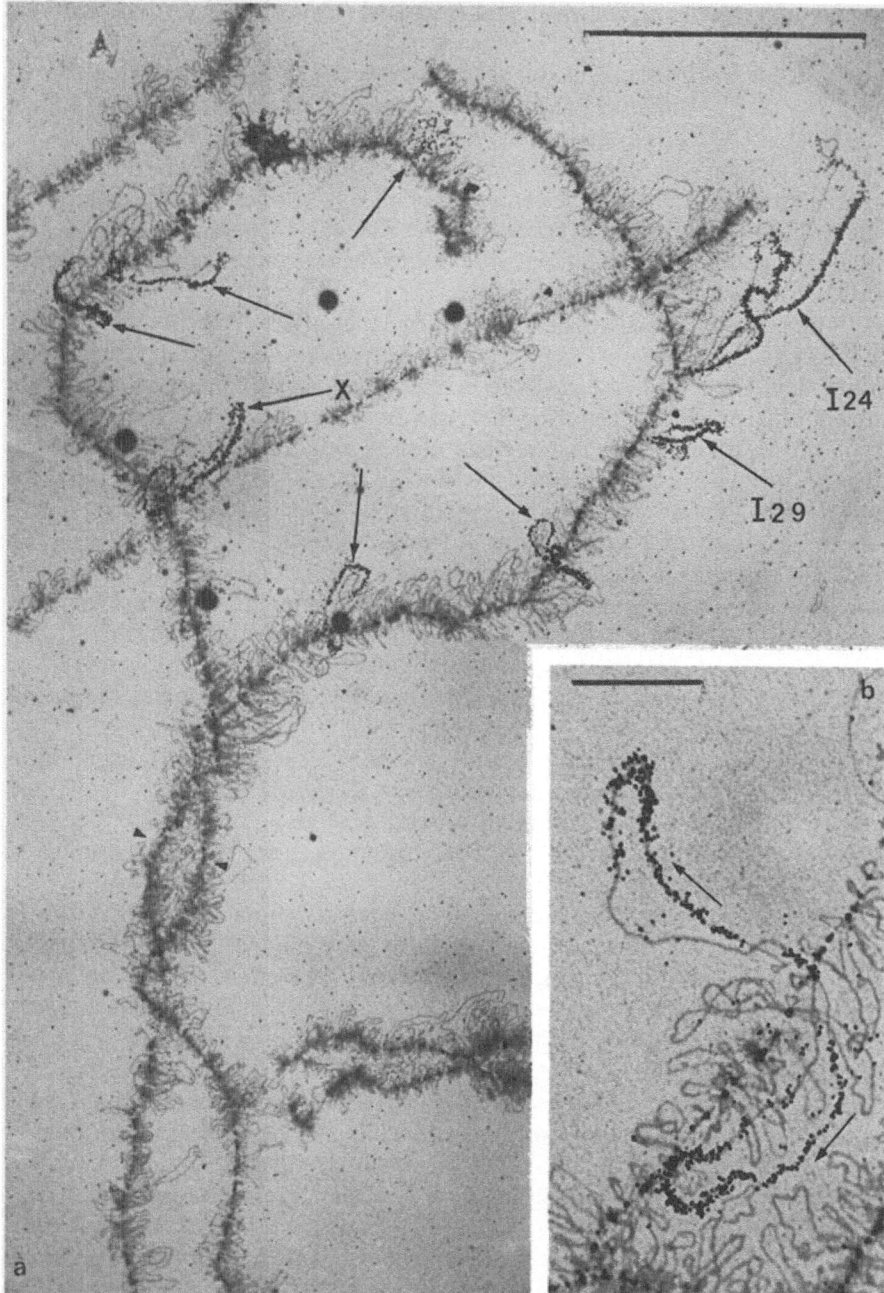

Fig. 5.3 a, b. Autoradiographs of lampbrush chromosomes of *Triturus c. carnifex* after hybridization in situ with denatured ^{3}H-labelled DNA nick-translated from a cloned 4.5 kbp histone gene cluster of *Strongylocentrotus purpuratus* containing the coding regions for H2B, H4 and H1 and intervening spacer sequences. **a** Part of bivalent I, whose centromeres are indicated by *arrowheads;* the eight labelled loop pairs are marked by *arrows*, two of which, I_{24} and I_{29}, are referred to in the text; all of them lie within the heteromorphic regions of the longer arms of bivalent I.

transcripts of other contiguous sequences, but becoming detached before the termination of transcription. I will return to consider an alternative interpretation later.

The constancy of labelling pattern of an identifiable loop in all preparations from a single female settled a question that for several years had remained unanswered. The extension/retraction theory (Callan and Lloyd 1960, Gall and Callan 1962, Snow and Callan 1969) had proposed that a loop's axis is continually extending from, and at its other end retracting into, the chromomere at its base all the while RNA synthesis is in progress. However a constant labelling pattern proves that a loop's axis remains stationary while it is being transcribed (Old et al. 1977, Macgregor and Andrews 1977, Macgregor 1980).

A few months after publication of the Old et al. (1977) paper we learned (personal communication, from Dr. L. H. Kedes 1977) that there are extensive simple sequence repeats of the form poly(C-T)·poly(G-A) in spacers between histone genes in all echinoderm species studied (Schaffner et al. 1978, Sures et al. 1978). We therefore (Callan et al. 1980, Callan 1982) set up more histone in situ DNA/ RNA hybrids, using lampbrush chromosome preparations from two other newts, five echinoderm probes known to contain the poly(C-T)·poly(G-A) repeat, one probe containing the entire histone-coding sequence of *Drosophila,* and three echinoderm probes whose entire sequences were known, and known not to contain the repeat. We also used two synthetic poly(C-T)·poly(G-A) DNAs and a synthetic single-stranded poly(C-T) DNA as probes.

All five echinoderm probes known to contain the repeat, the *Drosophila* probe, and all three synthetic probes, labelled all of the loops that had been identified as labelling with the three echinoderm probes used previously. The autoradiographs revealed yet more individual-specific differences between newts. Thus, for example, none of the four newts used in the first series of hybridizations showed any labelled loops on bivalent II, but of the two newts used for the second series, one proved to be homozygous for labelled loops close to the centromeres of bivalent II, the other heterozygous (Fig. 5.4c, d). Another and particularly striking individual-specific character concerns loop I_{24}. In both of the newts used for the second batch of in situ hybrids this loop was labelled throughout its length, whereas in all four of the newts used for the first batch the partial labelling of this loop was a regular and characteristic feature. The labelled length of I_{24} was much the same in all six individuals, some 100–150 µm. Either the unlabelled part of loop I_{24} seen in all preparations made from the first four newts includes transcripts of sequences that are absent from the later two newts, or these transcripts are present, but organized as another TU on a neighbouring lateral loop, with its own insertions in the chromosome axis.

All three synthetic probes labelled, though to a lesser degree, a few loops on bivalent I over and above those that had been identified in the first experimental series, and recognized again in the second series where probes including the

---

Bar = 100 µm. **b** The loop pair, corresponding to that marked with a *cross* in **a,** from another preparation; *arrows* indicate the directions of transcription; notice the unlabelled regions, and the cluster of silver grains where transcription would appear to terminate. *Bar* = 20 µm. (From Old et al. 1977)

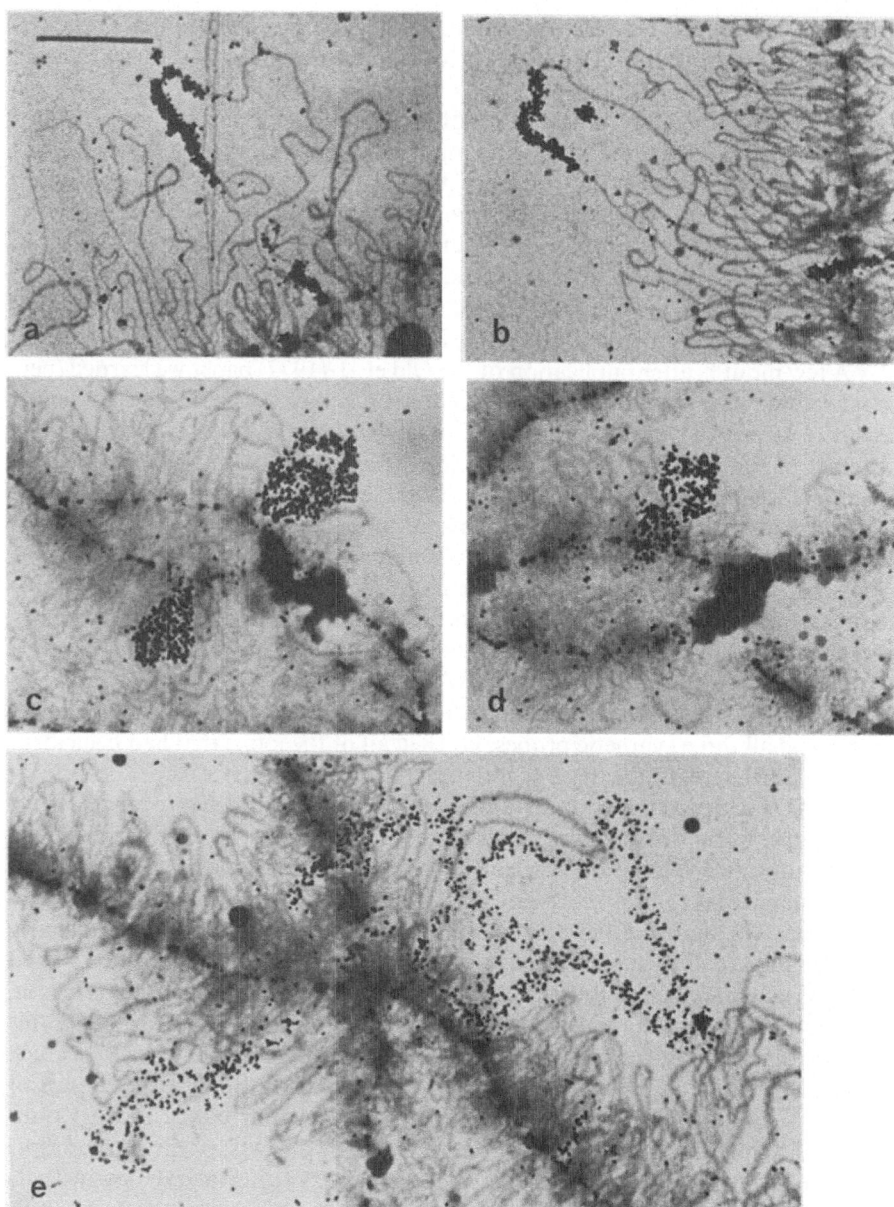

Fig. 5.4 a–e. Autoradiographs of lampbrush chromosomes of *Triturus c. carnifex* after hybridization in situ with denatured DNA probes. **a** and **b** The same loop, I_{29}, in the heteromorphic region of chromosome I, in two preparations; a short labelled part flanked by longer unlabelled parts was characteristic of this loop in this newt; the probe DNA was nick-translated from a clone that included the complete 6 kbp histone gene cluster of *Psammechinus miliaris*. **c** The middle regions of bivalent II; the probe used was a synthetic poly(C-T)·poly(G-A) DNA; in this newt a pair of loops on both homologues were regularly labelled. **d**, as **c**, but from a newt with labelled loops on one homologue only. **e** Loop I_{24} (see Fig. 5.3 a and text) labelled throughout its length with a probe nick-translated from a cloned DNA that included the coding region for H2B, without flanking spacer sequences, of *Strongylocentrotus purpuratus;* one loop is intact, the other broken. *Bar* = 20 µm. (From Callan 1982)

poly(C-T)·poly(G-A) repeat had been used. Although different degrees of loop-matrix compaction can influence an estimate of relative labelling intensity, it is evident that there are more loops containing transcripts from the simple repeat sequence than those that had first been recognized. However these loops presumably contain fewer of them.

The outcome of hybridizing with the three echinoderm probes known to lack the poly(C-T)·poly(G-A) repeat was altogether different, in that only one loop was regularly labelled, I_{24}, and over its entire length. The autoradiographs resulting from these three probes produced, after similar exposure, a lower density of silver grains over loop I_{24} than was the case with the probes containing the simple sequence repeat, and instead of showing a steep rise of labelling intensity that quickly reached a plateau, showed a gradient of labelling intensity throughout the length of the loop (Fig. 5.4e). This suggested that hybridizable histone RNA sequences in the transcripts increase in number as transcription proceeds, and are not confined to a short length of loop axis near to where transcription initiates. There may well have been an error of interpretation in the Old et al. (1977) paper caused by over-exposure and saturation of the autoradiographs. Another error, more serious, disclosed by the second series of in situ hybridizations is that all the labelled loops carry transcripts from poly(C-T)·poly(G-A) stretches of DNA, but only one loop, I_{24}, carries histone-coding sequences as well. However further doubts concerning this apparent clarification have arisen as a consequence of an impressive series of studies on the histone genes of *Notophthalmus viridescens* recently carried out by J. G. Gall and his colleagues.

Stephenson et al. (1981 a) showed that within the nine kilobase (kb) histone gene cluster of *N. viridescens* the order of the coding sequences is H1-H3-H2B-H2A-H4, with intervening spacers, all but H2B being transcribed in one direction. Cloned newt DNAs containing the histone sequences were initially identified by their capacities to hybridize with various subclones of echinoderm (*Psammechinus miliaris*) DNA that were known to include histone sequences. In *N. viridescens* the histone gene clusters do not lie adjacent to one another; instead they are separated by long tracts of a tandemly repeated satellite DNA, called satellite I (Stephenson et al. 1981 b) sequenced and shown to be 222 bp long (Diaz et al. 1981).

Nick-translated ³H-labelled probes containing satellite I DNA were hybridized to lampbrush chromosomes of *N. viridescens*. When the preparations were digested with RNase and denatured with alkali before hybridization, axial regions including all the centromeres were intensely labelled; the regions of chromosome axis at the sphere loci (one on chromosome II, one on chromosome VI) were also labelled. When the preparations were not RNased or denatured before hybridization, loops close to the sphere loci were labelled throughout their lengths, with labelling intensity proportional to the amount of RNP matrix at any given place. There was also light labelling of the axial regions including the centromeres; this was shown to be the result of unintended denaturation of DNA in these regions, for preparations digested with DNase I before hybridization still showed labelled loops near the spheres, but the centromeric regions were unlabelled (Diaz et al. 1981).

The analysis was extended by in situ hybridization using RNA probes containing sequences complementary to separated, cloned, single strands of satellite

I DNA. Two H[3]-labelled cRNAs were synthesized from the single-stranded satellite DNAs, and applied to lampbrush chromosomes for RNA/RNA hybridization. Both probes labelled loops around the sphere loci. Diaz et al. (1981) were
able to recognize a pair of exceptionally long loops, with single matrix gradients,
frequently in the form of a double-loop bridge with their thin ends distal relative
to the centromere, near the sphere locus on one homologue of bivalent VI in preparations from one particular newt. This loop pair was regularly labelled by one
of the two probes, with a labelling gradient matching the distribution of loop matrix; it was not labelled by the other probe (Fig. 5.5 a–d). Thus its morphology,
and the results of hybridization with complementary single-stranded probes,
agree with one another and indicate that this TU has a constant transcriptional
polarity within the chromosome.

When making these preparations the lampbrush chromosomes had been dispersed in saline diluted to one-quarter the strength of that previously used for sap
dispersal. As a result loop matrix was less compact than usual, thereby improving
the resolution of TU polarity. There were numerous examples of loops near the
spheres that were only partially labelled with the single-stranded probes. By taking photographs of the preparations before covering them with autoradiographic
emulsion, Diaz et al. (1981) were able to demonstrate that the limits of the labelled
regions coincided with reversals of the polarity of transcription within the lengths
of single loops (Fig. 5.5 e, f). In such loops all TUs of one and the same polarity
were labelled, whereas those of opposite polarity were not. This confirms in an
elegant fashion the evidence from "Miller-spreads" that adjacent TUs may be of
similar polarity, i.e. head-to-tail, or of reverse polarity, either head-to-head or
tail-to-tail, and that in regard to transcription of satellite I DNA wherever reversals of polarity occur they indicate that opposite strands of the DNA are being
transcribed.

Labelling with one of the single-stranded satellite probes was much more extensive, in the sense that greater lengths of loops were labelled, than were labelled
with the other probe. Knowing that in *N. viridescens* the 9 kb histone gene clusters
(3 μm of DNA if fully extended) are separated from one another by variable
amounts (up to at least 100 kb) of satellite I (Stephenson et al. 1981 b) Diaz et al.
(1981) suggested that this difference in the extent of labelling by the two satellite
I probes could be explained if transcription initiated at a promoter site at the 5′
end of a histone-coding sequence and continued without interruption through

---

→

Fig. 5.5. a, c and **e** Phase contrast photographs of lampbrush chromosomes of *Notophthalmus
viridescens* dispersed in dilute saline and centrifuged, before in situ hybridization; they include
sphere loci. **b, d** and **f** Autoradiographs of the same regions after in situ hybridization with single-
stranded RNA probes prepared from cloned satellite I DNA. **a** and **c** Double-loop bridges on
one chromosome VI at the fused sphere locus in two oocytes from a certain newt; **b** shows that
transcripts on this double-loop bridge hybridize with one satellite RNA probe, whereas **d** shows
that they do not hybridize with the complementary probe. **e** Long loops can be seen with multiple
transcription units of diverse polarities beside the sphere locus of chromosome II and **f** shows
the outcome of hybridization with the same probe as that used in **d**; two short transcription units
on one loop are of the same polarity and are labelled, corresponding to those marked with *arrowheads* in **e**, whereas the much longer transcription units of opposite polarity do not hybridize
with this probe. *Bars* = 20 μm. (From Diaz et al. 1981)

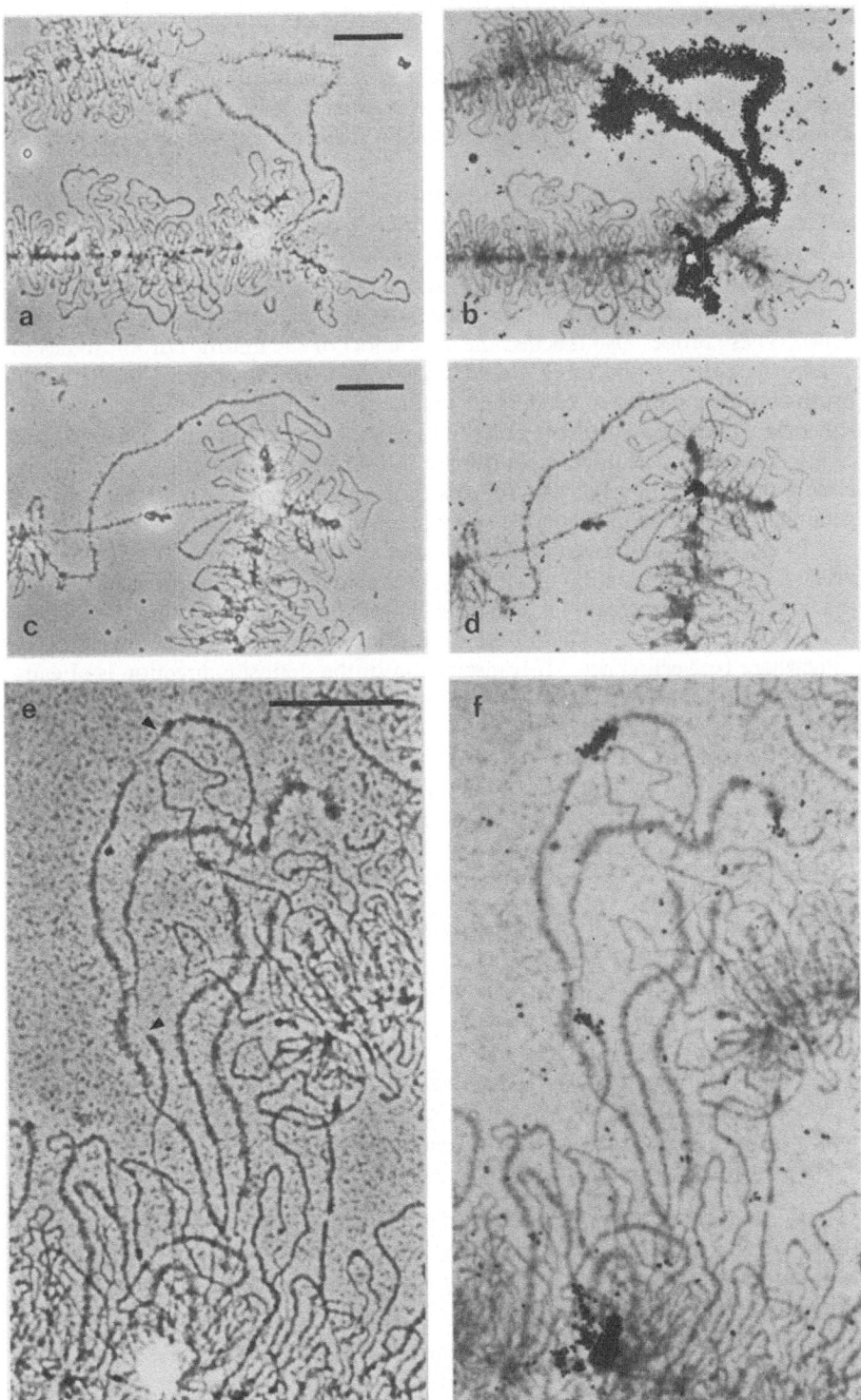

Fig. 5.5 a–f

neighbouring spacer and satellite sequences, "read-through", with four promoters available for initiation in one direction, but only one in the other. Such read-through transcription, where termination signals are not recognized, had been proposed by Varley et al. (1980 b) as a peculiar feature of transcription on lampbrush chromosomes, and would explain the great lengths of some of their TUs.

Meanwhile Gall et al. (1981) had shown by in situ hybridization to denatured mitotic chromosomes of *N. viridescens* of cRNA probes synthesized from single-stranded subclones of the histone gene cluster, the probes containing coding sequences and associated spacers but without satellite I sequences, and of DNA probes nick-translated from similar subclones, that the sphere loci alone were labelled. These probes also labelled the sphere loci of the mitotic chromosomes of *Triturus c. carnifex* and of *T. alpestris apuanus*. When hybridized to the lampbrush chromosomes of *N. viridescens* similar cRNA probes labelled loops at the sphere loci like those labelled by satellite I probes (Fig. 5.7 a). These observations clearly suggested that there is coordinate transcription of histone genes and satellite I on loops alongside the spheres, an association already established for genomic DNA at the molecular level by Stephenson et al. (1981 a, b).

In loops containing several TUs Diaz et al. (1981) and Gall et al. (1983) envisaged that the termination of a TU including satellite I sequences occurs within the limits of a downstream histone gene cluster if a promoter within that cluster is already being used for the initiation of transcription in the same direction, or somewhere between clusters if transcription in the opposite direction is already in progress from a promoter site downstream, i.e. two convergent, or tail-to-tail TUs.

So the three main features of the hypothesis are that all TUs on the histone loops initiate at promoters at the 5' end of any of the histone genes, that transcription proceeds without interruption past the 3' end of the gene(s), continues through internal spacer(s) and on beyond the limit of the cluster through satellite I sequences. All three features of the hypothesis have been tested and substantiated by Diaz and Gall (1985) in a masterly series of in situ hybridization experiments where cRNAs, synthesized from five pairs of single-stranded RNA subclones from within the *N. viridescens* histone gene clusters, have been used as probes. The subclones are displayed in Fig. 5.6 a. For simplicity the DNA strand that serves as a template for the transcription of histone genes H1, H3, H2A and H4 is arbitrarily designated strand A, while the complementary strand from which the H2B gene is transcribed is strand B. The cRNA probes are designated by the sequences that they contain, i.e. an A probe contains sequences that will hybridize with RNA transcribed from genomic strand A, and vice versa.

The predictions as stated by Diaz and Gall (1985) are as follows:

"(a) TUs that contain transcripts from the A strand should be more common than those with B transcripts, since there are 4 histone gene promoters in the A strand but only one in the B." All five comparisons between complementary A and B probes supported this prediction.

"(b) All sequences in strand A downstream from the H1 promoter or in strand B downstream from the H2B promoter should be transcribed regardless of whether they are from coding regions or internal spacers." Probe 51–7A contains only the internal spacer sequence between the H2A and H4 genes, but it labelled as extensively (Fig. 5.7 b) as A strand probes containing sequences(s) coding for mRNA(s). Probe 51–3B contains the spacer between sequences complementary to those transcribed by genes H1 and H3, and lies downstream from the promoter for H2B. As expected, it labelled relatively few TUs, just like probe 51–9B that contains the sequence coding for H2B mRNA.

"(c) No sequences in strand A upstream from the H1 promoter or in strand B upstream from the H2B promoter should be transcribed. Exceptions will occur if transcription continues without interruption into or through a second histone cluster." Probe 51–12B labelled no loops at the sphere loci. Probe 51–7B also did not label loops at the sphere loci, apart from a pair of loops in one preparation in which the label was restricted to a short section at the thick end of a long TU. This was precisely predicted as a possible exception.

"(d) Any TUs labelled by probes indicated in (b) should be labelled with uniform intensity over their whole length. This is because sequences immediately downstream from a promoter will be represented once and only once in every transcript of that TU. Again, read-through of a second cluster will produce exceptions, in this case TUs with stepwise label." Unlike TUs labelled with probes containing satellite I sequences, in which labelling gradients are a feature, TUs labelled with probes lacking satellite I sequences were labelled uniformly. There were examples of the predicted exceptions. With probe 51–9A a pair of loops, probably sisters, were labelled less intensely over the first three-fourths of their length than other TUs in the same preparation, and then followed a sharp transition to higher labelling intensity; this pattern would occur if a downstream promoter had been used for initiation, e.g. that for H2A, and transcription continued on through a second histone cluster. Sister loops with TUs showing similar stepwise labelling patterns were also seen in a preparation hybridized with probe 51–3A (Fig. 5.7 c). Diaz and Gall made two other predictions, but they require more complex analysis, and for lack of sufficient data will not be discussed here.

In short, the hypothesis has been fully supported by these experiments, and Fig. 5.6 b, c explains by diagrams how some of the types of matrix distribution found on the histone loops, and how they are labelled or not labelled by various probes, arise. In their discussion Diaz and Gall draw some further conclusions from their results. TUs are practically always labelled throughout their lengths with these single-stranded probes; if transcription sometimes initiated within satellite I sequences, the thin ends of the resultant TUs would not label with such probes. There are alternative explanations for the very few (only two) exceptional partially labelled TUs that were seen.

In the earlier study of satellite I transcription Diaz et al. (1981) and Gall et al. (1983) had found a strong bias in favour of one strand over the other, indeed this had largely prompted the hypothesis and experiments described above. It has now been demonstrated beyond doubt that the satellite strand more frequently transcribed is continuous with that which includes the four histone gene template sequences in a histone cluster. However the results have also shown a superim-

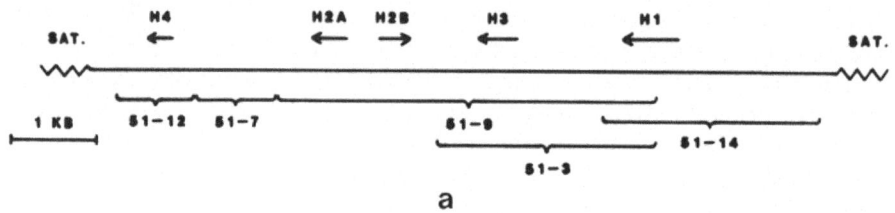

a

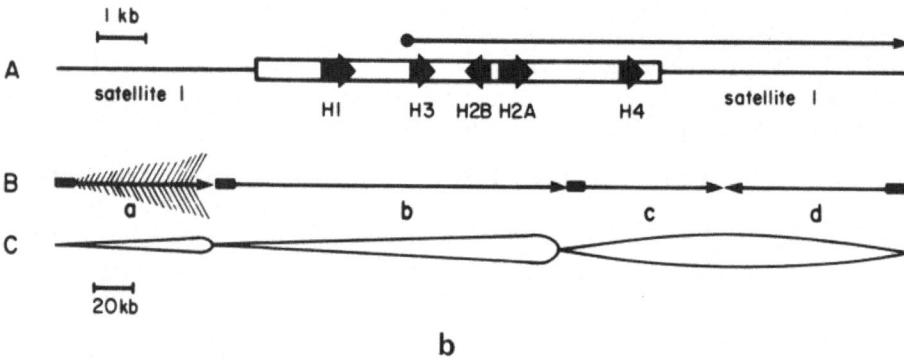

b

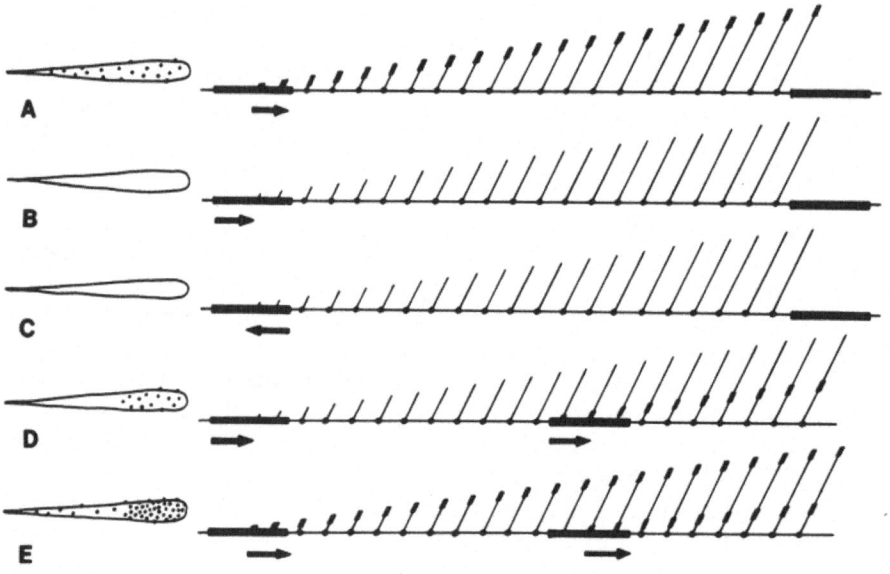

c

Fig. 5.6 a–c

posed complexity in that different histone gene promoters are utilized at different frequencies, with H1 preponderating over H3, H2A and H4. Another consequential discovery is that read-through of a complete histone cluster by transcription that initiated upstream is a rare event for, inter alia, few examples of stepwise labelling were found. Thus different lengths of simple thin to thick matrix gradients in the histone loops must generally reflect different numbers of satellite I repeats between histone gene clusters.

A further provocative observation is that several cases were noted where two loops originating from a single sphere locus had similar and distinctive labelling patterns. Although loops can only be identified as sisters if they form a double-loop bridge, these examples may well represent sisters utilizing the same promoters, but long after they had been formed at the last premeiotic DNA replication.

The last matter discussed by Diaz and Gall concerns the processing of transcripts containing histone-coding sequences. Whereas satellite I transcripts far exceed histone transcripts in the oocyte nucleus of *N. viridescens,* none were detected in oocyte cytoplasm. On the other hand, and confirming the earlier observations of several other investigators that histone mRNAs are stored in oocyte cytoplasm in large amounts, Diaz and Gall found mRNAs (for H1 and H4) in the cytoplasm of enucleated oocytes. Diaz and Gall consider it probable that endonucleolytic cleavage removes the most terminal histone sequence from the 5' end of the giant transcript at an inverted repeat processing signal at the 3' end (established for H4), and *all* downstream sequences (be they of other histones, internal spacers

---

Fig. 5.6. a Simplified map of a 9 kb histone gene cluster from *Notophthalmus viridescens,* showing the positions and orientations of the mRNA coding regions, flanked by satellite I sequences indicated by *wavy lines.* Subclones used for preparing the single-stranded RNA probes described in the text are indicated by *brackets* and *clone numbers.* **b** Read-through transcription at the histone gene cluster: *A* transcription initiates at a histone gene promoter, in this example H3, and continues downstream into the satellite region; *B* four histone gene clusters (*rectangular boxes*) separated by tracts of satellite I; *arrows* represent possible directions of transcription, giving rise to four TUs (*a, b, c, d*). TUs *c* and *d* run into one another within a satellite tract between two clusters, TUd having originated at an H2B promoter, whereas TUs *a, b* and *c* are assumed to have originated at any of the other four promoters. Nascent transcripts are shown on TUa; *C* TUs represented in *B* as they would appear in the light microscope. **c** How the labelling patterns of histone TUs arise after in situ hybridization with single-stranded probes that contain no satellite I sequences; *solid bars* in loop axis DNA represent histone gene clusters, and in each case transcription is shown as originating in the middle of a cluster; the thicker regions on transcripts, when present, represent sequences that have hybridized with a probe and thus show up labelled in autoradiographs. The sequences contained by various probes are indicated by *arrows;* an *arrow pointing right* represents RNA sequences corresponding to DNA sequences on the transcribed strand, while an *arrow pointing left* represents sequences corresponding to those on the non-transcribed strand. *A* TU labelled uniformly throughout its length: the probe contains sequences downstream from the active promoter; *B* TU unlabelled: the probe contains sequences upstream from the active promoter; *C* TU unlabelled: the probe contains sequences downstream from the active promoter, but from the non-transcribed strand; *D* TU unlabelled at its thinner end, but labelled at its thicker end: the probe contains sequences upstream from the active promoter, but transcription reads through a neighbouring cluster downstream; *E* TU more heavily labelled at its thicker than its thinner end; the probe contains sequences downstream from the active promoter, but transcription reads through a neighbouring cluster downstream. (From Diaz and Gall 1985)

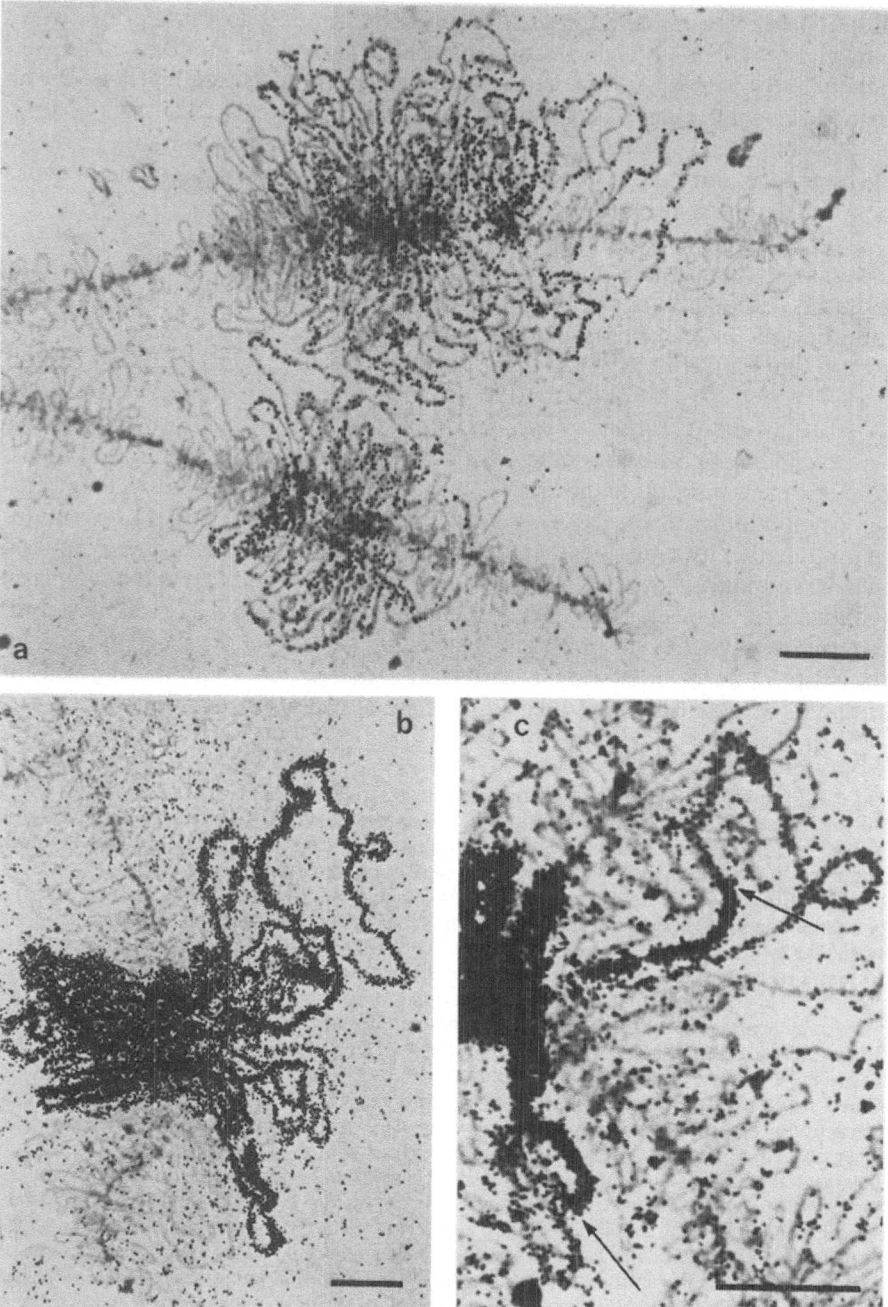

Fig. 5.7 a–c. Autoradiographs of regions of the lampbrush chromosomes of *Notophthalmus viridescens*, including the sphere loci of chromosome II, after in situ hybridization with ³H-labelled RNA probes copied from single-stranded clones shown in Fig. 5.6 a and designated as described in the text. **a** The probe was 51–9A and includes sequences complementary to transcripts beginning within the coding region of H1 and ending within the spacer downstream from the coding region of H2A; many long loops at the homologous sphere loci are labelled. **b** The probe

and satellite I repeats) are destroyed in the nucleus. In support of this, at first glance astonishing conclusion, they cite evidence that when synthetic transcripts from *Xenopus,* the chicken, and the sea urchin *Psammechinus,* with flanking sequences at boths ends of histone coding regions, are injected into *Xenopus* oocytes, whereas these are cleaved at the appropriate 3' terminus of the coding region, no processing occurs at the 5' end (Krieg and Melton 1984, Georgiev et al. 1984). This exposes in all its stark reality the C-value paradox, reviewed by Gall in 1981. Natural selection operates with severity on characteristics that determine the viability of whole organisms, including one characteristic that certainly does relate to C-value, namely cell size (see e.g. Horner and Macgregor 1983), but it seems to exert little control over the multiplication of DNA sequences, satellites in particular, that on presently available evidence appear to have little or no functional rôle.

I will end this review of Diaz and Gall's 1985 paper by quoting a provocative matter raised in their last four sentences. "In our view read-through transcription is a general feature of the lampbrush chromosome stage and is not a peculiarity of the histone genes. To what extent somatic transcription in the newt follows the same pattern remains to be seen. It is possible that the levels of the various factors involved in RNA transcription and processing could differ between oocytes and other cells. Or it may be that we can simply see more clearly in the lampbrush chromosome what occurs in other cell types as well".

In the light of this penetrating study by J. G. Gall and his colleagues, how are the observations of Old et al. (1977), later qualified by Callan et al. (1980) to be interpreted? The major discordance is the demonstration that in *T.c. carnifex* as well as in *N. viridescens,* the main concentrations of histone gene clusters lie close to the sphere loci, yet none of the probes used by Old et al. (1977) labelled loops near the spheres. Gall et al. (1981) considered the possibility that sea urchin histone sequences may differ too much from those of the newt to form stable hybrids, but discount this explanation because Stephenson et al. (1981 a) found good cross-reactions between three histone genes of *Psammechinus miliaris* and the corresponding genes of *N. viridescens.* They offer as a more likely explanation that the oocytes of *T.c. carnifex* used by Old et al. (1977) for in situ hybridization were at a stage in oogenesis (the onset of yolk accumulation) when the major histone sequences of *T.c. carnifex* were not being transcribed, whereas a minor group on chromosome I that happened to lie downstream from some other active gene was transcribed by read-through. Some support for this explanation comes from two observations. Gall et al. (1981) "... found extreme variation in the number and sizes of hybridizing loops at the sphere loci of *Notophthalmus*"; and Barsacchi

---

was 51–7A and contains sequences complementary to transcripts from part of the spacer beginning downstream from the coding region of H2A and ending upstream from the coding region of H4. The loops at a sphere locus are strongly labelled, though the probe contained no sequences complementary to those coding for histones. **c** The probe was 51–3A and contains sequences complementary to transcripts beginning within the coding region of H1 and ending within the spacer downstream from H3; *arrows* point to abrupt transitions from low to high levels of labelling in two long transcription units in loops that may be sisters, explicable if transcription initiated in both at a promoter for H3 and continued on through the next histone cluster downstream. *Bars* = 20 µm. **a** From Gall et al. (1981); **b** and **c** from Diaz and Gall (1985)

Pilone and Humphries (1975) noticed that in the same species the loops adjacent to the spheres become especially conspicuous as oocytes approach maturity, for they remain extended when most other loops have regressed. No doubt this uncertainty will soon be cleared up.

It is appropriate at this point to reconsider how loops that are labelled over only part or parts of their lengths are to be interpreted. For the histone loops, as has been discussed in the preceding paragraphs, J. G. Gall and his colleagues have given a convincing answer, reviewed by Gall et al. (1983). The multiple TUs within single loops are clearly related; regardless of their polarity, they all contain histone and satellite I transcripts. The partially labelled loops observed by Old et al. (1977) and by Macgregor and Andrews (1977) had been interpreted by these investigators as a consequence of transcript processing before the completion of transcription. Diaz et al. (1981) questioned this interpretation, suggesting instead that reversals of TU polarity within single loops could explain the observations. My own impression now is that Diaz et al.'s (1981) claim is justified for many of the partially labelled loops that have been observed, and in particular a short labelled region in the middle of a loop is hard to account for in any other way. However where a labelled region extends from one end of a loop, and where the matrix gradient, beyond the labelled region appears to continue with the same polarity as that exhibited by silver grains in the labelled region (Fig. 5.3 b), such cases I would consider more likely to indicate the processing of transcripts.

Whether this be so or not, the probes used by Old et al. (1977) and by Macgregor and Andrews (1977) included labelled, double-stranded, repetitive DNA sequences, so after their denaturation complementary strands should be available for hybridization. If the transcripts in multiple TUs on single loops include related repetitive sequences, then hybridization should occur irrespective of the polarity of transcription, resulting in such loops being labelled throughout their lengths. The fact that partially labelled loops were nevertheless observed shows that, at least in regard to these particular repetitive sequences, some transcripts include them but others do not. In this respect, then, they differ from the histone loops. However other loops with multiple TUs no doubt do resemble the histone loops in their repetitive sequence organization, an example being the giant loops on chromosome II of *N. viridescens*.

To date, no transcripts of unique coding sequences have been identified by in situ hybridization to lampbrush chromosomes. However one attempt to do so (Callan and Old 1980) should be considered in this chapter because it revealed another potential source of error in the method. Two *Xenopus* globin cDNAs and one human γ-globin cDNA were cloned in a plasmid vector, and nick-translated to give three probes. All three probes labelled throughout their lengths, and with a gradient in conformity with a simple uniform matrix gradient, a single pair of loops near the middle of chromosome IX of *T.c. carnifex*. However whereas there might have been sufficient sequence homology between *Xenopus* and *Triturus* globin genes to give detectable heterologous hybrids, evidence was already available that mammalian globin sequences do not cross-hybridize with *Triturus* DNA. Now the plasmids containing globin inserts all had one feature in common; they were constructed with G.C homopolymer tails, and when nick-translated the probes necessarily contained ^{3}H-labelled poly(C)·poly(G) stretches as well as globin sequences. Callan and Old therefore hybridized synthetic poly(C)·poly(G) probes with *T.c. carnifex* lampbrush chromosomes, and found that once again a single pair of loops near the middle of chromosome IX became labelled. So this attempt had failed to locate transcripts of globin sequences, but had revealed that one pair of loops is transcribing from a remarkably monotonous DNA. Perhaps even more remarkable is the fact that this loop was present on both chromosomes IX in four unrelated females of *T.c. carnifex*.

It will be of great interest to learn the outcome of in situ hybridization of cloned single copy structural gene sequences to amphibian lampbrush chromosomes. There is good prospect that this question will be attacked and resolved in the near future, for in the discussion of their results Diaz and Gall (1985) make the valid point that "... the fact that histone probes do give detectable label with reasonable exposure times (days) implies that probes of the sort we use could detect *any* single copy sequence transcribed on a lampbrush chromosome. The fact that histone sequences are repetitive in the genome is irrelevant to their detectability within a single TU".

Jamrich et al. (1983) have recently succeeded in making lampbrush preparations from oocytes of *Xenopus laevis* of a quality adequate for autoradiographic analysis. The technique used was that described by Gall et al. (1981). For many years the lampbrush chromosomes of *Xenopus* have proved to be refractory material, primarily because the nuclear sap remains extremely rigid in the salines generally used for making preparations. Müller (1974) obtained sap dispersal by adding formaldehyde to a phosphate-buffered saline of 60 mM overall concentration, and he found it essential to work strictly within a pH range of 6.8 to 7.0. The individual chromosomes proved to be much less easily recognized than those of most newts, there are more of them (n = 18), centromeres were not evident, and few structures served as reliable "landmarks".

From the standpoint of molecular biology, however, the relatively small size of the *Xenopus* genome, 3 pg, is distinctly advantageous, its genome is being intensively dissected in several laboratories, many of its DNA sequences have already been established as clones, the biochemistry of oogenesis in *Xenopus* is firmly based, and its oocytes have proved to be ideal vehicles for studying the behaviour and fate of injected nucleic acids and cell nuclei (for a general review of this latter topic, see Gurdon and Melton 1981).

The distinct virtue of using *Xenopus* for in situ hybridization is that several different fully sequenced cloned genes and components of genes can be applied as probes to lampbrush chromosomes of the same species from which the clones were originally derived. The pitfalls arising from the use of heterologous probes have been illustrated by the results of hybridizing histone sequences derived from echinoderms to newt chromosomes; trustworthy information only came when newt histone sequences were used instead.

In the centrifuged lampbrush preparations made by Jamrich et al. (1983) the lateral loops appear to be of about 10 µm average length, and they display recognizable matrix gradients. ^{3}H-labelled RNAs complementary to cloned, strand-separated repetitive *Xenopus* DNA sequences were used as probes.

A satellite sequence of 77–79 bp, present in about 10^5 copies per genome, provided two complementary probe cRNAs. When one of these was hybridized to mitotic preparations, both the terminal regions of all chromosomes were labelled to about the same degree, the hybridization being RNA/DNA. When this same probe was hybridized to lampbrush chromosomes (RNA/RNA) a cluster of loops on one end of one chromosome was intensely labelled after only 4 days exposure (Fig. 5.8 a). The complementary probe RNA, of similar specific activity, also labelled what is probably the same single cluster of terminal loops after similar autoradiographic exposure, though with much lower intensity (Fig. 5.8 b). When autoradiographs were exposed for 30 days, loops at the terminal regions of all the chromosomes were labelled, and with both probes. So loops at all the chromosome ends contain this satellite DNA sequence, and both strands are transcribed during the lampbrush stage, though with greater intensity in one terminal region of one chromosome than elsewhere.

An unrelated, tandemly repeated sequence 741 bp long, present in about 4×10^4 copies per genome, similarly provided two complementary probe cRNAs. When hybridized to mitotic preparations, single intercalary sites on about half of the chromosomes were labelled, all to about the same degree. When these two probes were hybridized to lampbrush chromosomes, both labelled a few pairs of loops, only one to four per preparation, at intercalary sites. So only a small proportion of this repetitive sequence, again both strands, is transcribed by lampbrush chromosomes. Most of the sites where this sequence is present are not transcribed, so presumably these remain condensed in chromomeres.

A third clone used by Jamrich et al. (1983) contained a widely interspersed sequence present at about 10^3 copies per genome, and again provided two complementary cRNA probes. These produced generalized labelling of mitotic chromosomes, and likewise both probes labelled several 100 loops widespread on the lampbrush chromosomes (Fig. 5.8 c), indicating that a considerable fraction of this repetitive sequence is transcribed.

These preliminary studies with *Xenopus* confirm the results of in situ hybridization to newt lampbrush chromosomes in showing that both satellite and moderately repetitive sequences are transcribed. The fact that only some of the repetitive sequences are transcribed, others not, strongly suggests that the regulation of their transcription is coincidental, i.e. regulated by initiation at sites upstream of some of them, not of others.

The well-labelled autoradiographs produced when cRNAs of the moderately repetitive interspersed sequence were used as probes demonstrates the extraordinary sensitivity of the technique, for if hundreds of loops become labelled though

---

Fig. 5.8 a–c. Autoradiographs of lampbrush bivalents of *Xenopus laevis* after in situ hybridization with ^{3}H-labelled RNA probes copied from cloned, strand-separated repetitive DNA sequences of *Xenopus*. **a** The probe was cRNA complementary to one strand of a 77–79 bp repeat sequence with about 10^5 copies per genome; loops near one end of the bivalent are intensely labelled. **b** The probe was cRNA complementary to the opposite strand of the repeat sequence used in **a**; loops near one end of what is probably the same bivalent as in **a** are much less heavily labelled, though the specific activities of the two probes, and autoradiographic exposure times, were similar. **c** The probe was cRNA complementary to one strand of a repeat sequence with about 10^3 copies per genome; many loops are labelled by this probe, which had a specific activity of the same order as those used for **a** and **b**, but this autoradiograph was exposed five times longer. The large objects of irregular shape are nucleoli. *Bar* = 5 μm. **a** Courtesy of Dr. M. Jamrich; **b** and **c** from Jamrich et al. (1983)

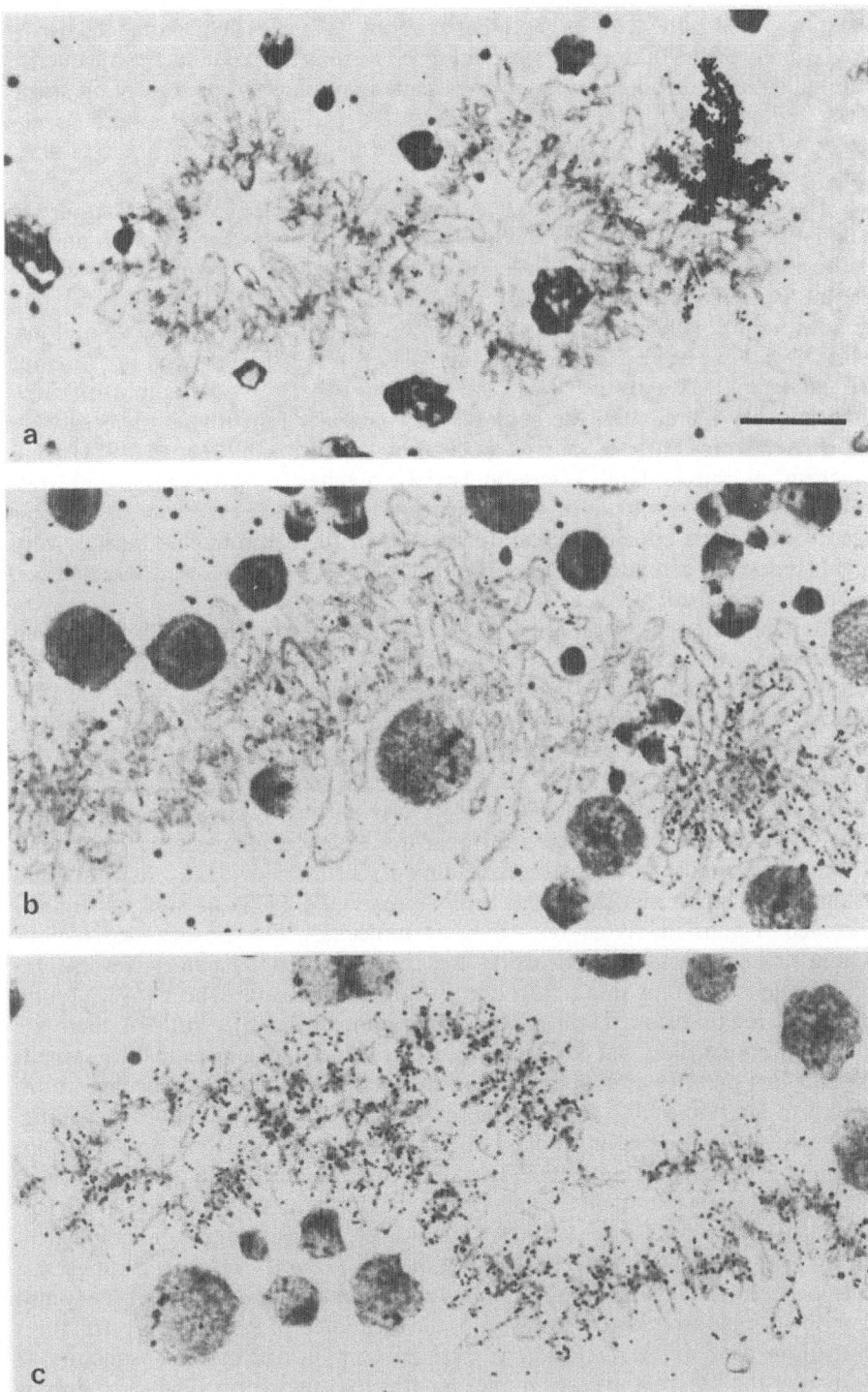

Fig. 5.8 a–c

the transcribed sequence is only present at about 10^3 copies per genome there cannot be many copies of the sequence per loop. The sensitivity results from sequence "amplification" through identical transcripts lying in close proximity on single TUs, and supports the contention of Diaz and Gall (1985) that multiple transcripts of unique coding sequences will likewise be detectable by in situ hybridization to lampbrush chromosomes.

It is generally accepted that DNA repeat sequences are a universal feature of eukaryotic genomes, and they have been shown to be interspersed with non-repetitive sequences in so many diverse species, largely as an outcome of the studies of R. J. Britten and E. H. Davidson and their colleagues, that interspersion may well, with the possible exception of some lower eukaryotes, be equally ubiquitous. This subject has been reviewed in detail by Lewin (1980) and I will not attempt to deal with it here. However I will consider how other work on lampbrush chromosomes has illuminated the subject. Intermittently repetitive sequences were shown to be transcribed in *Xenopus* oocytes by Davidson and Hough (1971) using biochemical criteria, and likewise in *Triturus* oocytes by Sommerville and Malcolm (1976). The results of in situ hybridization described earlier in this chapter accord with these conclusions, and have shown furthermore that various very highly repetitive (satellite) sequences, and also extremely simple repeat sequences, are also transcribed.

Sommerville and Scheer (1982) have demonstrated that in hnRNA extracted from hand-isolated germinal vesicles of previtellogenic oocytes (therefore essentially uncontaminated by cytoplasmic components or by transcripts from 18 + 28 S rDNA) of four Amphibia, complementary repeat sequences are present in abundance. The species used covered a 3 to 78 pg range of C-values, and were discussed in Chap. 4; Scheer and Sommerville (1982) showed that the oocyte nuclei of all four species contained hnRNA molecules of roughly similar lengths. When hnRNA molecules of *Triturus* are spread for electron microscopy under non-denaturing conditions extensive secondary structure is evident, "hairpin" and "stem-loop" structures (Malcolm and Sommerville 1977) as well as double-stranded regions involving two or more molecules. Double-stranded regions within single molecules demonstrate that there are complementary (reverse) repeat sequences within single transcripts. Sommerville and Scheer (1982) found that after renaturation followed by renaturation of *Triturus* hnRNA there are many intramolecular, but few intermolecular duplex structures in EM spreads made a few minutes after renaturation, whereas most of the molecules are involved in intermolecular duplex structures when left for 24 h before spreading. The number averages for duplex length in untreated hnRNA are: *Xenopus*, 0.11 µm; *Triturus*, 0.26 µm; *Amphiuma*, 0.28 µm; *Necturus*, 0.86 µm, so although the hnRNA molecules of all four species are of about the same average length (see Chap. 4), i.e. are not correlated with C-value, the average lengths of complementary repeat sequences within the series *Xenopus–Triturus–Necturus* do suggest a relationship with C-value (though *Amphiuma* is somewhat out of step). They also describe comparative kinetic analyses of the formation of RNA duplex structures in the four species; interpretation of these are complicated by various factors, in particular the different average duplex lengths that they form. They also discuss comparative kinetic analyses of the renaturation of *Triturus* tissue culture and oocyte hnRNAs; these are similar, which implies that the transcription of complementary repeat sequences is not restricted to oocytes.

Sommerville and Scheer (1982) made two further observations. When the single-stranded regions are removed from double-stranded RNA of *Triturus* oocytes, and the double-stranded RNA is then denatured and hybridized with homologous (*Triturus*) or heterologous (*Necturus*) denatured DNA, the heterologous reaction is quantitatively less than 5% that of the homologous reaction. So the complementary repeat sequences in these two genera are decidedly different and do not show evolutionary conservation. The other significant observation is the wide dispersion of the complementary repeat sequences amongst the nascent transcripts of *Triturus* lampbrush chromosomes. Labelled double-stranded *Triturus* hnRNA, similarly devoid of single-stranded regions, was denatured and hybridized in situ with *Triturus* lampbrush chromosomes. Autoradiographs showed scarcely any loops to be unlabelled.

The correlation between the lengths of the TUs of lampbrush chromosomes and C-values mentioned towards the end of Chap. 4 and earlier in the present chapter needs now to be considered along with the perplexing range and diversity of repetitive sequences transcribed on these TUs. There has been a great deal of argument concerning the possible functions of interspersed repetitive DNA sequences, most of which is beyond the scope of this monograph. However I wish to consider one aspect that is directly relevant. The overwhelming majority of repetitive sequences have no coding functions, and unlike coding sequences, and some exceptional satellite sequences (Macgregor and Sessions 1985), they have not been conserved during evolution, i.e. they show divergence when taxonomically related species are compared. A particularly appropriate example is provided by 15 species of North American salamanders, all belonging to the genus *Plethodon,* studied by Mizuno and Macgregor (1974). All the species have similar karyotypes, but their C-values range from 18 to 69 pg. Details of their phylogenetic relationships, the work of R. Highton, are given in the introduction to Mizuno and Macgregor's paper. In brief, between 60% to 90% of the repetitive sequences are held in common by closely related eastern species. In more diverse eastern species groups between 40% and 60% of these sequences are held in common, but less than 10% are shared by eastern and western species groups.

In view of such divergence what then can be the functional significance, if any, of these repetitive sequences? They may have played an evolutionary rôle by interfering with orderly synapsis in hybrids between different populations, thereby reinforcing geographic isolation, but why are they transcribed by lampbrush chromosomes? According to the evidence reviewed by Davidson (1976) at least some of these sequences are accumulated in oocytes, though not in their germinal vesicles, they persist beyond fertilization and are inherited by embryos. Why?

The repetitive sequences mentioned above are covalently linked to polyadenylated RNAs. They have been studied in particular detail by Anderson et al. (1982) who showed by denaturation and renaturation that about 68% of the mass of polyadenylated RNA in the cytoplasm of late ovarian oocytes of *Xenopus* include interspersed repeat sequences which Richter et al. (1984) have shown to be untranslatable, whereas in *Xenopus* tadpoles the proportion has dropped to less than 15%. To quote from Davidson and Posakony (1982): "The interspersed transcripts stored in mature eggs are especially interesting from the standpoint of possible function because they represent the major part of the mass of the egg poly(A)RNA and result from accumulation during a long period of synthesis during oogenesis. They are evidently destined for use during embryonic development". The linkage of these particular repetitive sequences to polyadenylated RNA (and whether there is a comparable accumulation of repetitive sequences not so linked is an open question) led Davidson and Posakony (1982) to consider various possible functions, but they end their review by stating that: "Despite their ubiquity, their quantitative prominence, their apparent developmental regulation and the amount of interest they have aroused, the repetitive sequence transcripts of animal cells remain a phenomenon in search of a physiological meaning".

The situation in sea urchin oocytes and embryos described by Posakony et al. (1983) shows great similarity to that in *Xenopus* and strengthens the evidence because it has been obtained by a different approach; the use of DNA clones complementary to polyadenylated RNA of *Strongylocentrotus* eggs. Various cloned repeat probes identified stable transcripts in egg cytoplasm ranging in length from 3 to 15 kb, similar to the unstable nuclear RNAs of embryos, but considerably longer than egg or embryo mRNAs. Moreover the experiments revealed that each of the repeat sequences when used as probes reacted with a multiplicity of transcripts each derived from a different component of the genome. This observation is in good accord with the evidence from in situ hybridization of cloned repeat sequence probes to transcripts on many different loops of newt lampbrush chromosomes. Posakony et al. (1983) state that: "... the shared sequence elements of these interspersed transcript families could provide a structural basis for recognition by other molecules, and thus potentiate various biological processes in which the RNAs of different families participate as a set".

While this may at least in part explain the synthesis and storage of interspersed repeat sequences covalently linked to polyadenylated RNAs in oocytes of *Xenopus* and *Strongylocentrotus,* one still has to account for the positive correlation between C-values and TU lengths, TUs that in the oocytes of Amphibia carry closely packed transcripts during most of oogenesis. In facing up to this problem, a theory proposed long ago by Brachet (1933, 1937) in worth recalling. Brachet was impressed by the way in which RNA «acide pentose nucléique» content diminishes, and DNA «acide thymonucléique» increases, during the early development of sea urchins and frogs, and he proposed that in such eggs, «oeufs à synthése partielle», nuclear DNA is synthesized at the expense of cytoplasmic RNA. When more became known about the chemistry of DNA and RNA and how they are synthesized, Brachet's theory in its original form could no longer be upheld.

However it has since been established that ribonucleotides can be converted into precursors required for DNA synthesis by the enzyme ribonucleotide reductase. Noronha et al. (1972) demonstrated that this enzyme is absent from unfertilized eggs of sea urchins; it first appears on fertilization, rises to peak concentration 5 to 6 h later, and continues at a high level during several further hours of development. Its synthesis is not affected by actinomycin D, so it must be encoded by maternal mRNA, masked prior to fertilization. This discovery was confirmed by De Petrocellis and Rossi (1976). Further confirmation has come from Standart et al. (1985) who have shown that in oocytes of the clam *Spisula* and the sea urchin *Arbacia* there is an abundant masked mRNA that encodes the smaller subunit protein of ribonucleotide reductase, and they think it probable that the larger subunit is stockpiled as already translated protein in the unfertilized egg. Synthesis of the smaller subunit is triggered by fertilization.

In Amphibia during cleavage, and their cleavages are rapid, the demand for DNA precursors must quickly reach a very high level, and the higher the C-value the higher the demand. It strikes me that the existence of a reductive pathway for deoxynucleotide synthesis may well account for the correlation between C-value and lampbrush TU length, and at the same time explain the incongruous nature of most of the repetitive sequences that are transcribed by lampbrush chromosomes. Any sequence of ribonucleotides will suffice if their function is to provide a pool of precursors for the synthesis of DNA during early embryogenesis, and the larger the C-value the larger the demand for such precursors.

I do not wish to imply that repetitive sequences in eukaryotic genomes play no part in the refined regulation of gene expression during development and differentiation, as first proposed by Britten and Davidson (1969). However the enormously different proportions of repetitive DNA sequences in the genomes of related organisms, the basis of the C-value paradox, and the instability of these sequences in evolutionary terms, hardly accords with the idea that the majority of such sequences exercise a precise regulatory function. Brachet's proposal of long ago does not explain the C-value paradox, but it could explain how embryos, when pushed to do so, succeed in synthesizing DNA at extremely high rates.

Some doubts have been raised as to whether any unique coding sequences for structural genes are transcribed during the lampbrush stage of oogenesis. In part these doubts arose because whereas there is unquestioned evidence that tRNA, 5 S RNA and 18 + 28 S RNA accumulate in *Xenopus* oocyte cytoplasm (for references, see Sommerville 1977), the amount of total polyadenylated RNA per oocyte rapidly reaches a plateau at the beginning of oocyte growth and remains at this level thereafter (Rosbash and Ford 1974, Dolecki and Smith 1979). Indeed Rosbash and Ford claimed that the synthesis of polyadenylated RNA in *Xenopus* oocytes occurs before the lampbrush stage is reached, but this is certainly incorrect because lampbrush TUs are quickly established once oocytes have entered diplotene and started to grow (see Chap. 4).

Golden et al. (1980) went on to demonstrate that what is true of oocyte polyadenylated RNA sequences in toto is also true of specific sequences. They constructed ten cDNA clones from *Xenopus* ovary RNA and nine cDNA clones from *Xenopus* tadpole RNA; three of the latter proved to be mitochondrial. When hybridized to oocyte polyadenylated RNA, sequences complementary to 2 of the tadpole cDNA clones were not detected, 13 nuclear-derived cDNAs indicated no changes in abundance of their complementary sequences during oocyte growth, whereas sequences complementary to the 3 mitochondrial cDNAs all showed large scale increase. If polyadenylated RNAs were stable throughout *Xenopus* oogenesis, as claimed by Ford et al. (1977), this would indeed imply that when lampbrush chromosomes are transcribing at their highest rate their transcripts do not include sequences contributing to the population of polyadenylated RNA. However this proposition would seem most unlikely, for oocytes are actively translating and accumulating proteins coded by mRNAs on polysomes, and this would normally be accompanied by mRNA degradation. In fact polyadenylated RNAs are transcribed and pass to the cytoplasm throughout the lampbrush stage (Dolecki and Smith 1979, Anderson et al. 1982, and Dr. E. H. Davidson personal communication, 1983) and although their relative abundancies and total amounts do not change during oogenesis, they are maintained at steady state levels by synthesis and turnover.

Among the observations of Golden et al. (1980) was the discovery that two of their cDNA clones constructed from *Xenopus* tadpole RNA did not hybridize at a detectable level to oocyte polyadenylated RNA, implying that these two sequences are not transcribed in oocytes though they are in tadpoles. Schäfer et al. (1982) have likewise failed to detect hybridization of one vitellogenin and three adult globin cDNA clones from *Xenopus* to oocyte polyadenylated RNA. Assuming that failure to detect means absence, *Xenopus* lampbrush chromosomes, despite their many TUs, do not transcribe all the coding sequences in the *Xenopus* genome. These observations are in conformity with the widely held view that maternal mRNAs, whether they be translated in oocytes and in embryos, or are unmasked and only translated after fertilization, do not play a dominant rôle in the embryonic development of Amphibia much beyond gastrulation; by this stage the embryo's own transcriptional machinery takes control. The phenotypes of reciprocal hybrids in animals in general resemble one another closely, and this would not be so if maternal mRNAs persisted and predominated in later stages of embryogenesis.

CHAPTER 6
Oocytes Without Lampbrush Chromosomes

In order to discuss this topic, the first requirement is a definition of what is intended by the term lampbrush chromosome. Lampbrush chromosomes are restricted to the diplotene stage of meiosis. They consist of intermittently distributed compact regions of chromatin, the chromomeres, that are for the most part inactive in RNA synthesis, and likewise intermittently distributed regions of extended DNP, the lateral loops, on which RNA transcription occurs. In organisms with relatively high C-values the lateral loops are visible in the light microscope because in them the units involved in transcription are unusually long, and the RNA transcripts with their associated ubiquitous "core" proteins are either exceptionally densely packed or, if the transcripts are relatively sparse, bulky aggregates of accessory proteins have accumulated amongst them. As will become apparent later in this chapter, though this may appear to be an acceptable working definition, and it is certainly valid for the lampbrush chromosomes of Amphibia, it is not entirely satisfactory when other organisms are being considered. I will return to the vexing definition question in Chap. 7.

Lampbrush chromosomes have been observed in the oocytes of a wide range of animals. Among the more recent literature on animals not discussed in detail elsewhere in this monograph are the following references: *Bithynia tentaculata,* Mollusca, Gastropoda (Bottke 1973), *Sepia officinalis,* Mollusca, Cephalopoda (Ribbert and Kunz 1969), *Brachydanio rerio,* Pisces, Teleostei (Baumeister 1973), *Gallus domesticus,* Aves (Koecke and Müller 1965, Ahmad 1970, Wylie 1972, Dr. N. J. Hutchison, personal communication). This list is certainly not exhaustive, but taken in conjunction with animals that are discussed elsewhere it gives some indication of how widespread lampbrush chromosomes are likely to prove to be. Nevertheless they are not ubiquitous.

The existence of animals whose oocytes do not contain lampbrush chromosomes demonstrates that the lampbrush stage is not an essential state of organization that the chromosomes must traverse in order to complete female meiosis, and the widespread absence of a lampbrush stage in plant meiosis argues likewise, nor is it essential for the synthesis and storage of the materials that equip an ovum for development. In some animals without lampbrush chromosomes the oocyte nucleus does not engage in RNA synthesis or if it does, to a negligible extent, and alternative mechanisms have evolved that supply their oocytes with all the materials they need. The variety and the phyletic distribution of these alternative mechanisms justifies the claim that, in evolutionary terms, they are secondary acquisitions. I will first describe the situation in insects.

The ovaries of insects are of two major contrasting forms, panoistic (without nurse cells) and meroistic (with nurse cells). In the more primitive panoistic form, present in Thysanura, Odonata, Plecoptera, Orthoptera, Isoptera (and also, exceptionally, in Siphonaptera amongst holometabolous insects), each ovariole consists of a graded series of single oocytes enclosed by an epithelial layer of somatic cells, the oogonia and least advanced oocytes occupying the tip of the ovariole (the germarium), the most advanced oocytes lying adjacent to the duct through which they will be discharged as fertilized eggs.

There are two forms of meroistic ovaries, polytrophic and telotrophic. In telotrophic ovaries, present in Hemiptera and Coleoptera-Polyphaga, the germarium at the tip of each ovariole contains nurse cell nuclei and presumptive oocyte nuclei in a syncytial cytoplasm, and cytoplasmic tubes confluent with this syncytium run down the ovariole to make open cytoplasmic junctions with a graded series of oocytes, the most mature of which lie furthest from the ovariole tip.

In polytrophic ovaries, present in Neuroptera, Dermaptera, Coleoptera-Adephaga, Hymenoptera, Lepidoptera and Diptera, each ovariole consists of a graded series of chambers, each chamber containing a number of nurse cells and a single oocyte. The simplest condition is found in *Forficula* (Dermaptera) where each chamber contains a single oocyte and a single nurse cell, the sister products of one mitosis with a persistent open cytoplasmic bridge between them. In other insects with polytrophic ovaries each ovariole chamber contains more nurse cells than the regular single oocyte. The best studied example is *Drosophila* (Koch et al. 1967). *Drosophila* has 16 cells per chamber, all of them derived by mitotic divisions from a single cell. Persistent cytoplasmic bridges connect these 16 cells with one another, the spatial distribution of the bridges being determined by the disposition of the mitotic spindles that gave rise to them. One of the 16 cells becomes an oocyte, the other 15 go through cycles of endomitosis and become polyploid nurse cells. A similar condition exists in other Diptera, such as *Musca* and *Calliphora*. When the ovaries of these Diptera are incubated with ^{3}H-uridine, followed by autoradiography, the nurse cell nuclei (and also the nuclei of the ovariole epithelium) show intense incorporation. The RNA synthesized by the nurse cell nuclei passes to their cytoplasm and thence, via the cytoplasmic bridges, to the cytoplasm of the oocyte (R.C. King and Burnett 1959, Sirlin and Jacob 1960, on *Drosophila,* Bier 1963a, b, on *Musca*). The RNA synthesized by the ovariole epithelial cell nuclei remains confined within these cells and is not transferred to the oocyte. In these autoradiographs the oocyte nucleus, though relatively large, is little labelled (Mahowald and Tiefert 1970, on *Drosophila*) and, what label there is, is confined to a small region, the "karyosome" or "karyosphere" (Bauer 1933) within which the much condensed oocyte chromosomes are compacted. According to Bier et al. (1967) the oocyte chromosomes of *Musca* and *Calliphora* contract immediately following pachytene, and the karyosome is the only Feulgen-positive portion of the oocyte nucleus. That the karyosome in *Drosophila* oocytes includes all of the chromosomes, and not merely those parts that are heterochromatic, has been demonstrated by Chandley (1966); she labelled the ovarian chromosomes with ^{3}H-thymidine during the pre-meiotic S phase, and in autoradiographs over ovarioles that had been left for a period to allow these labelled cells to develop into oocytes, labelling was confined to the karyosome.

Although there is a trace of ^{3}H-uridine incorporation by the karyosome, by no stretch of the imagination can the chromosomes within it be considered lampbrush in their organization. Almost all of the RNA present in the oocytes of *Drosophila, Musca* and *Calliphora* originates from nurse cells, not from the oocyte's own nucleus. In *Calliphora erythrocephala* the endomitoses that occur in the nurse cell nuclei sometimes, though exceptionally, lead to the formation of synapsed polytene chromosomes comparable in organization to the more familiar polytene chromosomes of the salivary glands of larval Diptera such as *Drosophila* and *Chironomus* (Bier 1960). Bier et al. (1969) drew attention to the fact that these nurse cell polytene chromosomes are unlike the polytene chromosomes present in other differentiated tissues of Diptera in that ^{3}H-uridine incorporation is widespread throughout their lengths, not discontinuous and restricted to relatively few regions. This is a significant observation, for it is in keeping with the widespread distribution of sites of RNA synthesis throughout the lengths of lampbrush chromosomes. It emphasizes the perhaps surprising feature of oogenesis that in order to equip an oocyte with the materials it needs so as to function appropriately when fertilized, demands appear to be made on RNA transcripts originating from a large number of sites that are liberally disposed throughout an animal's genome.

Bier et al. (1967) describe oogenesis in several beetles belonging to the group Coleoptera-Adephaga, all of which have polytrophic ovaries. In *Carabus, Pterostichus* and *Abax* there is a brief phase of chromosome extension following pachytene when the chromosomes become dispersed within the oocyte nucleus. During this phase ^{3}H-uridine is incorporated throughout the oocyte nucleus, but except for its nucleolus the level of incorporation is much less than in the endopolyploid nurse cell nuclei. Associated with the oocyte's chromosomes are several „Binnenkörper", which fuse together as the oocyte matures, ending up as a single spherical mass. These Binnenkörper show some similarity to the spheres of amphibian lampbrush chromosomes in being smooth in outline, vacuolated, in lacking RNA and incorporating no ^{3}H-uridine. While the chromosomes are maximally extended inside isolated oocyte nuclei they are scarcely visible by phase contrast, which Bier et al. (1967) ascribe to deficiency of matrix, i.e. low rate of RNA transcription. However during the early stages of chromosome contraction, which sets in rapidly, the diplotene bivalents when isolated in saline appear to be genuine lampbrush chromosomes, though their lateral loops are very short, presumably because transcript density is low and most axial DNP nucleosomal. The contracting oocyte chromosomes compact together to form a karyosome applied to the surface of a single Binnenkörper. Even when the oocyte chromosomes are densely compacted, a low rate of RNA synthesis remains detectable in the karyosome, but just as in the ovaries of Diptera the overwhelming majority of the RNA in these beetles' oocytes is synthesized in the nurse cell nuclei and transported to the oocyte's cytoplasm.

Bier et al. (1967) have also studied oogenesis in dytiscid beetles, which like carabid beetles belong to the group Coleoptera-Adephaga. In dytiscid beetles oogenesis is complicated by the presence in the oocyte nucleus of a massive quantity of extrachromosomal rDNA (in *Dytiscus* 67 pg, see Macgregor 1982, for references) as well as genomic DNA. RNA passes to the oocyte's cytoplasm from two sources, the nurse cells and the oocyte's own nucleus; the latter contribution must be for the most part rRNA, for although the oocyte's chromosomes undergo an initial post-pachytene phase of extension they remain visible as fine Feulgen-positive strands and as oogenesis proceeds they compact to form a karyosome. Bier et al. (1967) did not observe a lampbrush organization of the genome at any stage in dytiscid oogenesis, and this has been confirmed by Macgregor (1982).

In another group of insects with polytrophic ovarioles, the Lepidoptera, Pollack and Telfer (1969) demonstrated that in *Hyalophora cecropia* the oocyte is provided with the great majority of its RNA by nurse cells, there being only a slight incorporation of ^{3}H-uridine in the oocyte's own nucleus prior to vitellogenesis, and none at all thereafter. Traut (1975) similarly found a brief early period in the oogenesis of *Ephestia* when ^{3}H-uridine is incorporated by the oocyte's chromosomes, which later become inactive. Weith and Traut (1980) studied the achiasmate oocyte chromosomes of *Ephestia* using the Miller-spreading technique. In the light microscope the *Ephestia* chromosomes at pachytene show a chromomeric organization, and when spread for electron microscopy the chromomeres disperse to form loops of transcriptionally inactive chromatin projecting from synaptonemal complexes. At a slightly later stage, though prior to vitellogenesis, there is a brief burst of activity, when large numbers of nascent RNP fibrils are present on lateral loops of chromatin. This phase, though brief, is therefore essentially a lampbrush stage.

There is little information on the status of the oocyte's chromosomes in insects with telotrophic ovarioles. Macgregor and Stebbings (1970) found that in *Notonecta* there is scarcely any synthesis of RNA by the oocyte nucleus, and Macgregor (1982) has seen no lampbrush chromosomes in these nuclei.

In insects with panoistic ovarioles and large oocytes, where the oocyte's own nucleus is responsible for the synthesis of all the RNA in the oocyte, one would expect to find lampbrush chromosomes. This expectation has not gone unchallenged, for Macgregor (1982) considers the evidence for lampbrush chromosomes in the cricket *Achaeta* unconvincing, and he has not seen lampbrush chromosomes in a cockroach *Periplaneta americana*. However Kunz (1967a, b, 1969a, b) and Bier et al. (1967) have provided reasonable evidence for the existence of lampbrush chromosomes in several species of Orthoptera, including *Locusta, Tettigonia,* the cricket *Achaeta* (= *Gryllus*) and another cockroach *Blatta orientalis*. In all these animals the lateral loops are much shorter than those of amphibian lampbrush chromosomes, and presumptive chromomeres are for the most part close to or below the resolving limit of the light microscope.

In animals belonging to several other invertebrate groups nurse cells accompany oocytes whether they complete most of their growth while floating free in coelomic fluid or while remaining attached to an ovary. In such animals little is known about the degree to which the oocyte's own chromosomes engage in transcription, i.e. whether they pass through a recognizable lampbrush stage.

A situation akin to that found in insects is also present in reptiles. In the more primitive reptiles, chelonians and crocodiles, the oocytes thoughout their development are surrounded by a single layer of follicle cells, all of which are of uniform appearance. In more advanced reptiles, lizards and snakes, whereas young and post-vitellogenic oocytes are similarly surrounded by a single layer of follicle cells, during intermediate previtellogenic stages of oocyte development the follicular epithelium is multi-layered and complex. A detailed and comparative account of the organization of the follicular epithelium in reptiles (and also in birds, all of which appear to have the simpler form as present in primitive reptiles) was provided by Loyez (1906). More recent accounts of the complex follicular epithelium present in lizards have been given by Neaves (1971), Andreuccetti et al. (1978), and see also Macgregor (1982).

The complex, multi-layered follicular epithelium develops from the simple form by division of the follicle cells followed by differentiation; the cells of the outer layer remain small, the inner layer of cells are of intermediate size, while the middle layer is occupied by larger "pyriform" cells, thought to be the final differentiated state of the cells which occupy the inner layer. The pyriform cells function as nurse cells; they have cytoplasmic protrusions that extend as narrow bridges through the outermost layer of the oocyte (the zona pellucida or zona radiata) so that direct cytoplasmic continuity is established between them and the oocyte.

Loyez (1906) studied stained sections of ovaries, and was familiar with the observations of Rückert (1892) on shark oocytes. She termed lampbrush chromosomes «chromosomes à filaments barbélés» or «chromosomes à filaments plumeux». She found such chromosomes in the young oocytes of three species of tortoises and also, to judge from her drawings, young oocytes of the nile crocodile. In the gecko, *Platydactylus muralis,* she also found lampbrush chromosomes distributed throughout the germinal vesicle in oocytes ranging from 200 to 800 μm in diam, likewise in the slow-worm *Anguis fragilis.* However in five species of lizards, *Lacerta muralis, L. stirpium, L. viridis, L. vivipara* and *Uromastix achantinurus,* lampbrush chromosomes were only present, if at all, in very young oocytes just entering diplotene; soon afterwards chromosome compaction occurred, and persisted throughout the remainder of oocyte development. Loyez stated emphatically that lampbrush chromosomes are absent from three species of snakes, *Tropidonotus viperinus, T. natrix* and *Vipera aspis;* the chromosomes become compact immediately after pachytene and occupy only a small central region of the germinal vesicle right from the beginning of diplotene.

Macgregor (1982), working with germinal vesicle nuclei of *Testudo hermani* isolated in saline, confirms Loyez's statement that tortoise oocytes contain lampbrush chromosomes; and that if a lampbrush stage is present in *Lacerta viridis* it is short and on a much reduced scale, for already in young pre-vitellogenic oocytes the chromosomes and nucleoli are confined to a small central region of the germinal vesicle, a regular morphological indicator that the chromosomes are inert. Within another group of reptiles, the Amphisbaenia, Macgregor and Klosterman (1979) found well-developed lampbrush chromosomes in two species of *Bipes,* this being true both of macro- and micro-bivalents. Yet *Bipes,* like lizards and snakes, has pyriform cells in a complex follicular epithelium that act as nurse cells to the oocyte.

On this evidence it would seem that there has been an evolutionary trend in reptiles, as in insects, whereby RNA-synthesizing roles originally performed by lampbrush chromosomes have come to be assumed by nurse cells. However the retention by *Platydactylus, Anguis* and *Bipes* of a lampbrush stage, despite their acquisition of nurse cells, shows that the two mechanisms can function together, and it would be of interest to know whether their respective rôles are complementary to one another, or overlap.

Early evidence (Balfour 1878, Retzius 1912, quoted and illustrated by Wilson 1925, p. 337) gave the impression that a similar situation exists in cartilaginous fish, for the illustrations show cells in a multi-layered follicular epithelium with cytoplasmic extensions penetrating through the zona radiata and coming into confluence with oocyte cytoplasm. However J. M. Dodd (personal communication 1984), who has examined the follicular epithelium surrounding oocytes of *Scyliorhinus* and *Hydrolagus,* has shown by electron microscopy that in sections cut perpendicular to the oocyte surface the follicular epithelium consists of only one cell layer, the multi-layered epithelium illustrated by earlier workers appearing so because the sections examined had been cut semi-tangentially. The apparent cytoplasmic extensions from follicle cells are in reality channels between follicle cells; yolk exudes through pores in the blood capillary walls surrounding the follicular epithelium, becomes partially broken down en route through these channels, and is engulfed by the oocyte by pinocytosis.

Although classical and also recent textbooks of cytology and developmental biology (e.g. Wilson 1925, Balinsky 1975) state that lampbrush chromosomes occur in the oocytes of animals that have large yolky eggs, the question whether they are also present in animals with small eggs lacking nurse cells has generally been left open. For many years it has been tacitly assumed that they are present in both, but that the degree of chromosome "decondensation" may be less in animals with small eggs than in those with large. This view was challenged by Bachvarova (1981) in biochemical studies on mouse eggs, to which I will return, and has been strongly supported by Anderson et al. (1982) from calculations based on the rates of synthesis and accumulation of polyadenylated RNA in *Xenopus* oocytes.

If the definition of a lampbrush chromosome given at the beginning of this chapter is accepted, a problem is presented by a comparison between two species of echinoderms. Starfishes in general have large yolky eggs 0.5 mm or more in diameter, and *Echinaster sepositus* is a typical example. Delobel (1971) has described the morphologies of the lampbrush chromosomes manually isolated from *Echinaster* oocytes and observed by light microscopy. There are 22 bivalents, and the chromosomes are small, the largest being only 70 μm long. Some chromomeres are visible but the existence of others can only be inferred from the common points of origin of sister lateral loops. Most of the loops are only 2 μm or so long though there are several "giant" loops some 15 to 20 μm long that serve for chromosome identification and which are sufficiently well characterized that Delobel could distinguish between *Echinaster* from Roscoff and *Echinaster* from Banyuls.

Sea urchins in general have much smaller eggs, 100 to 150 μm diam, and *Strongylocentrotus purpuratus* is a typical example. E. H. Davidson (personal communication 1984, and see Anderson et al. 1982) has informed me that *Strongylocentrotus* oocytes do not contain lampbrush chromosomes. These sea urchin oocytes are not fed by nurse cells, so for the provision of RNA during oogenesis they must be as dependent on synthesis carried out in their own germinal vesicles as are the oocytes of *Echinaster*. The evidence for Davidson's statement comes from unpublished work of B. Hough-Evans, O. L. Miller Jr. and E. H. Davidson, who made "Miller-spreads" of nuclear contents from small, medium and large oocytes. O. L. Miller (personal communication 1984) showed me some photographs of these spread preparations. Whereas the transcripts of rDNP are much like those of other animals, i.e. densely packed on repeated units with spacers in between, transcripts on non-ribosomal DNP are sparsely distributed, and there appear to be no significant differences in transcript density between preparations from oocytes of different sizes. If non-ribosomal transcripts are always sparse in these oocytes they would not accumulate sufficient matrix for loops to be visible in the light microscope, moreover the nucleosomal organization known to be present in these transcription units (TUs) would in any case make them appear very short indeed. O. L. Miller estimates that the longest TUs in *Strongylocentrotus* oocytes are of the order 40 kb (13.6 μm of B-conformation DNA, but only 2 μm of compact nucleosomal DNP). Assuming that *Echinaster* lampbrush loops are heavily transcribed and their axes fully extended, the longest are in fair accord with the B-conformation DNA lengths of *Strongylocentrotus* oocyte TUs estimated by Miller.

Busby and Bakken (1979) have studied the TUs of *Strongylocentrotus* gastrulae in Miller-spreads, and estimate that they have a mean length of 27 ± 16 kb, i.e. not significantly shorter than the TUs of oocytes. Thomas et al. (1982) have analyzed in detail a cloned genomic region of *Strongylocentrotus*, 17 kb long, that includes a 9.5 kb segment transcribed in oocytes. This segment contains a single copy, presumed coding sequence 2 kb long, in several allelic forms in different individuals, while the remainder consists of repeated sequences. A transcript of similar length and sequence as that present in the oocyte is likewise present among the nuclear RNAs synthesized in blastulae and gastrulae. If this similarity in TU length between oocyte and embryonic cells proves to be widespread, it will come as a surprise, for hitherto it has been generally assumed that the TUs on the chromosomes of oocytes are a great deal longer, by orders of magnitude, than those of the nuclei of the cells of developing embryos.

However the main point at issue here is why should *Strongylocentrotus* oocytes transcribe non-nucleolar RNA so much less actively than *Echinaster*? Their chromosome organization is presumably similar, for the growing oocytes of both are at diplotene, and doubtless the TUs of *Strongylocentrotus* are arranged on loops of chromatin despite the sparsity of their transcripts. Moreover the mechanisms responsible for the early development of their eggs are likely to be similar, for the embryology of all echinoderms is essentially uniform as far as the dipleurula stage, by which time an alimentary canal with gut and anus has developed, and the ectodermal cilia have aggregated to form a closed loop; it is only thereafter that the special larval features of the various classes differentiate.

The explanation must surely lie in the different sizes of their eggs, for in terms of volume *Echinaster* eggs are about a hundred times larger than those of *Strongylocentrotus*. Some of this difference in size is attributable to yolk, but assuming that other aspects of oogenesis in these two echinoderms, including its duration, are reasonably commensurate, in particular that the proportions of polyadenylated RNAs exported to the cytoplasm relative to unstable RNAs coordinately transcribed, but turning over within their oocyte nuclei are comparable, that the polyadenylated RNAs are degraded at similar rates, and that when their eggs are mature they require to have accumulated comparable concentrations and varieties of RNAs and/or proteins translated from them, then *Echinaster* requires rates of transcription that are much more rapid, possibly by two orders of magnitude, than does *Strongylocentrotus*. Consequently *Echinaster* oocyte nuclei contain lampbrush chromosomes whose loops are visible in the light microscope, whereas *Strongylocentrotus* does not.

This argument has been convincingly presented by Anderson et al. (1982). It is based on quantitative estimates of rates of non-nucleolar and non-mitochondrial RNA synthesis, turnover, and entry to the cytoplasm in oocytes of *Xenopus*, and calculations based on these estimates and those of other authors cited. In essence, these calculations show than on average each kind of polyadenylated RNA enters the cytoplasm of a rapidly growing oocyte at about 45 copies per minute throughout the lampbrush phase. The lampbrush chromosomes contain the 4C quantity of DNA, so for a single copy gene this represents 11 copies per minute per gene, and thus a new transcript initiation every 5.5 s. This is close to the maximum possible rate of initiation, based on the transcript elongation rate of 10 to 15 nucleotides per second measured by Anderson and Smith (1978), of one initiation every 7 s. In other words all TUs in *Xenopus* require to be transcribing as fast as possible, i.e. with polymerases at maximum packing density. For good measure, and from further calculations, the estimated number of different polyadenylated RNAs being synthesized in *Xenopus* oocytes, of mean size 2 kb, is about 2×10^4, which is in reasonable accord with a rough estimate of the total number of TUs in the haploid chromosomes complement.

The difficulty arising from the definition of a lampbrush chromosome that was mentioned at the beginning of this chapter is now apparent; at what level of transcript density can, or should, a line of distinction be drawn? That this is not an entirely specious semantic problem is demonstrated with particular clarity by observations on spermatocytes, and I will return to consider it again in Chap. 8. Here I want to take account of mammalian oocytes, for in 1967 T. G. Baker and Franchi (1967 a, b) gave some evidence for concluding that human oocytes in primordial follicles contain lampbrush chromosomes. That evidence is no longer convincing.

The first demonstration of the correct sequence of stages in the prophase of the first meiotic division was provided by de Winiwarter's (1901) observations on the ovaries of newborn rabbits, and several of the now familiar terms used to describe meiotic prophase (leptotene, synapsis, pachytene and diplotene) owe their origin to de Winiwarter. In half-day old rabbits most of the oocytes are at leptotene, but towards the centre of the ovary a few oocytes show synapsis in progress. In rabbits 1.5 to 2.5 days old oocytes at leptotene are confined to the outer regions of the ovary, whereas oocytes in the interior of the ovary, the majority, are engaged in synapsis. In rabbits 4 to 5 days old some oocytes in the inner regions of the ovary are at pachytene, more peripherally lie oocytes engaged in synapsis, and more peripherally still lie some oocytes at leptotene. In rabbits 10 to 12 days old the innermost oocytes are at diplotene, while in 28-day-old rabbits the majority of the oocytes are at diplotene, a few at pachytene, and more at earlier stages. These observations established the meiotic prophase sequence in oocytes as being leptotene, synapsis (zygotene), pachytene and diplotene, and gave some idea of its time course. The observations of de Winiwarter are now known to apply to oogenesis in eutherian mammals generally (see T. G. Baker et al. 1969, for references). Oocytes remain essentially unchanged in size until the end of pachytene; thereafter each becomes surrounded by a single layer of follicle cells (the primordial follicle stage), grows during early diplotene, then growth of the oocyte ceases, but the follicle enlarges by the secretion of a fluid-filled antrum separating the now multi-layered follicle from the oocyte within.

Oocytes that have reached diplotene remain at this stage until ovulation, when they complete the first meiotic division. Rabbits and some other rodents are exceptional amongst eutherian mammals in that oocytes at diplotene pass on to a "dictyate" or "dictyotene" stage in which the chromatin is so greatly dispersed that chromosomes cannot or can scarcely be seen by light microscopy; the oocyte nuclei appear "empty". In many other eutherian mammals, including humans, there is no dictyate stage, so that diplotene bivalents remain visible right through meiotic prophase.

T. G. Baker and Franchi (1967a) studied foetal and postnatal human ovaries by light and electron microscopy. They found that oocytes enter the prophase of meiosis at variable times from as early as the second month after conception, and by the time of birth most oocytes have reached diplotene, i.e. significantly earlier than in the rabbit. They identified leptotene by the presence of single "cores" associated with condensing chromatin, pachytene by the presence of tripartite cores (synaptonemal complexes) associated with synapsed bivalents, and diplotene by the absence of tripartite synaptonemal complexes, but whose lateral elements, now split lengthwise, remain associated with sister chromatids that are no longer synapsed with their homologues. The sister chromatids invest these double cores as fibrillar and granular material. T. G. Baker and Franchi (1967b) consider that this investing material, mostly fibrillar close to the double cores, mostly granular at the periphery, with objects of similar morphology present in pairs on either side of the double cores in some regions, represents the lateral loops of lampbrush chromosomes. They state that when sections were stained with methyl green pyronine the investing material proved to be pyronine-positive, and from this finding assumed that the investing material contains RNP.

T. G. Baker et al. (1969) carried out an autoradiographic study of ^{3}H-uridine incorporation by the ovaries of rats and monkeys. The nuclei of both primordial oocytes (prior to diplotene) and of growing oocytes (during diplotene) incorporate ^{3}H-uridine, but incorporation ceases when oocytes have attained their maximum size. These observations clearly establish that at these stages the oocytes are transcribing RNA, but the paper gives no indication that there is a rapid and massive increase in the rate of RNA synthesis as oocytes pass from pachytene to diplotene, which is characteristic of oocyte chromosomes with a well-developed lampbrush phase. Moreover the double cores of the chromosomes seen by T. G. Baker and Franchi in diplotene oocytes of both rats and monkeys are not present in the lampbrush chromosomes of Amphibia. T. G. Baker et al. (1969) also demonstrated incorporation of ^{3}H-uridine in rat oocytes at the dictyate phase, which

T. G. Baker and Franchi (1967 b) had proposed might correspond even more precisely with a lampbrush condition because of the extreme state of dispersal of the chromosomes. However here again entry to the dictyate phase was not accompanied by a notable increase in the rate of incorporation of ^{3}H-uridine.

In a more recent autoradiographic study of the incorporation of ^{3}H-uridine by mouse ovaries, in a paper where the literature has been extensively reviewed, Bakken and McClanahan (1978) have shown that there is an increase in the rate of both chromosomal and nucleolar RNA synthesis as oocytes pass from pachytene to diplotene to dictyotene and reach their maximum size, but that once the dictyotene oocytes have become surrounded by somatic cells, forming primordial follicles, the incorporation of label decreases. Although this decrease could have resulted from an increase in endogenous precursor pool size or diminished permeability of the follicle to exogenous ^{3}H-uridine, on the evidence of Miller-spreads the rate of chromosomal RNA transcription in both diplotene and dictyotene oocytes is very low, with few transcription gradients identifiable and these comprising only three or at the most four RNP fibrils widely spaced apart (Bakken and Hamkalo 1978, and Dr. A. H. Bakken, personal communication).

Uncertainty regarding the status of RNA synthesis in mouse oocytes has been dispelled by Bachvarova (1981) in a quantitative biochemical study. The primordial mouse oocyte is 20 μm in diam and grows to its full size, 80 μm diam, in 14 days. Throughout growth there is a constant rate of accumulation of rRNA and heterogeneous RNA of ≤ 36 S, most of which is polyadenylated. Polyadenylated RNA represents only a 2% product of heterogeneous nuclear (hn) RNA (> 36 S) and is relatively stable, whereas the rest of the hnRNA is unstable, with a half-life of about 20 min. The quantity of polyadenylated RNA in a mouse oocyte at the end of growth is 3000 times less than that present in a mature *Xenopus* oocyte which means, if some allowance be made for the volume of inert yolk in *Xenopus*, that the concentration in the cytoplasm of polyadenylated RNA is of the same order of magnitude in the oocytes of both animals. Mouse and *Xenopus* have similar C-values, but the rate of transcription of hnRNA in mouse oocytes is 40 times less than the rate in *Xenopus* oocytes when the lampbrush phase is fully established, and this without making allowance for their different environmental temperatures. Thus Bachvarova recognizes that because the rate of polymerase passage along a mouse oocyte TU is likely to be higher than that along a *Xenopus* oocyte TU, 40-fold will be an underestimate of the difference in polymerase packing densities between these two organisms, so the evidence of sparsely distributed RNP transcripts in Miller-spreads of mouse oocytes described by Bakken and Hamkalo (1978) is fully substantiated.

If lampbrush chromosomes are not even present in mouse oocytes, whose period of growth is particularly short, it is most improbable that they occur in any mammalian oocytes at any stage, and the significance of the dispersion of the chromosomes during the "dictyate" or "dictyotene" stage is problematic; perhaps the nuclei lose H1 histone and supra-nucleosomal compaction cannot be maintained. Bachvarova was unable to detect any turnover of the 2% conserved fraction of the total hRNA, mostly polyadenylated RNA accumulating in the cytoplasm, within 2 days, but she cites evidence that it is not entirely stable, it has a half-life of 8–12 days, and suffers considerable degradation between the end of growth and ovulation.

The last observation prompts the speculation that the absence of lampbrush chromosomes from mammalian oocytes may be associated not only with the small size of mammalian eggs, but with a feature of development that is peculiar to mammals. Though mammalian eggs are small, their cleavage and developmental rates are exceptionally slow. Balinsky (1975) makes the point that several hours elapse between successive cleavages of blastomeres at mammalian body temperatures, whereas intervals of less than 1 h between cleavages are usual in echinoderms, fish and amphibians at much lower ambient temperatures. Laskey (1983) alludes to this spectacular diversity; by the time a mouse egg has divided once, a *Xenopus* embryo consists of more than 20,000 cells. Moreover the blastocyst stage in mammals, when the inner cells mass becomes differentiated from the trophoblast, is only reached several days after fertilization, and several more days pass before the mammalian counterpart of gastrulation occurs.

Although there is experimental evidence that at the blastocyst stage the inner cell mass is committed to give rise to an embryo, whereas the trophoblast is committed to an extraembryonic role, implantation in the wall of the uterus (Gardner 1972), the inner cell mass is known, in particular from the embryo fusion experiments of A. K. Tarkowski, B. Mintz (cited by Balinsky 1975), A. McLaren and collaborators (McLaren 1976), to have a remarkable capacity for regulation. To quote from Balinsky (1975, p. 268) "... the evidence is strongly in favour of the view that the inner mass cells are completely devoid of any determination in respect to developing the various tissues and organs of the embryo". In other words, no control over development is exercised by the spatial segregation of materials derived from the synthetic activities of the maternal genome. To quote likewise from Davidson (1976, p. 306) "... no evidence for polarity or other forms of morphogenetic localization exist... Cytoplasmic localization may be completely absent in the mammalian developmental scheme". Instead development, right from the beginning, is presumably under the sole control of the zygote nucleus. In this respect marsupial and placental mammals may be unique within the animal kingdom.

The comparative times of onset of RNA transcription in *Xenopus* and mouse embryos supports this proposition. According to Newport and Kirschner (1982) the first 12 cleavages of *Xenopus* eggs are rapid and substantially synchronous; G_1- and G_2-phases are missing from the cell cycles, and no transcription of RNA occurs until there is a dramatic switch-on at the mid-blastula stage, first confined to polymerase III activity, after about 4000 nuclei have been produced. This transition is accompanied by an extension of cell cycle duration, G_1- and G_2-phases appear, and the S-phase lengthens. In the mouse embryo, on the contrary, nonnucleolar RNA synthesis is already detectable by autoradiography at the 2-cell stage; it becomes more intense at the 4-cell stage, by which time rRNA synthesis has already begun (Bernstein and Mukherjee 1972). Hughes et al. (1979) have confirmed these observations by studying Miller-spreads from 2-, 4- and 8-cell stages, with convincing photographs of non-nucleolar TUs showing nascent RNP fibrils well over 1 μm long.

I have reviewed the evidence that the oocytes of animals from several different groups have dispensed with lampbrush chromosomes, at least according to the definition given at the beginning of this chapter, in the course of evolution. Mammals may have dispensed not only with lampbrush chromosomes, but also with maternally-derived morphogenetic substances that must number amongst the products of their activities in other animals groups. The absence of lampbrush chromosomes in animals that have small eggs, and no nurse cells supplying their oocytes, appears to be merely a matter of scale. It is most improbable that the varieties, fates and functions of transcripts synthesized in the oocytes of sea urchins, with small eggs, are significantly different from those synthesized in the oocytes of starfish, with larger eggs. Thus, with the possible exception of mammals, what can be learned more readily about the disposition and nature of DNA sequences that are transcribed in oocytes that have lampbrush chromosomes, e.g. by in situ hybridization, and about the accumulation and distribution of proteins in oocytes with large germinal vesicles, e.g. by immunological techniques, is likely to have general validity, and not excluding oocytes whose own chromosomes are transcriptionally inactive.

CHAPTER 7

Spermatocytes with Lampbrush Chromosomes

The fuzzy outlines of pachytene and diplotene chromosomes in the spermatocytes of many animals have for long invited comparison with the lampbrush chromosomes of oocytes. Three early papers (Hsu 1948, Srivastava 1951, 1954) made the claim, based on morphological similarity, that the lateral extensions from the chromosomes to be seen in fixed and stained preparations of spermatocytes of grasshoppers during meiotic prophase correspond to the lateral "fibres" projecting from lampbrush chromosomes. At first I thought so too (Callan 1957).

However the morphological similarity is in part deceptive. When preparations of grasshopper testes are stained with Feulgen the lateral "fuzz" is clearly Feulgen-positive, whereas in amphibian lampbrush preparations similarly stained only the chromomeres in the chromosome axes are Feulgen-positive; although the axes of the lateral loops contain DNA, the dispersion of this DNA is much greater than that present in the lateral projections from grasshopper's pachytene and diplotene chromosomes.

Despite the considerable degree of compaction of DNA in the lateral fuzz of the prophase chromosomes of the spermatocytes of many grasshoppers, Henderson (1964) has shown that in *Schistocerca, Cyrtacanthracis,* and *Chorthippus* the autosomes engage in RNA synthesis, at about the same rate, throughout meiotic prophase, that synthesis ceases during metaphase and anaphase of the two meiotic divisions, and starts up again for a brief period in young spermatids prior to their differentiation into spermatozoa. Throughout meiosis the compact X-chromosome is inactive in transcription.

Henderson (1971) made a case for considering the chromosomes of male grasshoppers as having a lampbrush-type organization during pachytene and diplotene. However he pointed out that they differ in one obvious respect from typical lampbrush chromosomes in that the sister chromatids are usually visible as separate entities at diplotene, whereas the paired lateral loops of oocyte lampbrush chromosomes are generally the only morphological evidence that DNA replication had occurred prior to their extension.

I think it worth pointing out here, and I will return to the matter later in this chapter, that the mode of compaction of chromosomes during meiosis is of itself responsible for the appearance of lateral loops (see Kezer 1970), rather than being a type of organization designed to facilitate transcription. At all events there are two other reasons why the adjective "lampbrush" ought not to be freely used to characterize the male prophase chromosomes of grasshoppers or amphibians. One is that the rate of RNA transcription is not markedly higher at diplotene than at earlier stages, in fact if anything it is lower; during oogenesis on the contrary, in organisms with lampbrush chromosomes, there is a spectacular increase in the rate of RNA transcription from the beginning of diplotene onwards, until the lateral loops retract. The other reason is that there is no massive accumulation of protein in association with the RNA transcribed during diplotene in male grasshoppers or amphibians, whereas the lateral loops of typical lampbrush chromosomes owe their visibility in the light microscope to protein, not to either of the nucleic acids. The accumulation of protein associated with lampbrush chromosomes in oocytes begins at the onset of diplotene.

Similar considerations apply to the meiotic chromosomes of mammalian spermatocytes. Using a modified C-banding technique, Klášterská et al. (1976) claim to have demonstrated that in both mouse (*Mus*) and monkey (*Macaca*) the late pachytene chromosomes have a lampbrush organization and that they proceed on to a diffuse diplotene where the chromosomes cannot be seen in fixed and stained preparations by light microscopy. Klášterská (1976) lists a number of plants as well as animals in which diplotene is diffuse, and implies (Klášterská 1978) that this diffuse stage represents a still greater dispersion of the chromosomes, based on a lampbrush organization, that was already present at pachytene.

However the contention that the diffuse diplotene stage in mammalian spermatocyte meiosis represents a lampbrush stage is not supported by other observations. In autoradiographic studies of ^{3}H-uridine incorporation by mouse spermatocytes, Monesi (1965, 1967) established that there is a low rate of incorporation during premeiotic S through to early pachytene; there follows an increase in incorporation that reaches a peak at mid-pachytene, after which there is a progressive decline through late pachytene, diplotene and diakinesis. There is always a problem, when making comparisons of the incorporation of radioactive precursors by different cells in autoradiographs, that different pool sizes may account for apparent differences in rates of incorporation. However in the case of mouse spermatocytes Kierszenbaum and Tres (1974a) confirmed Monesi's finding that there is an apparent peak of RNA synthesis reached at mid-pachytene; they went on to demonstrate (Kierszenbaum and Tres 1974b) that the peak is genuine by examining Miller-spreads of spermatocytes by electron microscopy. At mid-pachytene loops of chromatin bearing RNP transcripts extend from synaptonemal complexes. The nascent RNP consists of chains of granules each of about 20 nm diam, the chains range in length from 350 to some 900 nm, but no transcripts were found to be anywhere near as close-packed, nor as long, as those present in amphibian lampbrush chromosomes. Yet this is the stage when transcription in mouse spermatocytes is at its peak; transcripts are much more sparse both in earlier and in later (i.e. diplotene) stages.

Comings and Okada (1970) provided the first demonstration that a lateral loop organization is responsible for the chromomeric compaction of chromatin in spermatocytes at pachytene. They spread cells from the testes of mouse, quail, frog and crayfish (*Cambarus*) at an air/distilled water interface, in a study primarily concerned with establishing the nature of the various components of the synaptonemal complex. Lateral, central and transverse elements of the complex are digested by trypsin. They retain their structural integrity after digestion with

DNase, but the lateral elements dissociate from one another when DNase digestion is followed by exposure to 0.2 M HCl, and considerably more disintegration
occurs when DNase digestion is followed by exposure to 8 M urea. From these
observations Comings and Okada concluded that all three components of the
complex contain basic protein. The better to study the distribution of chromatin
relative to the synaptonemal complex, Comings and Okada (1970) chose *Cambarus* as a test object because its chromosomes are many and small. In *Cambarus*
bivalents at pachytene spread at an air/water interface the chromomeres disperse;
chromatin fibres project from the lateral elements of synaptonemal complexes
and are looped back upon themselves, the longest loops measuring 7 μm. On the
basis of this observation Comings and Okada stated: "The lampbrush type chromosome is not restricted to oogenesis but is a configuration common to spermatogenesis as well".

Keyl (1975) has similarly spread zygotene, pachytene and diplotene chromosomes of *Chironomus* spermatocytes on an air/water interface, the water being
alkaline (pH 8.8), but of low ionic strength. In unpaired regions of chromosomes
at zygotene the chromatin consists in its entirety of loops, with adjacent or nearly
adjacent origins in a common "axis". Keyl states that in places favourable for
viewing, this axis is a double strand, each strand being about 6 nm wide. The
loops are all of about the same thickness, 30 to 50 nm, each consisting of a loosely
wound 12 nm fibre enveloped in a matrix which Keyl thinks may be involved in
the winding. Keyl (1975) emphasizes that this matrix is not comparable to the
RNP matrix (i.e. transcripts) of a lampbrush chromosome, for it is distributed
uniformly throughout the length of each loop. Matrix is absent from the thinner
chromosome axis between neighbouring loops; this axis is so thin that it cannot
include the lateral element of a synaptonemal complex. Keyl thinks that
synaptonemal complexes have been disaggregated in his surface-spread preparations, unlike those of Coming and Okada, and his illustrations support this contention.

Keyl (1975) found no evidence of transcription either at pachytene or diplotene. He considers that the loops originate individually by the total disaggrega-
tion of chromomeres that were present prior to spreading; there are certainly no
signs of chromomeres where the bases of the loops are attached to the axial filament. Keyl (1975) concludes that the lateral loop organization shown by these
spermatocyte chromosomes is the result of a mode of chromatin packing that enables synapsis and crossing over to occur, and that transcription by lampbrush
chromosomes in oocytes must of necessity take place on lateral loops because of
this mode of packing.

Weith and Traut (1980) have made Miller-spreads of spermatocytes (chiasmate) and oocytes (achiasmate) at pachytene in the meal moth *Ephestia*. In fixed
and stained preparations examined by light microscopy the pachytene chromosomes display a chromomeric organization, whereas in Miller-spreads no chromomeres remain intact and their chromatin is dispersed as loops with their bases
adjacent to one another in the lateral elements of synaptonemal complexes. The
looped-out chromatin is nucleosomal, and in spermatocytes the great majority of
the loops carry no transcripts; in the few regions where transcripts are arranged
in a length gradient they cover only parts of the loops. Weith and Traut support

the conclusion of Keyl, describing these chromosomes as lampbrushlike, the loops having "... evolved primarily not for transcription but as an adaptation to meiosis". In oocytes of *Ephestia* the organization is much the same as in spermatocytes, except that in pre-vitellogenic oocytes of adults there is a burst of transcriptional activity on the loops.

Rattner et al. (1980, 1981) have studied Miller-spreads of the chromosomes of spermatocytes of the silkworm moth *Bombyx* and they too have demonstrated that the chromatin is organized as loops ranging from 5 to 25 μm in length, about 7000 per haploid genome, projecting from "axes" which run throughout the length of each chromatid. An important observation is that in chromosomes with as yet incomplete lateral elements of what will later become synaptonemal complexes, therefore from leptotene nuclei, the loops are already present in regions not yet furnished with these lateral elements. The bases of the loops are well-defined, there are no chromomeres, and they are spaced along an axis at intervals of some 150 to 200 nm. Nucleosomal organization is evident both in interloop regions of axis and in the loops, including loops showing low levels of transcription. In later prophase the lateral elements of the synaptonemal complexes are completed, and presumably reinforce the interloop regions of axis; at all events during pachytene the loops appear to project from the lateral elements. Thus the loop organization precedes the laying down of material that forms the lateral elements of synaptonemal complexes; this organization is therefore directly responsible for the primary compaction of chromatids, for in their native state the loops are condensed upon themselves to form chromomeres. Miller-spreads of meiotic metaphase chromosomes of *Bombyx* show loops projecting from the margins of these even more compact chromosomes, so the reflected loop organization that is responsible for the primary compaction of chromatin evidently persists throughout meiosis.

Thus far in this chapter I have considered the evidence that the chromosomes of the spermatocytes of a wide range of animals are organized in such a way that when their chromatin disperses in nature or under appropriate experimental conditions it does so by the lateral extension of loops, the bases of which remain connected to one another and form a linear series. I have also implied that unless a high rate of RNA transcription occurs on such loops, these structures do not fall within the compass of a reasonable definition of a lampbrush chromosome. It is convenient at this point to reconsider the question of this definition.

Until about 1970 the distinctive attribute of a lampbrush chromosome was generally considered to be the manner in which its DNP is organized, i.e. a compact chromomeric chromosome axis, with laterally projecting loops. Then followed the evidence that the chromosomes of spermatocytes as well as oocytes are organized in a similar fashion. More recently still compelling evidence has been obtained in support of the view that lateral loop organization is also a widespread feature of eukaryotic chromosomes at interphase in somatic cells (see e.g. Benyajati and Worcel 1976, on *Drosophila;* Hancock and Hughes 1982, on mouse; and review by Hancock 1982).

In the meantime the long-entrenched idea that the chromosomes of both somatic and germ-line cells undergo compaction during mitotic prophase by forming hierarchies of coils has been shown to be invalid. While there is plenty of evidence (see e.g. Rudak and Callan 1976) that the highest level of compaction occurs by the coiling of a fibre that is for the most part of uniform width, the lower levels of compaction that produce this fibre are achieved by supercoiling, involving histone H1 interactions (Finch and Klug 1976, Thoma et al. 1979), of looped domains of DNP that have their two ends neighbouring each other and anchored to a non-histone protein "scaffold" (see e.g. Paulson and Laemmli 1977, Marsden and Laemmli 1979, and Mullinger and Johnson 1980).

According to the evidence of Hancock and Hughes (1982) and other observations reviewed by Hancock (1982), in the interphase nuclei of somatic cells these chromosome scaffolds are extended and integral with the "peripheral lamina" of the nuclear "skeleton" or "matrix", from which loops of chromatin, active in transcription, project towards the interior of the nucleus. Whether the chromosome scaffold and the peripheral lamina of the nuclear matrix contain identical proteins has yet to be established. In any event, so far as concerns a definition appropriate for lampbrush chromosomes, it is clear that loops of DNP projecting laterally from an otherwise condensed chromosome axis can no longer be considered distinctive.

Two features common to all lampbrush chromosomes are both negative attributes; they are unlike somatic chromosomes in altogether lacking a chromosome scaffold of non-histone protein, and they are not anchored to the nuclear membrane. Meiotic prophase chromosomes during leptotene and zygotene develop a special scaffold with a special function, the lateral element of a synaptonemal complex. This scaffold persists through pachytene; however except for animals with achiasmate meiosis it disintegrates at the onset of diplotene. Diplotene chromosomes, both of plants and animals and in both male and female meiosis, show a remarkable diversity between related species in the extent to which their chromatin is dispersed in vivo, with loops extended laterally whether or not they are transcribed. Absence from the constraints imposed by a scaffold could explain the diplotene chromosome's freedom to disperse, and as was mentioned in Chap. 6 the degree of dispersal may then depend on the availability of histone H1 to promote supra-nucleosomal condensation.

Evidently, then, we are still faced with the question of finding an acceptable definition for a lampbrush chromosome. In my opinion it must of necessity have some regard for the intensity of transcription, for the exceptional lengths of transcription units (though whether in reality they are exceptional has yet to be established), and for the degree to which proteins accumulate alongside transcripts. As none of these attributes can be expressed in limiting terms, the working definition given at the beginning of Chap. 6, imprecise though it is, will have to suffice.

I want now to consider the remarkable status of the Y-chromosome of *Drosophila* during male meiosis. A review of the first 10 years of research in this field was published by Hess in 1971, and was followed by another in 1981. More recently still Hennig (1984) has surveyed this topic, and by including in his survey much background information about spermatogenesis in *Drosophila* he has exposed several problems that have arisen in attempts to explain Y-chromosome function.

Meyer et al. (1961), for the most part studying thin sections through primary spermatocytes of *Drosophila melanogaster* by electron microscopy, found dispersed chromatin and a nucleolus of normal appearance, but they also discovered aggregates of various unusual structures that they designated „Tubuli" and „reticuläres Material". The tubular structures, sparse in young spermatocytes and lying near the nucleolus, but rapidly increasing in quantity as the spermatocytes grow, consist of closely packed, convoluted tubes each some 30 to 40 nm wide and possibly several microns long, with indications of a regular arrangement when viewed in favourably cut sections. At this stage the masses of tubular material can be seen in living spermatocytes by phase contrast microscopy. By diakinesis the tubular structures have dispersed. They do not reappear at later stages of meiosis; they were not present in spermatocytes at earliest prophase; they are thus confined to the period of spermatocyte growth, i.e. to what would be termed diplotene if male *Drosophila* had a normal chiasmate meiosis. The reticular material also makes its first appearance in young growing spermatocytes, though later than the tubular structures. It consists of interlacing, irregularly bent strands about 20 nm wide. The reticular material is much more loosely distributed than the compact masses of tubular structures; like the tubular structures, the reticular material disperses by diakinesis, and is confined to the period of spermatocyte growth.

Meyer et al. (1961) mention other, less easily recognizable, unusual structures in the spermatocyte nuclei of *Drosophila melanogaster;* these need not be considered here, for they are better characterized in some other species of *Drosophila*. The crucial discovery of Meyer et al. (1961) was that all these unusual structures are absent from the spermatocytes of *D. melanogaster* males that lack a Y-chromosome. It has for long been known that X/O males of *D. melanogaster* develop normally, but are sterile (Bridges 1916); spermatozoa are formed, but they are immobile. Later Stern (1929) studied *D. melanogaster* males with genotypes that lacked various parts of the Y-chromosome, and showed that several "fertility factors" (a series of seven were later established by Brosseau 1960) lie on both the long and short arms of the Y; all must be present if a male is to produce functional sperm. Meyer et al. (1961) associated these observations, and on this account termed the unusual objects that they had found in *D. melanogaster* spermatocytes „phasenspezifische Funktionsstrukturen", representing the Y-chromosome or parts of this chromosome in a metabolically active state. Except for its nucleolus organizer and involvement in ribosome synthesis, the Y-chromosome of *D. melanogaster* is metabolically inert (and heterochromatic) in somatic cells.

Another correlation established by Meyer et al. (1961) was that in spermatocytes lacking the tubular structures, crystals of protein appear, first in their nuclei and later in their cytoplasm, and persist through the rest of spermiogenesis. They suggested that this material might accumulate as a consequence of a breach in a synthetic pathway, a pathway that in animals of normal phenotype involves activity of the Y-chromsome. These crystals are reminiscent of the needle-shaped protein crystals sometimes found in newt and frog oocyte nuclei when for unknown reasons the rate of RNA transcription is particularly low (Callan, unpublished).

Although activity of the Y-chromosome in the primary spermatocyte of *D. melanogaster* was clearly shown to be involved in some manner in the normal process of spermiogenesis, how this activity confers motility on the spermatozoa re-

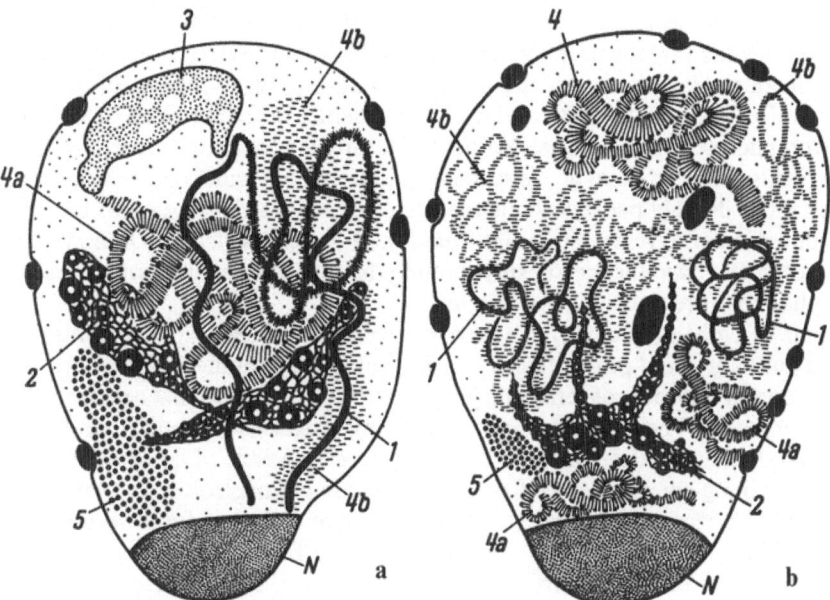

Fig. 7.1 a, b. The Y-chromosomal structures visible in phase contrast in spermatocyte nuclei of *Drosophila hydei* (**a**) and *D. neohydei* (**b**). *N* nucleolus; *1* dense axes of the threads; *2* clubs; *3* pseudonucleolus of *D. hydei;* *4* distal coil of *D. neohydei;* *4a* tubular ribbons; *4b* diffuse matrix surrounding and extending beyond the dense axes of the threads; *5* accumulation of granules. (From Hess and Meyer 1963)

mained (and still remains) unclear. The fully differentiated spermatozoa of X/O males possess all the cytoplasmic organelles that are present in the functional spermatozoa of X/Y males, though Meyer et al. (1961) mention that these various organelles differ from normality in their size and arrangement.

Hess and Meyer (1963) and Meyer (1963) extended this study to other species of *Drosophila,* 54 being described by Hess (1967 b). Two of these, *D. hydei* and *D. neohydei,* provided particularly valuable information. These two species have exceptionally large pear-shaped spermatocyte nuclei (up to 40 µm long × 26 µm wide) and also striking unusual structures (the phasenspezifische Funktionsstrukturen) whose gross morphology can be readily studied in living cells, mounted in physiological saline, by phase contrast. The observations made on this material led to the conclusion that the unusual structures are parts of the Y-chromosome engaged in RNA synthesis, and that they correspond in organization to the lateral loops of amphibian lampbrush chromosomes.

The spermatocytes of both species have a single nucleolus, closely applied to the nuclear membrane at the narrower end of the nucleus. Four structures of characteristic morphology are particularly evident in both species (Fig. 7.1). A pair of threads (Fäden) in *D. hydei* project from the nucleolus as dense structures, each about 20 µm long and surrounded by a less dense matrix; beyond this compact region the distal segments of the threads are diffuse and convoluted. In *D. neohydei* the pair of threads also consist of dense and diffuse segments; the dense segments cannot be seen inserted in the nucleolus, though in young spermatocytes of both species the threads first appear as short diffuse loops in immediate proximity to the nucleolus, thereafter the threads extend into distal regions of the nucleus and differentiate into dense and diffuse segments. In thin sections examined by electron microscopy the diffuse segment of a thread appears as a twisted skein of 5 nm fibrils. The proximal dense segment (about 0.3 μm wide) consists of similar fibrils more densely packed, while the surrounding matrix appears as convoluted tubular material resembling that found in *D. melanogaster*. In *D. neohydei*, though not in *D. hydei*, the matrix surrounding the threads also contains reticular material like that present in *D. melanogaster*.

A pair of clubs („Keulen") are of poor overall visibility in phase contrast, about 10 μm long and 4 μm wide in *D. hydei*, longer and thinner in *D. neohydei*, but they contain numbers of refractile granules up to 1 μm wide that are conspicuous in living cells. The clubs of *D. neohydei* being longer and thinner than those of *D. hydei*, their granules tend to be lined up in single rows. Like the threads, the clubs in young spermatocytes first appear beside the nucleolus, and later extend distally. In occasional spermatocytes the pair of clubs are fused to one another. The clubs have their limits sharply defined; in fine structure they appear as a loose network of 20 nm fibrils. The granules with which they are associated have a uniform texture, apart from a central vacuole in large examples, and 30 to 40 nm grains at the periphery that appear to be shed and become spread through the volume of the nucleus.

The "pseudonucleolus" of *D. hydei* is characteristic of this species. It generally lies well away from the genuine nucleolus, near the other end of the nucleus, but not attached to the nuclear membrane. It is similar in size to the nucleolus, some 10 μm long, but instead of being uniformly rounded the pseudonucleolus has two short conical projections from the main mass that are directed towards the interior of the nucleus. The fact that the pseudonucleolus has two such projecting cones is an indication that it consists of two structures fused together, and in exceptional spermatocytes of *D. hydei* there are two separate pseudonucleoli, each with a single projecting cone. In thin sections examined by electron microscopy the fine structure of the pseudonucleolus is unlike that of the nucleolus. Its boundary is sharply defined and its texture dense though spongelike, with vacuoles and canals whose limits are similarly sharply defined. Spermatocytes of *D. neohydei* contain no pseudonucleoli, but in the corresponding region of the nucleus lies a convoluted band of material („distaler Knäuel" or distal coil). The width of the band making up this structure is about 0.5 μm, and in thin sections it appears as aggregates of 10 nm tubules. It shows no sign of being a paired structure.

The fourth unusual structure in these spermatocyte nuclei is likewise a convoluted band of diffuse material, with refractile granules along its borders, and lying closer to the nucleolus. This object gives no sign of being a double structure in *D. hydei*, but it frequently takes the form of two separate convoluted bands in *D. neohydei*. In thin sections these diffuse convoluted bands („Tubuli-Bänder" or tubular ribbons) can be seen to consist of closely packed tubules, 20 to 30 nm wide in *D. hydei*, only 10 nm wide in *D. neohydei*.

Meyer (1963) gave cytochemical evidence that the structures peculiar to the spermatocyte growth stage in *D. hydei* and *D. neohydei* contain RNA, that they are Feulgen-negative when fully differentiated though they originate from a region near the nucleolus in young spermatocytes which at this stage is Feulgen-positive, and that trypsin is more effective than pepsin in digesting them. This, together with the morphological evidence, led Meyer to the conclusion that all these structures are comparable to the lateral loops of amphibian lampbrush chromosomes, specifically that the clubs are comparable to the giant granular loops, and the pseudonucleus to the giant fusing loops, of oocytes of *Triturus c. cristatus*. However because these structures are not individually reflected on themselves and hence are not, strictly speaking, loops, he makes the suggestion that a better comparison can be made with a univalent lampbrush chromosome (the *Drosophila* Y-chromosome has no homologue) in which extensive double-loop bridges have been formed. In this context it should be mentioned that chromomeres have not been seen in *Drosophila* spermatocytes.

Hess and Meyer (1963) discussed in greater detail the cytological evidence then available that these structures represent phase-specific differentiations of the Y-chromosome in *D. hydei* and *D. neohydei;* this evidence is compelling. The nuclei of spermatocytes of X/O males of *D. hydei* contain no structures that are in any way comparable to those present in X/Y spermatocytes (Fig. 7.2 a, c), though they do contain dense aggregates of various kinds including tubules that are, however, much wider than normal (40 to 50 nm) and are not arranged in bands. Unlike X/O males of *D. melanogaster,* the testes of X/O males of *D. hydei* contain no spermatozoa, though they do produce a few early stages of spermatids (Prof. O. Hess, personal communication). X/Y/Y males of *D. hydei* contain four threads, not the normal two, that are clearly identifiable by their dense segments which lie proximal to the nucleolus (Fig. 7.2 b). They may contain four clubs, though fusion between these structures can occur and results in clubs that are larger than normal. They may contain two pseudonucleoli, though more frequently only one, and that this is a consequence of fusion is clearly shown by the presence of four conical projections instead of the normal two. The spermatozoa of X/Y/Y *D. hydei* are twice as long as normal, 12 to 14 mm instead of the 6 to 7 mm long spermatozoa of X/Y *D. hydei,* though Hennig (1984) considers that the normal length has been underestimated.

Absolutely critical evidence for direct involvement of the Y-chromosome in determining the morphologies of the phase-specific structures in the spermatocytes of *Drosophila* was provided by an examination of male hybrids. F$_1$ hybrids (*D. hydei* ♀ × *D. neohydei* ♂) show the intranuclear structures typical of *D. neohydei* (absence of a pseudonucleolus, instead presence of a distal coil, dense regions of the threads not visibly connected with the nucleolus), while the reciprocal hybrids (*D. neohydei* ♀ × *D. hydei* ♂) show the intranuclear structures typical of *D. hydei* (presence of a pseudonucleolus, dense regions of the threads projecting from the nucleolus). So the species-specific morphologies of the peculiar structures in the spermatocytes would at first sight appear to depend simply and directly on the constitution of the Y-chromosome. In the light of these observations Hess and Meyer (1963) considered the possibility that the Y-chromo-

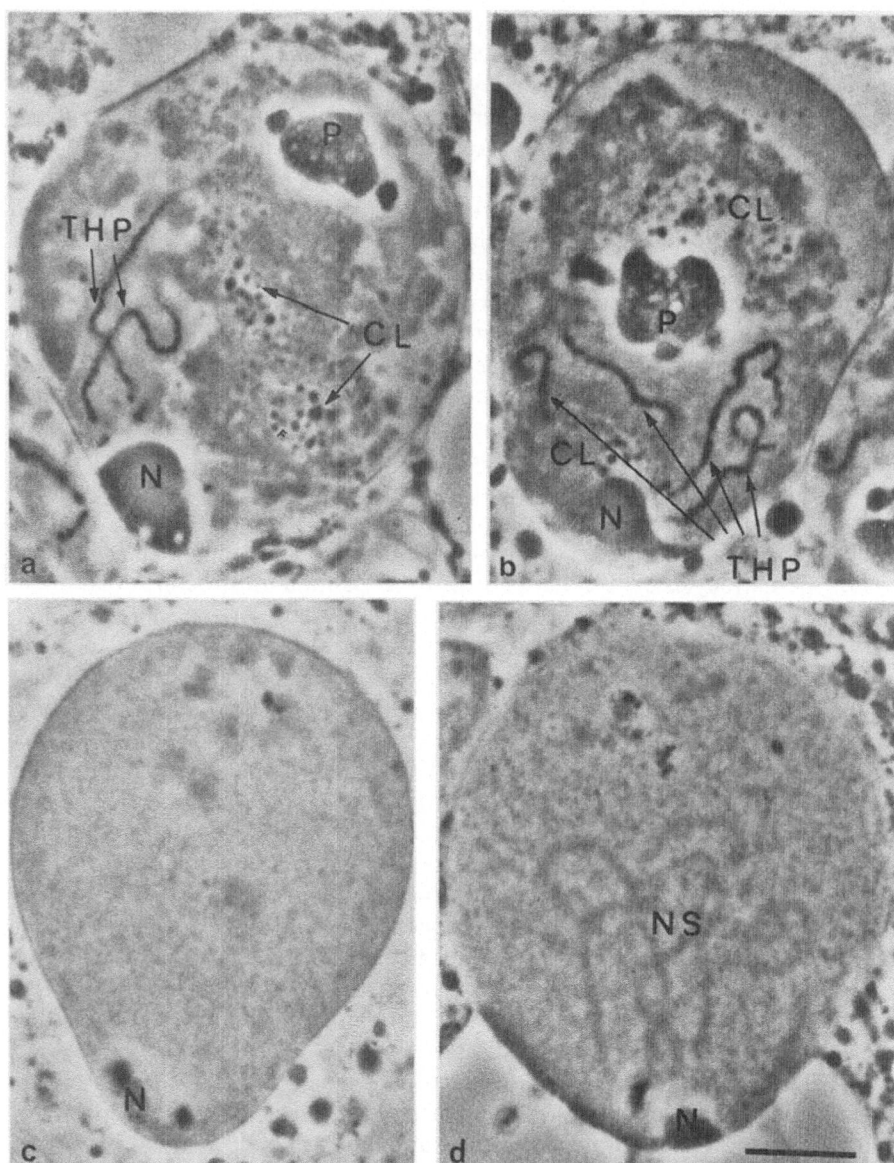

Fig. 7.2 a–d. Phase contrast photographs of living spermatocyte nuclei of *Drosophila hydei*. **a** X/Y male; **b** X/Y/Y male; **c** XO male; **d** male with X and a fragment of the Y including the noose locus in the short arm. *N* nucleolus; *THP* dense proximal parts of the threads; *CL* clubs; *P* pseudonucleolus; *NS* noose loops. *Bar* = 10 µm. **a** and **b** From Hess (1965 a); **c** courtesy of Prof. O. Hess; **d** from Hess (1968 b)

some, despite its being heterochromatic in somatic cells, might code for specific proteins in spermatocytes that are, in turn, responsible for the specific morphologies of the intranuclear structures. They did not exclude other possible interpretations, which will be discussed later in this chapter.

One of the next questions to be tackled concerned the distribution and order of the phase-specific structures on the Y-chromosome in spermatocytes of *D. hydei* which, for the sake of convenience, will from now on be referred to as "Y-loops" (Hess 1965b, 1967a). The analysis made by Hess depended on reciprocal translocations produced by X-rays. The first experiment was straightforward. An X-rayed male was crossed with a normal female and amongst the offspring was a single male of "vermilion" phenotype. This mutant male was crossed with a normal female, and the F_1 generation backcrossed to build up a stock homozygous for vermilion. The stock was fertile, and cytological study showed that it contained a reciprocal translocation between one of the autosomes and the Y-chromosome as well as the vermilion mutation. One of the break points that produced the translocation lay within the heterochromatin close to the centromere of an acrocentric autosome, the other at about two-thirds the length of the long arm of the Y as measured from its telomere. Of the two translocation chromosomes, "Y-A" consists of the short arm, centromere and proximal one-third of the long arm of the Y plus all the euchromatin of the autosome, while "A-Y" consists of the proximal heterochromatin and centromere of the autosome plus the distal two-thirds of the Y. As a result of different modes of segregation at meiosis, various kinds of deficiency and duplication males, as well as males with the balanced genotypes, are produced by this translocation stock. Those containing both Y-A and A-Y, i.e. balanced, contain the normal set of Y-loops in their spermatocytes. A class lacking threads and pseudonucleolus lack A-Y. A class lacking clubs and tubular ribbons lack Y-A. A class with duplicate threads and pseudonucleoli, but otherwise normal, contain duplicate A-Y and a single Y-A. A class with duplicate clubs and probably duplicate tubular ribbons also, but otherwise normal, contain duplicate Y-A and a single A-Y. Finally there is a class containing all the Y-loops in duplicate, and these contain duplicate A-Y and Y-A translocation chromosomes. From these observations it is clear that the loci for threads and pseudonucleolus lie in the distal two-thirds of the long arm of the Y, while the loci for clubs and tubular ribbons lie in the proximal one-third of the long arm and/or the short arm of the Y chromosome.

The later series of experiments, designed to locate more precisely the loci involved in the formation of the Y-loops, required a complex genetic analysis. The normal X-chromosome of *D. hydei* has a median centromere. One arm is euchromatic except for a short proximal heterochromatic segment, while the other arm is wholly heterochromatic. There exists an attached-X stock of *D. hydei,* in females of which both arms of a single X-chromosome are euchromatic except for short proximal heterochromatic segments, the heterochromatic arms of both the original X-chromosomes have been lost, including their nucleolus organizers, and the karyotype of the female includes a Y-chromosome. Females of this stock carry the recessive sex-linked markers "white" (eye colour) and "light" (body colour) and are fertile. Attached-X females were X-rayed and crossed with wild-type males. The normal offspring from such a cross are white-eyed attached-X daughters and wild-type sons. However there are exceptional white-eyed sons, and these generally arise as a result of a reciprocal translocation, in which one break point lies in the proximal heterochromatic segment of one arm of the attached X-chromosome, the other somewhere in the Y with which it was associated during female meiosis. In this manner, by so-called detachment, X-Y translocations are produced. Such exceptional males were mated with attached-X females, and seven different "detachment stocks" were successfully established in mass culture. The males in these stocks are fertile because their karyotypes include a normal Y (derived from the non-irradiated male in the original cross) as well as an X-Y translocation chromosome. In order to study the loop-forming potentialities of the X-Y translocations (and therefore of the different Y fragments), white-eyed detachment males from the seven different stocks were crossed with normal wild-type females. The great majority of the male progeny have wild-type eye colour, and must therefore be of normal X/Y karyotype. However classical primary non-disjunction (Bridges 1916) occurs in a few oocyte meioses (the frequency can be raised tenfold by X-raying this parent) and as a result patroclinous (white-eyed) males are produced whose karyotypes contain an X-Y detachment chromosome, and no normal Y. These males are sterile, but they can be reconstituted at will by repeating the cross that gave rise to them, thereby permitting precise cytological appraisal.

I will not go into the details of each and every experiment, but will summarize Hess' findings, incorporating some more recent information (Prof. O. Hess, personal communication 1984). He divided the long arm of the *09bD. hydei* Y-chromosome into ten segments of equal length, the first segment lying proximal to the centromere, the tenth distal. At the junction between the tenth and ninth segments, dense proximal regions of the threads are located; these lead directly on to the diffuse regions of the threads, which in their turn lead directly on to the conical projections of the pseudonucleolus. It is presumed that two fine strands lead back from the pseudonucleolus to the rest of the Y-chromosome, which lies embedded in the nucleolus. This then represents a compound loop. Segments 8 to 3 do not include loop loci, but at the junction between segments 2 and 1 lies a pair of tubular ribbons, these lead directly to the clubs, and it is presumed that two fine strands lead back from the clubs to the rest of the Y-chromosome. This then represents another compound loop. Another discovery concerns the short arm of the Y; one X-Y translocation consisted of the short arm of the Y and its centromere, plus the X, with little or none of the long arm of the Y intervening between the X and the centromere. Spermatocytes with this translocation and no other parts of the Y showed a hitherto unnoticed pair of diffuse loops referred to as „Schlingen", nooses (Fig. 7.2 d). These structures are present in spermatocytes with normal Y-chromosomes, but they are so diffuse and intermingled with other loops, particularly the tubular ribbons, as to go undetected. Their locus therefore lies in the short arm of the Y. The distribution of these various loops along the Y-chromosome of *D. hydei* is shown in Fig. 7.3.

With these translocation stocks available, Hess (1967 a, 1968 a) was able to undertake a complementation analysis of the Y-loops' involvement in spermatogenesis in *D. hydei*. The results can be concisely summarized: each and every one of the Y-loops must be present for functional spermatozoa to be formed.

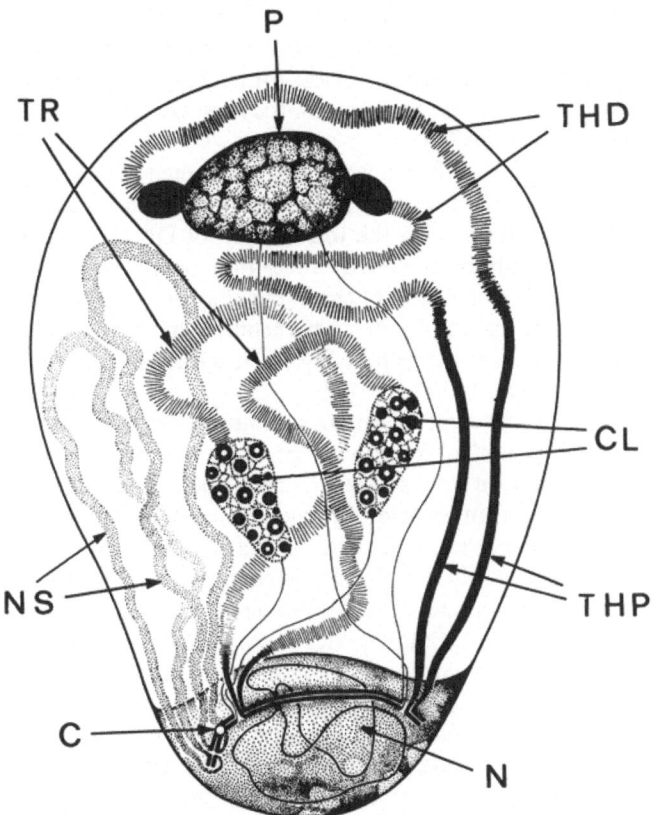

Fig. 7.3. The distribution of lampbrush loops along the Y-chromosome in a spermatocyte nucleus of *Drosophila hydei,* from Hess and Meyer (1968). The condensed parts of the Y are shown embedded in the nucleolus. The nucleolus organizer locus was not known at the time, and is shown at a hypothetical site in the long arm of the Y, with fine strands representing the transcriptionally active rDNP. It has now been established that there are two nucleolus organizers on the Y of *D. hydei* (see text). *N* nucleolus; *C* centromere; *NS* nooses; *TR* tubular ribbons; *P* pseudonucleolus; *CL* clubs; *THP* dense proximal parts of the threads; *THD* diffuse distal parts of the threads

Deficiencies for particular Y-loops, or combinations of Y-loops, lead to the arrest of spermiogenesis at various stages. Thus with only the nooses present the meiotic divisions occur and spermatids are formed, but they do not differentiate further. At the other extreme, in males that lack only the pseudonucleolus, spermiogenesis proceeds to completion, a few spermatozoa may even be motile, but nevertheless such males are sterile. These conclusions have been fully supported by more recent and very sophisticated experiments where males with any chosen X-Y translocation can be regularly produced in large numbers (Hess 1976, and personal communication 1984). In all those Y-fragment combinations where spermiogenesis proceeds to some degree, all major structural components of the developing spermatozoa are formed in at least some spermatids (Meyer 1968), and from this Hess argues that in all probability the Y-loops do not code for proteins or enzymes that are directly involved in the production of these organelles. An alternative proposal is that the Y-loops are regulatory genes of specific types (Beermann 1965, Hess 1966), possibly "... structural devices for packaging and storing gene products which are synthesized during the spermatocyte stage but used only later during spermiogenesis" (Hess 1967a).

Hess (1967a) also drew attention to the length of DNA assumed to be present and fully extended in each of those Y-loops of *D. hydei* that can be measured, some 50 μm, and compared this figure with the total length of DNA in the Y-chromosome, calculated to lie within the range 2.5 to 12.5 mm. So all the Y-loops taken together represent only a small fraction of the Y-chromosome's DNA, but each is roughly comparable to the length of DNA per chromomere of an average band of a polytene chromosome. This length is also well within the range of lengths of amphibian oocytes' lateral loops.

In 1970 Hess described the outcome of further complementation tests in which a situation akin to "variegation" of Y-loop morphology was exposed. Two self-maintaining "tester" combination stocks were first constructed. One consists of attached-Y females with a Y-A translocation instead of a normal Y (the Y-A chromosome includes the loci for nooses, clubs and tubular ribbons) and males carrying the same Y-A translocation plus an X-Y translocation that includes the complementary loci for threads and pseudonucleoli, necessary for fertility, and no free Y. X-chromosomes of both males and females carry the recessive markers white and light. The other tester combination stock is converse to the one just described, i.e. attached-X females with the A-Y translocation and males carrying the same A-Y translocation plus a complementary X-Y translocation.

Males to be tested, carrying particular X-ray-induced X-Y translocations, owe their fertility in stocks to the presence of an intact Y. When such males are crossed with females of one or other of the combination stocks, all F_1 males carry 2 Y fragments, Y-A or A-Y derived from the mother, and a particular X-Y derived from the father. Their spermatocyte nuclei, capacity to complete spermiogenesis, and fertility are then examined.

The results can be summarized as follows. Those X-Y translocations that confer full fertility carry, as is to be expected, normally expressed Y-loop loci that include all those that are absent from Y-A, or alternatively from A-Y. Certain specific X-Y translocations however produce male offspring, some of whose spermatocytes show incomplete expression (extension) of particular Y-loops, while other spermatocytes show full expression of these same loops. Such males are of reduced fertility. This condition may be considered as variegation within an individual. Other specific X-Y translocations produce male offspring, some of which show the full expression of their Y-loops in all spermatocytes, and are fertile, while their brothers, with the same X-Y translocation chromosome, show uniform degrees of sub-standard Y-loop expression in all their spermatocytes, coupled with defective spermiogenesis. This condition may be considered as sibling variegation. A remarkable feature of these X-Y translocations leading to variegation is that they are unstable. The variegation is only evident in the F_1 males from test-crosses made soon after X-ray induction of the particular translocation. As more and more generations intervene between induction and testing, the X-Y translocation stabilizes so that expression of a particular Y-loop locus

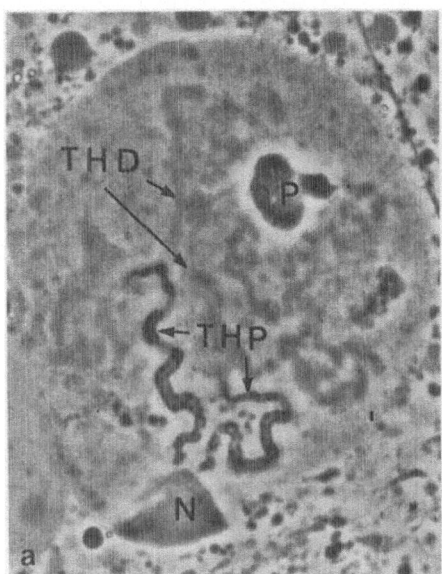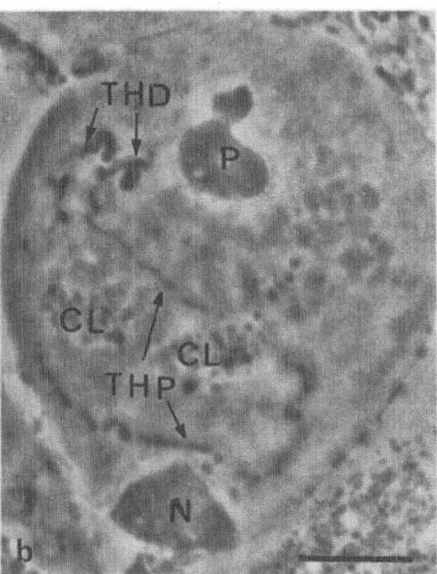

Fig. 7.4 a, b. Phase contrast photographs of living spermatocyte nuclei of *Drosophila hydei*. **a** "Tube-proximal" and **b** "tube-distal" mutations. *N* nucleolus; *P* pseudonucleolus; *CL* clubs; *THP* proximal parts of the threads; *THD* distal parts of the threads. *Bar* = 10 μm. (Courtesy of Prof. O. Hess)

is completely lost (repressed), or, in one case only, there is reversion to uniform and normal Y-loop expression. A satisfying explanation of these results is not yet available, though Hess makes the point that variegation in general in *Drosophila* is known to occur when chromosome rearrangements place heterochromatic regions in juxtaposition to euchromatic regions that are not normally so positioned, and vice versa, hence the term "position effect variegation" (Lewis 1950, W. C. Baker 1968). Some of these X-Y translocations are probably of this nature.

In the course of his induction of chromosome translocations in *D. hydei* by X-rays, two remarkable Y-loop mutations were discovered by Hess (1965 a). One of these, designated "tube-proximal", resulted from the irradiation of a normal male and after mating was found in several of his sons. In the mutant the normally dense segments of the two threads, the part that lies proximal to the nucleolus, was transformed into a broad tube tapering at its distal end (Fig. 7.4 a). Beyong this region the diffuse segment was of normal morphology. Males with such mutant Y-loops in their spermatocytes were fertile.

The other mutant resulted from the irradiation of an attached-X female. After mating, amongst her progeny were males with a fragment of a broken attached-X chromosome, a normal Y and a mutant Y. In this mutant, designated "tube-distal", the proximal segment of each thread was normal, but the normally diffuse segment was transformed into a contorted narrow tube with sharp outlines (Fig. 7.4 b). After further crosses, males were produced that contained only the mutant Y, and like "tube-proximal" these also proved to be fertile.

Hess went on to construct X/Y/Y males of *D. hydei*, one of the Y chromosomes being the normal wild type, the other carrying the tube-proximal or tube-distal mutant. In the spermatocytes of such males both normal and mutant threads were present. These observations therefore established that there are at least two different genetic components of the threads, and furthermore that both components differentiate autonomously, i.e. they express intrinsic specific features of these loci on the Y-chromosome, and do not owe their characteristic morphologies to mutations that may have occurred elsewhere in the genome. The morphologies of these mutant threads have been shown to be under the control of factors that lie within loop-forming sites themselves, or at least immediately adjacent to them, for among X-Y translocation chromosomes produced by "detachment" from attached-X females whose Y-chromosome carried one or other mutant thread, some have been recovered with extremely short distal fragments of the long arm of the Y. Despite the small size of these fragments, the thread loop loci that they include display their characteristic mutant forms in spermatocytes (Prof. O. Hess, personal communication). The interpretation of the uniform morphologies of these loop mutants over lengths of tens of microns presented Hess with conceptual problems just as do some of the lateral loops of amphibian oocyte lampbrush chromosomes; I will leave the question open meantime, except for returning to Hess' speculation that the overall functions of the Y-loops might be to store and protect from breakdown mRNAs synthesized by the spermatocytes but translated later, during spermiogenesis. This idea was given in greater detail by Hess (1968 b), and by Hess and Meyer in an extensive review published in 1968.

One line of evidence that the lateral loops of amphibian lampbrush chromosomes extend because they are transcribing RNA, and their loops become visible in the light microscope because of the accumulation of a matrix of close-packed RNP transcripts, came from observation of the dramatic effects produced by actinomycin D. RNA synthesis is suppressed, RNP transcripts are shed and the loops retract (Izawa et al. 1963). The process is reversible, for when oocytes recover from poisoning by actinomycin D, the loops begin again to transcribe RNA, and in so doing they extend (Snow and Callan 1969). Meyer and Hess (1965) showed that actinomycin D (and C) has similar effects on the Y-loops of *D. hydei*. Injection of 0.5 µl of 20 µg ml^{-1} actinomycin dissolved in *Drosophila*-Ringer into a freshly emerged male causes almost immediate disappearance of the distal diffuse segment of the threads; the proximal compact segment becomes vacuolated and disintegrates about 35 h after injection. The tubular ribbons also disintegrate; in the electron microscope the tubules are still evident, but their distribution is haphazard. The clubs and pseudonucleoli suffer less damage and residues remain visible in the light microscope, but their characteristic morphologies are largely lost. Some 40 h after injection of actinomycin the Y-loops begin to regenerate, and by 90 to 120 h the spermatocyte nuclei have regained their normal appearance.

In order to answer the question whether the loop regeneration is genuinely autonomous (there existed an alternative possibility that loop matrix already shed from the Y-loops might return to its various sites of origin during regeneration) X/Y/Y males were constructed, one Y-chromosome being normal wild type, the other carrying the mutant tube-proximal. Males of this constitution were injected with actinomycin C, and the destruction of their Y-loops was observed. When

other injected males were given enough time for Y-loop regeneration to have occurred, normal wild-type and mutant threads had reappeared in their spermatocytes, not threads having some intermediate morphology. So the autonomous determination of these Y-loops' morphologies was again established.

Hess (unpublished) has isolated further Y-chromosome mutants that individually alter the phenotypes of all the other Y-loops. Some of these do not impair
male fertility, i.e. they behave like the tube-proximal or tube-distal mutants but
others, which can be generally characterized as inhibiting normal Y-loop extension, cause sterility. Still more Y-chromosome mutants have been induced by
breeding from *D. hydei* males whose food contained the mutagen ethyl methyl sul-
phonate, EMS (Leoncini 1977). This mutagen has a direct and destructive effect
on the Y-loops of males; after 24 h exposure to EMS in the food the Y-loops (and
nucleolus) of spermatocytes are much reduced in size or absent altogether. The
mutant male progeny from EMS-treated males fall into four classes: sterile males
with normal Y-loops; sterile males with morphologically altered Y-loops; sterile
males with one or more Y-loops missing; and temperature-sensitive males, fertile
at 18 °C, but sterile at 26 °C. In the spermatocytes of the fourth class all Y-loops
develop normally at both permissive and restrictive temperatures. One other mutant was found in which the Y-loops develop normally at 18 °C, but do not develop at all at 26 °C; yet even when cultured at the lower, apparently permissive
temperature, such males are sterile. With these further mutants available a genetic
fine structure analysis of the Y-chromosome of *D. hydei,* using deletion/comple-
mentation experiments, has been undertaken by Hackstein et al. (1982). The main
conclusion resulting from this study is that not more than one complementation
group can be detected within each Y-loop site.

In order to study the organization of the Y-loops in greater detail Meyer and
Hennig (1974) spread spermatocytes of *D. hydei* on an air/water interface, the pH
of the water being stabilized with minimal concentrations of electrolytes. The material at the interface was picked up on carbon-coated and glow-discharged grids,
fixed and dried after rinsing in Kodak Photoflo, stained before or after air-drying,
and examined in the electron microscope. To identify individual loops, advantage
was taken of the availability of males with various Y-deficiencies, and of X/O
males. At pH 6 the morphology of the loops corresponds more or less to their appearance in phase contrast, i.e. they are not greatly dispersed, with RNP close
packed around loop axis, but with a degree of compaction and distribution of
RNP characteristic of the different Y-loops. The nooses show simple thin to thick
gradients of matrix RNP, each noose is about 80 µm long in young spermatocytes, but Meyer and Hennig consider that its length when fully extended is some
150 µm.

At pH 9 the various Y-loops and their RNP transcripts are maximally extended, but can no longer be identified with assurance. RNP fibrils more than
10 µm long were seen attached to DNP axes, and if their compaction is of the
same order as that of rRNP fibrils visible in the same preparations Meyer and
Hennig (1974) supposed that giant transcripts more than 50 µm long are synthesized on single loops. They also noted that the distances between neighbouring
transcripts differ from one loop to another, and are in all cases greater than between RNP transcripts of ribosomal units. As was to be expected, all these struc-

tures are absent from the spermatocytes of X/O males, though transcripts from other components of the genome were observed.

Glätzer and Meyer (1981) used the Miller-spreading technique to obtain further information about the structure of the Y-loops of *D. hydei,* and discovered a peculiar feature of RNP transcripts on one of the Y-loops, the threads. Spreads were made at pH 6.3 and pH 9. They found transcription units (TUs) with well-defined polarity in spermatocytes of X/O males, in conformity with Hennig's (1967) autoradiographic observation that the autosomes contribute about 50% to the total non-nucleolar RNA synthesized by spermatocytes. Some of these TUs are tens of microns long, and Glätzer and Meyer (1981) do not hesitate to term them lampbrush loops (Fig. 7.7 b, c).

In nuclei weakly spread at pH 6.3 several different RNP configurations are of evident Y-loop origin, though only those derived from the threads could with assurance be assigned to a particular Y-loop. When viewed in phase contrast each thread appears as a dense axis embedded in more diffuse matrix. As seen in the electron microscope in spreads at pH 9 this diffuse material consists of linear RNP fibrils up to 20 µm long (Fig. 7.5 a). In more strongly dispersed preparations these long fibrils can be seen attached to axial DNP, but at the point of attachment there are bush- or mycel-like structures (Fig. 7.5 b), presumably formed by RNP fibrils that are folded back on themselves in such a complex way that they appear branched, with many free ends. These bushlike structures, when not dispersed, form the dense axial component of the thread that is visible in phase contrast. Similar bushlike RNP structures, but without associated linear fibrils, are also present in the distal regions of the threads where these make connection with the conical projections of the pseudonucleoli.

In spreads made at pH 9 a great variety of RNP configurations are apparent, ranging from long, smooth, linear fibrils with their attachments to axial DNP relatively close to one another (Fig. 7.6 a), simple linear beaded RNP fibrils, beaded fibrils for the most part linear but reflected and branched at their free ends, and at the extreme of complexity enormous aggregrates of RNP fibrils reflected back and forth, though each with a single point of attachment to axial DNP (Fig. 7.6 b). Comparison of spreads made at pH 6.3 and pH 9 lead Glätzer and Meyer (1981) to the conclusion that the native unit of organization in all these transcripts is a uniformly beaded RNP fibril, the individual beads being about 30 nm wide and therefore comparable with those described by Mott and Callan (1975) and by Malcolm and Sommerville (1974, 1977) in urodele lampbrush transcripts. The beaded fibrils can unwind to produce smooth linear or bushlike structures (Fig. 7.7 b, c) in strongly dispersed preparations, but in spreads at pH 6.3 where

---

Fig. 7.5 a, b. Miller-spreads at pH 9 showing the organization of transcripts on parts of the threads in *Drosophila hydei* spermatocytes. In **a,** at lower magnification and relatively little dispersal, the axial component of the thread is so dense that it obscures the DNP axis, while the diffuse component consists of extended linear transcripts; it is within this diffuse component that antigenically-demonstrable adventitious proteins accumulate (Glätzer 1984). In **b,** at higher magnification and with more dispersal, mycel-like structures are evident where the transcripts are attached to DNP axis (*arrowheads* at lower left). These mycel-like portions of the transcript, when compacted together, constitute the dense axial component of the threads. *Bar* in **a** = 5 µm; in **b** = 2 µm. (From Glätzer and Meyer 1981)

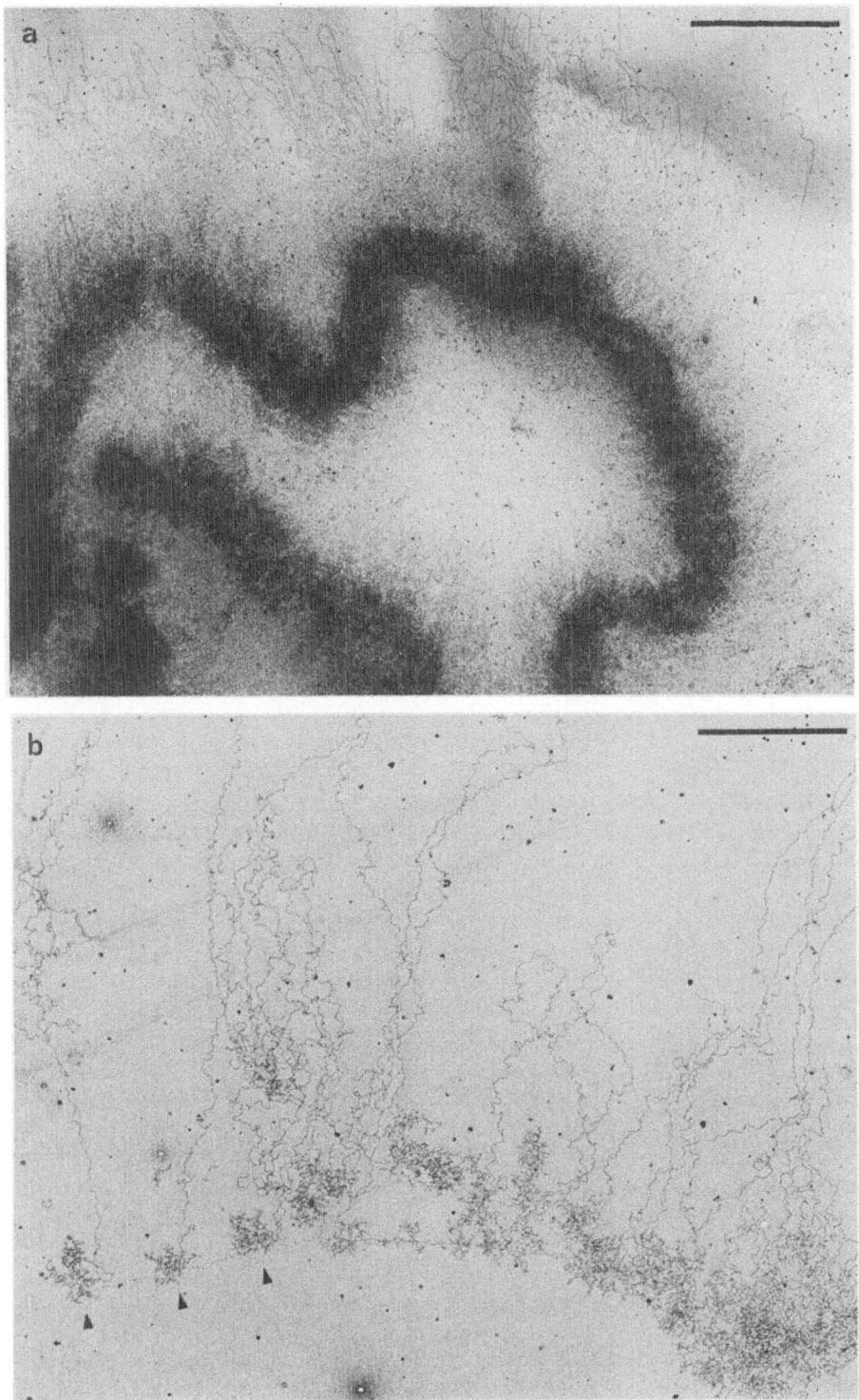

Fig. 7.5 a, b

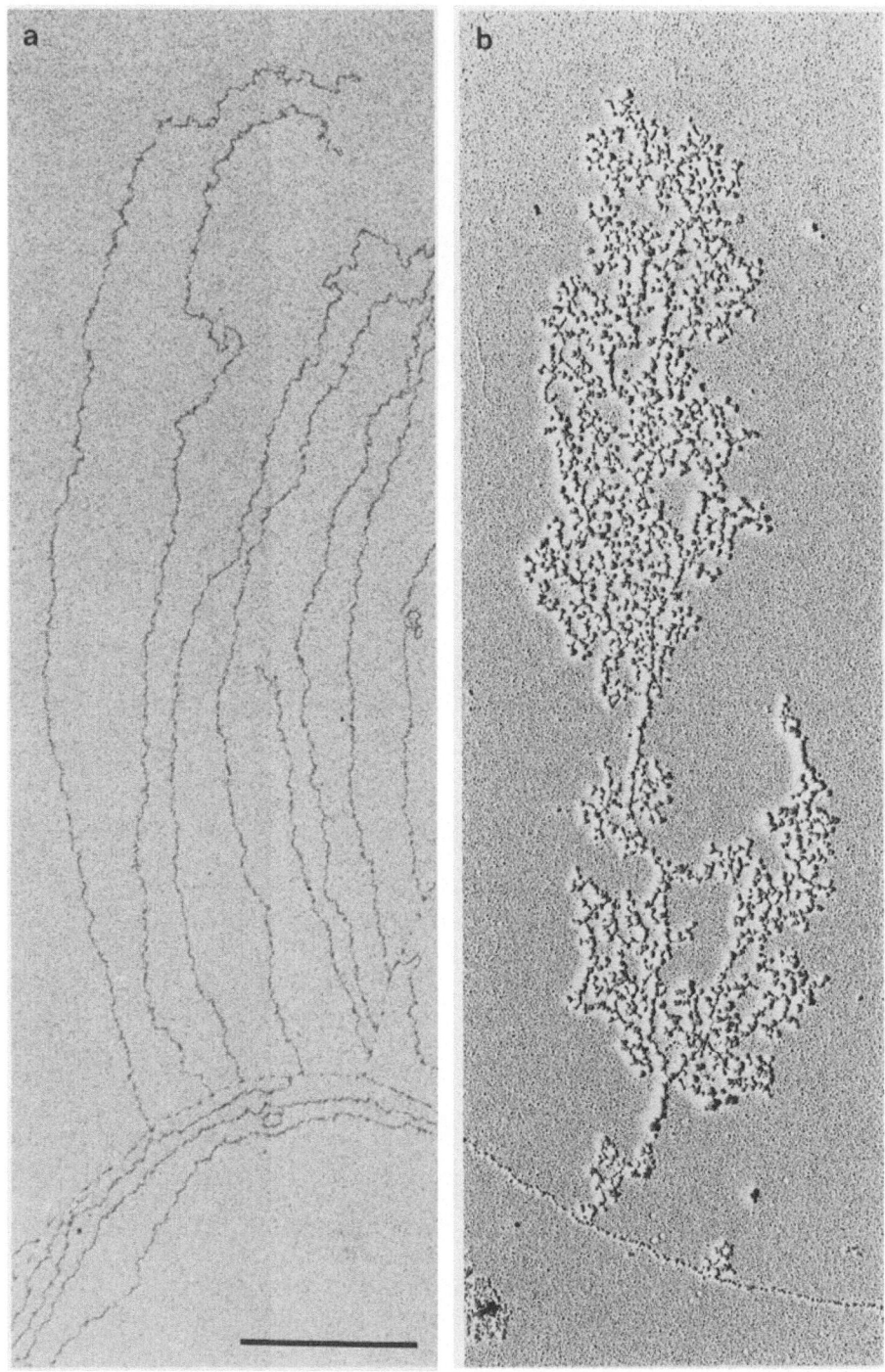

Fig. 7.6 a, b. Miller-spreads at pH 9 showing long transcripts in *Drosophila hydei* spermatocytes at the same magnification. In **a**, the transcripts are linear and their insertions on DNP are rel-

less dispersal has occurred and conditions are more lifelike the beaded fibrils are densely compacted on one another. Arguing in particular from the dual structure exhibited by the threads, Glätzer and Meyer (1981) suggest that the dense accumulations of giant transcripts around the axes of the threads may consist not only of transcripts from the Y-loops themselves, but may also include transcripts from other parts of the genome that are aggregated and "stored" on these loops. If this interpretation is correct it could account for some of the functional interactions between Y-loops and genes essential for fertility that are situated elsewhere in the genome. Glätzer and Meyer (1981) also considered the possibility that proteins, rather than RNP transcripts originating elsewhere in the genome, enter spermatocyte nuclei, accumulate amongst the Y-loop transcripts and so compact them that they generate aggregate structures of loop-specific morphology. If there are weak binding forces between such proteins and Y-loop transcripts they would tend to be eliminated at high pH, the proteins would disperse, and so permit the transcripts to unravel. In fact there is now evidence that immunologically-detectable proteins accumulate in the diffuse peripheral regions, not in the dense axes, of the threads (Glätzer 1984).

This raises two questions to which I will return: do loop-specific proteins exist, and if so where are they encoded; and to what extent are specific transcript sequences on the Y-loops ultimately responsible for determining their various morphologies? The X-ray-induced tube-proximal and tube-distal mutations studied by Hess (1965a) suggest that specific transcribed DNA sequences are of overriding importance, but if so such a mutated sequence must presumably lie near the initiation of transcription and impose the nature of its specificity on each RNA transcript as it continues its way along the DNA sequences included in the rest of the transcription unit; we are back to one of the most enigmatic questions presented by lampbrush chromosomes!

Grond et al. (1983) have made an ultrastructural study of the noose loops of D. hydei, making sure of their identiy by using males whose genotypes include only the short arm of the Y-chromosome. In thin section the noose loop RNP consists of granules of about 35 nm diam. In Miller-spreads each noose loop is evident as a single TU about 50 μm long, and it carries transcripts whose points of attachment to loop axis are on average 350 nm apart from one another. Close to their points of attachment to loop axis the transcripts appear as chains of granules, but more distally where greater dispersal has occurred they are folded back and forth to form complex branching structures (Fig. 7.7a). The DNP between transcripts is nucleosomal, and the total length of B-conformation DNA in the noose TU is estimated to be at least 260 kb (90 μm).

The granules on the noose-loop transcripts evidently unwind in spreads, but in this state they continue to display a great deal of secondary structure, possibly a consequence of annealing of complementary RNA sequences transcribed from

---

atively close together, separated by few nucleosomes. In **b,** which is an enormously long beaded transcript showing complex secondary structure, neighbouring transcripts lie far apart. In view of their lengths both **a** and **b** probably represent transcripts on Y-loops; precisely which Y-loops have not been determined, though threads and nooses can be excluded. *Bar* = 1 μm. (From Glätzer and Meyer 1981)

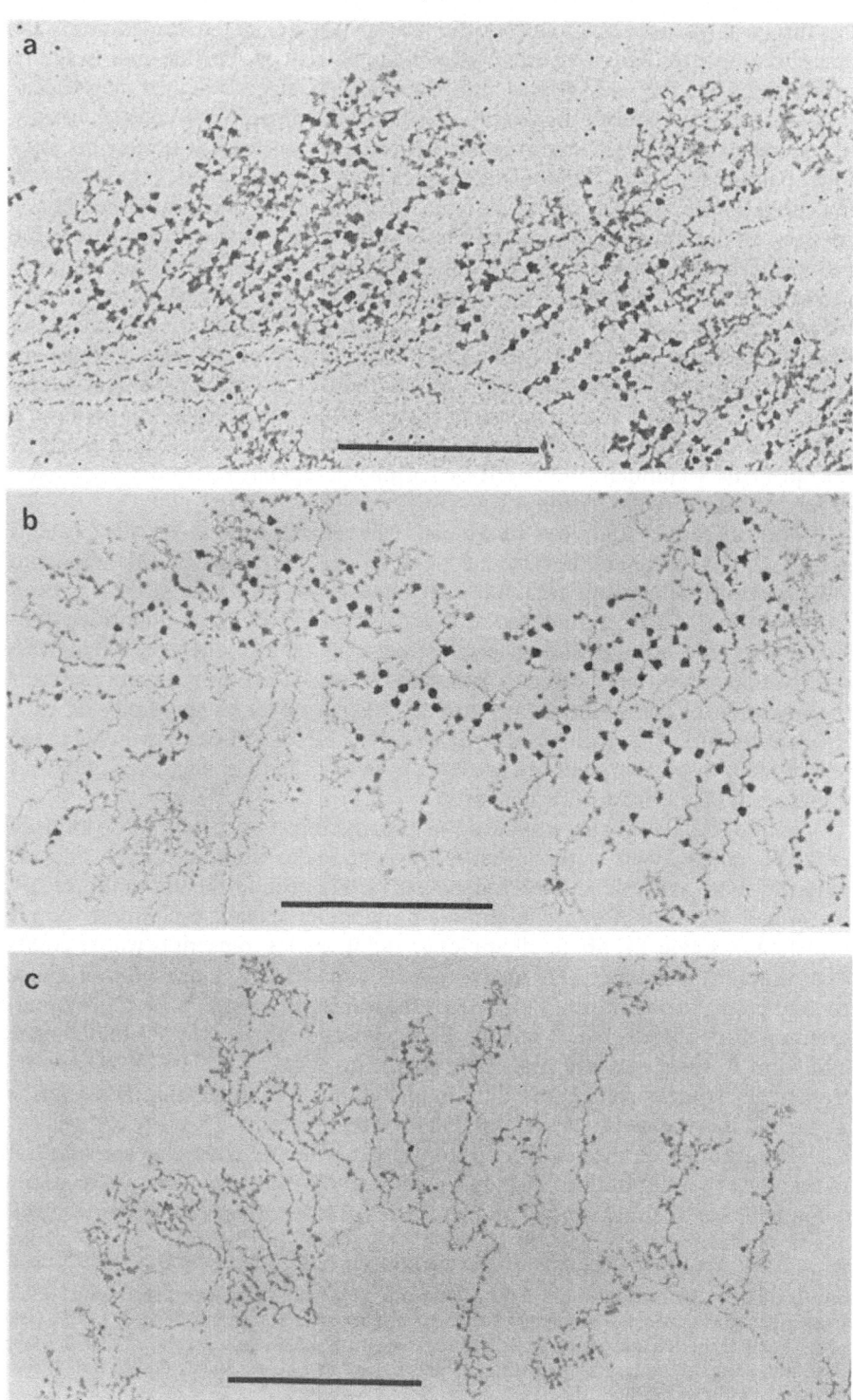

Fig. 7.7 a–c

inverted repeat sequences thought to be present in noose DNA (Vogt et al. 1982). Unlike the transcripts on the thread loops described by Glätzer and Meyer (1981) no mycel-like structures surround their points of attachment to axial DNP. Grond et al. (1983) describe the noose transcripts as being in other respects much like those of unremarkable or "normal" lampbrush loops, and find no evidence that transcripts from other genetic loci accumulate alongside them.

Grond et al. (1984) have continued to study the ultrastructure of the Y-loops of *D. hydei,* and have concluded that the tubular ribbons, like the nooses, have no extraneous material associated with their transcripts. However the threads, pseudonucleoli and clubs are different in that their transcripts, appearing granular in thin section like the RNP of the nooses, accumulate large quantities of interspersed accessory protein. Miller-spreads (de Loos et al. 1984) show that the giant transcripts on the axial DNP of the threads are more sparsely distributed than transcripts on the noose loops, on average 3 µm apart, while those of the pseudonucleoli may be even further apart, up to 20 µm; evidently one cannot take the presence of densely packed transcripts on TUs as a reliable criterion of lampbrush organization. An exceptional feature of the pseudonucleoli is that neighbouring transcripts are often strikingly different in size; de Loos et al. (1984) argue that this is not a result of unspecific "degradation" for the transcripts of nooses and threads show no such irregularity, and the interspersed small pseudonucleolar transcripts display size gradients as though secondary initiation of RNA synthesis takes place at several sites within the length of the entire TU. As was found by Glätzer and Meyer (1981), where the transcripts of the threads are attached to axial DNP extensive secondary structure is evident, but not elsewhere. By contrast the pseudonucleolus transcripts show a much branched secondary structure throughout their lengths; the transcript shown in Fig. 7.6 b is of this form.

The detailed cytogenetical study of the Y-loops of *Drosophila,* and the simplicity of the sytem when compared with the lampbrush chromosomes of amphibian oocytes, less than ten loops to be allocated functions as against thousands, has invited biochemical analysis. Moreover an additional advantage of the *Drosophila* system is that the transcriptional activities of the Y-loops, confined to meiotic prophase, are implicated in a well-documented developmental process that occurs immediately after they cease transcription, the differentiation of spermatids into spermatozoa. In *D. melanogaster* this differentiation is known to be autonomously determined by the genetic constitution of the diploid spermatocytes. Stern and Hadorn (1938) transplanted Anlagen of gonads from fertile donors into genetically sterile hosts, and during later development normal spermiogenesis proceeded in the transplants, leading to fertility; on the contrary, transplanted Anlagen of gonads from sterile donors into genetically fertile hosts did not produce functional sperm. These transplanted Anlagen contained germ-

---

Fig. 7.7 a–c. Miller-spreads at pH 9 of *Drosophila hydei* spermatocytes. **a** Transcripts on a small portion of the noose loops. **b** and **c** Parts of unidentified TUs in X/O spermatocytes. In **a** and **b** the transcripts have retained their native compact granular organization proximal to the DNP axis, whereas distally the granules have disaggregated to produce fibrillar RNP showing complex secondary structure; in **c** the dispersal is more extreme and the granular organization has been entirely lost. *Bars* = 1 µm. **a** From Grond et al. (1983); **b** and **c** courtesy of Dr. K. H. Glätzer

line cells and somatic cells of the same genetic constitution, so influence of the latter on the former could not be ruled out. However Marsh and Wieschaus (1978) transplanted primordial germ cells from X/Y embryos into X/O embryos, and as a result 5 of 14 host males were rendered fertile. This experiment proves that the action of the Y-chromosome in conferring fertility is strictly autonomous in the germ line; lack of the Y-chromosome in somatic tissues of the testis is not responsible for the infertility of X/O males. Furthermore Lindsley and Grell (1968) found that the absence of any chromosome or even any group of chromosomes from the normal complement in a spermatid nucleus has no influence on the progress of spermiogenesis; the genetic constitution of the diploid spermatocyte is what matters. Biochemical studies were begun by Hennig in 1967, working with *D. hydei.*

The *Drosophila* system does have certain drawbacks vis-à-vis amphibian oocytes. The spermatocytes are much smaller than amphibian oocytes, the spermatocyte chromosomes cannot be isolated in a lifelike state for study by light microscopy, including autoradiography, and the morphologies of the Y-loops are destroyed when spermatocytes are exposed to methods of fixation that are commonly used for studying chromosomes. However Hennig devised a squash method that serves for autoradiography. Testes are gently squashed in saline between coverslip and slide; the preparation is then plunged into absolute ethanol cooled to $-50°$ to $-70°C$ with solid CO_2, the coverslip flipped off the frozen preparation with a razor blade, and the slide replaced in ethanol. Such squash preparations can be taken down an ethanol series, diluted with 10% formalin in water, to water; the fixation is stable, and Y-loop morphologies are tolerably well preserved.

Like others before him, Hennig failed to detect DNA in the Y-loops by conventional methods, but DNase was found to break the threads and tubular ribbons in squashed and ruptured unfixed spermatocytes. The ^3H-thymidine incorporation gave equivocal results in autoradiographs because all the genome was labelled; however a time-course study established that the growth stage of spermatocytes, i.e. from the first appearance of the Y-loops to their disappearance at diakinesis, lasts for about 5 days; within this period the Y-loops take about 20 h to reach their full development.

Hennig confirmed the presence of RNA in the Y-loops by conventional methods, and by ^3H-uridine incorporation. RNA is synthesized throughout the spermatocyte growth stage, and ceases abruptly at diakinesis; it does not recur thereafter. However just as with ^3H-thymidine incorporation, ^3H-uridine incorporation by the rest of the genome complicates the interpretation of autoradiographs, and according to Hennig the autosomes are responsible for about half of the total non-nucleolar RNA synthesized by spermatocytes. Despite this difficulty Hennig (1967) demonstrated that the threads, conical projections from the pseudonucleoli, and clubs are sites of RNA synthesis, that this synthesis is suppressed by actinomycin C, and that labelled RNA remains associated with these structures for a limited time, not more than 20 h, considerably less than the overall duration of the Y-loops' fully differentiated state in spermatocytes.

Protein synthesis continues during the differentiation of spermatids into spermatozoa, for 2 to 3 days after RNA synthesis has ceased, implying that mRNAs remain available over this period. In discussing his observations Hennig (1967) followed Hess in proposing that the function of the Y-loops may be to protect such mRNAs from premature degradation by associating them with protein. And he posed the question as to why only the Y-chromosome develops lampbrush loops when this chromosome is responsible for only about half of the non-nucleolar RNA synthesized by spermatocytes. As mentioned already, this question has been resolved by Glätzer and Meyer (1981) and by Grond et al. (1983); transcriptionally active loops are not confined to those on the Y-chromosome, but they are not visible in the light microscope.

Although Giemsa staining is not a reliable cytochemical technique in the sense that the materials differentially stained cannot generally be identified, differential staining does at least give an indication of material diversity. Yamasaki (1977) subjected squash preparations of spermatocytes of *D. hydei* and two related species to Giemsa staining at pH 10, 5.8, and 2.8; differential staining was most striking at pH 10. In *D. hydei* the threads and nooses are virtually unstained; the matrix of the pseudonucleolus stains light blue, it is traversed by red staining channels that may represent the axial elements of this loop pair, and the conical projections also stain red; the axes of the tubular ribbons stain light blue and are embedded in red matrix (i.e. the reverse of the staining of the pseudonucleolus); the clubs stain red, their granules more intensely than the matrix; and the nucleolus stains blue-violet. When the squash preparations were extracted with RNase or hot trichloroacetic acid before staining at pH 10, none of the Y-loops were stained. From these results Giemsa is apparently staining RNA, and the differential staining presumably depends on different molecular associations between RNA and protein, either because different RNAs, or different proteins, or both together, are present on different loops.

Hennig (1968) demonstrated by RNA/DNA hybridization on filters that some rapidly labelled RNA extracted from the testes of *D. hydei,* but not from other male tissues, is complementary to Y-chromosomal DNA. More of this RNA is present in extracts from the testes of X/Y/Y males. From the hybridization conditions used for RNA/DNA annealing Hennig considered that the DNA sequences complementary to this RNA must be repetitive, and from saturation experiments he concluded that these hybridizable sequences only represent some 1% to 2% of the Y-chromosome's total DNA. Hennig stated that the simplest interpretation to be put on these observations is that the Y-chromosome synthesizes RNA coding for proteins required for sperm differentiation, though he did not rule out other possible interpretations.

Hennig et al. (1974) gave further information on the proportion of Y-chromosomal DNA transcribed in spermatocytes of *D. hydei*. Minimum estimates for the lengths of all the Y-loops of one chromatid total about 660 µm, and a further 400 µm is occupied by 18 S + 28 S rDNA, including non-transcribed spacers totalling about 100 µm; the Y-chromosome of *D. hydei* is unusual in that it includes two nucleolus organizers, one at the distal tip of its long arm, the other in the short arm (Schäfer and Kunz 1975). So the transcriptionally active Y-chromosomal DNA is about 1 mm overall, about $1/_{12}$th of all the DNA in the Y-chromatid (12 mm), and of this the Y-loops' contribution is more than one-half. There is, then, reasonable concordance between this and the earlier estimate based on

RNA/DNA hybridization. What is the nature of the residual, apparently non-transcribed $^{11}/_{12}$ths? On the evidence of in situ hybridization of cRNA transcribed in vitro from total DNA of *D. hydei* to mitotic chromosomes of this species, the Y-chromosome does not contain highly repetitive simple sequence (i.e. satellite) DNA; this is confined to the heterochromatic arm of the X-chromosome (the arm that is absent from attached-X females), and regions alongside the centromeres of all the autosomes. The latter part of the Hennig et al. (1974) paper described an analysis of proteins extracted from the testes of *D. hydei* and *D. neohydei* with various Y-chromosome and Y-fragment constitutions, and including X/O males, separated by electrophoresis on SDS-acrylamide gels. One particular protein was shown to be present in wild-type but absent from X/O testes of *D. hydei,* and that in wild-type testes this protein is confined to post-meiotic stages of spermatogenesis. In testes of *D. neohydei* this particular protein is absent, but another protein of nearly the same mobility is present, one that is absent from *D. hydei.* In the testes of X/Y/Y males that contain the Y-chromosomes of both species but whose genomes otherwise contain, for the most part, only *D. hydei* chromosomes, both of these proteins (molecular weights almost 300,000) are present. Hennig et al. (1974) considered that this finding ruled out the possibility that the Y-chromosome is only responsible for the regulation of synthesis of a protein that is encoded elsewhere in the genome; rather, that information carried by the Y-chromosome is directly responsible for its synthesis. However this conclusion has not gone unchallenged.

Ingman-Baker and Candido (1980) have made a detailed two-dimensional analysis of the gel patterns of labelled proteins from the testes of X/O, X/Y and X/Y/Y males of *D. melanogaster.* Approximately 1200 proteins were found in all three genotypes, of which 92 are structural proteins present in spermatozoa. No differences attributable to the Y-chromosome could be detected, and they conclude that the Y-chromosome does not code for any major proteins. This study was carried out on *D. melanogaster,* and the results are not necessarily valid for *D. hydei.* In fact, and unlike *D. melanogaster,* Hulsebos et al. (1983) have found that two major testis proteins of X/Y *D. hydei,* of molecular weights 155,000 (previously estimated to be 300,000 by Hennig et al. 1974, see above) and 35,000, are absent from the testes of X/O *D. hydei.* Nevertheless their evidence, from comparisons between mutant strains lacking various parts of the Y-chromosome, now suggests that the coding sequences for these proteins lie on the autosomes, and that part of the Y-chromosome, including the club and noose-loop loci, play a regulatory rather than a coding role in their production.

Kloetzel et al. (1981) have worked with *D. hydei,* and although they found no nuclear proteins that are present in X/Y, but absent from X/O spermatocytes, they have come to a similar conclusion regarding Y-loop function. Spermatocyte nuclei from X/Y and X/O males, and from males carrying the Y-A or A-Y translocations constructed and described by Hess (1965b), but with no free Y, were analyzed for their protein contents. As compared with the nuclei of X/O spermatocytes, seven proteins, of molecular weights about 45,000, 52,000, 54,000, 66,000, 80,000, and 170,000, were found to be enriched in the nuclei of spermatocytes containing two or more active X-loops. None of these proteins are absent from X/O spermatocytes, but proteins of molecular weights 35,000, 46,000, 58,000, and 110,000 are enriched vis-à-vis spermatocyte nuclei containing Y-loops. These findings support the view that the transcripts on the Y-loops do not themselves include coding sequences but rather regulate in some way the relative proportions of different nuclear proteins that are synthesized during spermatogenesis, the abnormal enrichment of four proteins in X/O males being the result of regulatory malfunction.

Of particular interest is the observation of Kloetzel et al. (1981) that when the RNP contents of spermatocyte nuclei are analyzed by differential centrifugation, a rapidly sedimenting heterogeneous nuclear (hn) RNP with an S-value >900 can be isolated. In X/Y males this fraction contains proteins as well as fast labelling RNA both enriched by a factor of about three as compared with X/O males. This hnRNP may well represent the main bulk of Y-loop matrix, for it includes in almost their entirety all seven of the major proteins that are enriched in X/Y males. Moreover this rapidly sedimenting hnRNP contains an unusually high proportion (25%) of polyadenylated RNA, is extremely stable in high salt concentration, and its protein composition remains almost unaltered if the crude hnRNP is digested with DNase prior to sucrose gradient analysis. Kloetzel et al. (1981) point out that in these several respects, and others that they specify, the proteins that are enriched in X/Y males of *D. hydei* resemble "nuclear matrix", i.e. non-histone proteins that are associated with hnRNA occurring in a wide variety of cell nuclei. They reach the conclusion that the function of the Y-loops is to store and protect from degradation, by their association with nuclear matrix proteins, mRNAs that are involved in the post-meiotic differentiation of spermatozoa. In essence this is precisely the same conclusion as was reached by Hess (1965a, 1967a) on altogether different evidence.

Early though not conclusive evidence that the Y-loops' DNAs contain repetitive sequences was given by Hennig (1968), Hennig et al. (1974) and Lifschytz (1975). This has now been substantiated by the production and use of cloned Y-loop DNA sequences. The first such clone was obtained by Lifschytz (1979); the clone is a constituent of "noose" loop DNA, and it exists as a middle repetitive sequence. More cloned DNAs homologous to the transcribed middle repetitive sequences of four of the Y-loops of *D. hydei* have recently been described by Lifschytz et al. (1983) and their study, given in particular detail, will be reviewed here. Out of several thousand plasmid clones harbouring ECoR1 fragments of male *D. hydei* DNA, eight were found that hybridized exclusively with male, but not female DNA. Several clones proved to contain more than one ECoR1 fragment; these were digested again with ECoR1, the fragments separated, tested for their preferential hybridization to total male DNA, and families of subclones of Y-linked middle repetitive DNAs were thus obtained. Representative subclones were chosen for hybridization after they had been nick-translated. For preliminary localization, DNA/DNA in situ hybridization to denatured mitotic metaphase chromosomes was carried out with five subclones. To increase the resolution within the length of the Y-chromosome of *D. hydei,* the stock carrying the Y-A and A-Y reciprocal translocations was used as test material because the two translocation chromosomes can be recognized in mitoses. Y-A includes the loci for nooses, clubs, and tubular ribbons in spermatocytes, whereas A-Y includes the loci for threads and pseudonucleolus. All clones hybridized to the Y-component of translocation Y-A, none to that of translocation A-Y.

When hybridized in situ to denatured squash preparations of the polytene chromosomes of *D. hydei* salivary glands, all subclones labelled the chromocentres of male, but not female polytene chromosomes, though the level of labelling was low (the Y-chromosome is underreplicated in salivary gland nuclei); two of the five subclones also hybridized to various euchromatic sites on all the other chromosomes.

All five subclones were hybridized to fractionated RNA samples extracted from total testes, nuclei of embryos and nuclei of spermatocytes. One of the subclones hybridized to none of these samples, and presumably represents a repetitive sequence in the Y-chromosome that is not transcribed even in spermatocytes. The other four subclones hybridized intensively to spermatocyte nuclear RNA, but negligibly or not at all to RNAs from other sources. These four positive subclones were hybridized in situ to squashed spermatocytes of wild-type *D. hydei,* and distinctively different distributions of silver grains resulted, in keeping with the regions characteristically occupied by the various Y-loops. Spermatocyte nuclei of X/O males were not labelled by any of the subclones, and spermatocyte cytoplasm of X/Y males was likewise unlabelled. In order to clinch the assignment of subclones to Y-loops, advantage was taken of the various X-Y translocations carried by different *D. hydei* stocks. To summarize, one subclone hybridized to the clubs, another to the tubular ribbons, and two subclones hybridized to the nooses. These latter two subclones were derived from different parental clones, and their sequences only show remote homology to one another. Another important fact established by Lifschytz et al. (1983) is that none of the Y-loop cloned DNAs hybridize with nuclei or cytoplasm of post-meiotic cells. The Y-loop transcript RNAs are strictly confined to spermatocyte nuclei, they are degraded during the meiotic divisions, and it is improbable that they are translated.

In discussing their observations Lifschytz et al. (1983) admit to the difficulty of assigning a sole regulatory or storage role for the Y-loops, but in noting that the various loops occupy characteristic positions in spermatocyte nuclei they suggest that this may provide a clue to their function. Following the proposals made by Cavalier-Smith (1978) to explain the significance of the lampbrush loops of amphibian oocytes, they think the Y-loops may play an indirect regulatory role by participating in the assembly of nuclear matrix in spermatocytes (cf. the suggestions made by Kloetzel et al. 1981), thereby establishing "... domains for gene activity as well as compartments for the storage of gene products". Such domains would thereafter require to be partitioned during the two meiotic divisions so that each of four spermatids receives its appropriate endowment. Once established, these domains would then ensure that coordinate assembly of the various organelles occurs during the differentiation of spermatids into spermatozoa. In essence this proposal likens a *Drosophila* spermatid to an egg with mosaic-type development, such as that of a tunicate.

Another cloned *D. hydei* sequence, with homologies to the clone prepared by Lifschytz (1979) and likewise present in noose-loop DNA, was prepared by Vogt et al. (1982). This clone and five others have been described by Vogt and Hennig (1983). Three of the cloned insert sequences proved to be present only in the Y-chromosome, and all of these hybridized in situ to transcripts of the noose loops in spermatocytes. The other three, although they hybridized in situ to the Y-chromosome in denatured mitotic preparations, did not hybridize to transcripts in spermatocytes, so presumably these sequences lie in inactive parts of the Y-chromosome; one of them also hybridized with a single band of autosome IV in preparations of polytene chromosomes, while the other two hybridized with autosomal pericentric heterochromatin.

Hennig et al. (1983), by microdissection of spermatocytes of *D. hydei* in which the only Y-loops present were the threads, succeeded in cloning four sequences whose complementary RNA hybridized in situ to the end of the long arm of the Y-chromosome (where the thread locus lies) in mitotic preparations, and in addition all of their cRNAs when hybridized to polytene chromosomes labelled various euchromatic bands on the autosomes. Restriction analysis of the six noose clones and the four thread clones mentioned above show that all these clones contain repeated sequences that are, however, not strictly homogeneous; they consist of families of related, but not identical sequences. In the case of the noose clones (Hennig 1984) there are subrepeats for the most part ten nucleotide pairs long, and these form the components of larger superimposed repeats. Hennig (1984) considers that this is probably a general feature of Y-chromosomal sequences of *D. hydei,* most of which have further copies located elsewhere in the genome. So far no protein-coding sequences have been found in any cloned inserts shown to have been derived from the Y-chromosome of *D. hydei.*

At this point we need to reconsider how the Y-loops' various morphologies are determined, for all the early experimental evidence had supported the view that they differentiate autonomously. An alternative possibility followed from the discovery by Lifschytz (1974, 1975) that X-ray mutations induced on the X-chromosome can, and evidently indirectly, influence the expression of the Y-loops. Three non-allelic male sterile mutations, XL2, XL4, and XL24, all cause disruption of spermatid elongation. In spermatocytes mutation XL2 causes reduction in size of all the Y-loops, greater at high (30 °C) than at low (18 °C) culture temperatures. The Y-loops all start to develop, but do not reach their normal dimensions. In the crosses set up to detect sex-linked male sterile mutations the original Y-chromosome had also been irradiated, but when this Y was substituted by a non-irradiated Y the same effect on the Y-loops was observed. Moreover the reduction in size of the Y-loops is the same whether or not an additional Y-chromosome or part of an additional Y-chromosome is present in XL2 spermatocytes. Lifschytz showed that XL2 lies close to the locus w and, most ingeniously, that it is a recessive mutation. There is an X-Y translocation in *D. hydei* where a small section of the X-chromosome, including the locus w, has been transferred to the Y-chromosome. The spermatocytes of males with XL2, a normal Y and this X-Y translocation form normal Y-loops and are fertile. Mutation XL2 does not act as a general repressor of the initiation of transcription on the Y-loops, for all the Y-loops start to differentiate normally. Rather it appears that the mutant acts indirectly by failing to supply some substance that the Y-loops need in order to complete their extension.

The two other male sterile mutants have diverse effects on the Y-loops. With XL4 the pseudonucleolus is most reduced in size, but the threads are almost normal and the other loops suffer to an intermediate degree. With XL24 the threads and nooses are more affected than the pseudonucleolus. These observations suggest that the three mutants represent different lesions in metabolic pathways all of which, however, are in some way involved in the Y-loop differentiations that in *D. hydei* males of normal karyotype confer full capacity for the development of spermatozoa from spermatids.

Hennig (1984) cites another example of a mutation, in this case autosomal, that affects Y-chromosome morphology in spermatocytes of *D. hydei*. In males homozygous for this recessive mutation the dense axial component of the threads is missing. Such males are sterile. When combined with the "tube-proximal" Y-mutant of Hess (1965a) the compact region of the threads, characteristic of this mutant, is drastically reduced in size. These males are also sterile.

These examples of male sterility that are not directly attributable to malfunction of the *D. hydei* Y-chromosome could be explained by assuming that the wild-type X or autosomal genes involved code for proteins which normally accumulate on Y-loops, in part determine their morphology, and by being so sequestered allow more of these proteins, relative to others, to be synthesized. By this I am implying relief from a negative feedback control of transcription or translation. The mutant genes cause sterility because they fail to code for these proteins, and this is manifested by a "phenotypic" alteration of Y-loop morphology.

There is also further evidence that appropriate interactions between genes of the autosomes, the X, and the Y-loops are needed to assure male fertility. As mentioned earlier in this chapter, Hess and Meyer (1963) showed that male F_1 hybrids between *D. hydei* and *D. neohydei* are viable and fertile. They contain a haploid set of autosome of both species, the X of one or the other, but in them the Y-chromosome expresses the complement of loops typical of the male parent.

Schäfer (1978, 1979), and also Hennig (1977), made F_1 hybrids between *D. hydei* ♀♀ × *D. neohydei* ♂♂, and repeatedly backcrossed to the maternal parent species, thereby obtaining males with nearly all or all of their autosomes, and their X-chromosome, derived from *D. hydei* and only their Y-chromosome from *D. neohydei*. In those backcross hybrid males containing only one pair of *D. hydei* autosomes, although one particular autosome pair disturbs spermiogenesis to a greater extent than the other five, at least some motile spermatozoa are formed in all such hybrids. However in those backcross male hybrids containing two or more pairs of *D. hydei* autosomes, spermiogenesis is disturbed to a degree roughly in accord with the number of pairs of *D. hydei* autosomes present, and the greater the disturbance the less the fertility. The most grave disturbances to spermiogenesis occur in backcross hybrid males where all but the Y-chromosomes are derived from *D. hydei*. These males are completely sterile. Schäfer (1978) describes their Y-loops as being of normal *D. neohydei* morphology, whereas Hennig (1977) describes them as being more compact and reminiscent of *D. hydei* Y-loops. In discussing his observations Schäfer makes the point that at least some spermatids in all these backcross hybrids contain all the normal structural components of the sperm organelles, but they tend to be improperly organized. In their disorder they resemble the disorganization caused by small deficiencies of the Y-chromosome in *D. melanogaster* and *D. hydei* described by Meyer (1968). Schäfer supposes that the Y-loops of *D. neohydei* are involved in cordinating the morphogenetic processes of spermatid differentiation but cannot do so in these backcross hybrids because the proteins at their disposal are encoded by *D. hydei* genes.

Immunological studies have recently supplied evidence that specific proteins are associated with particular Y-loops of *D. hydei*. One study is that of Hulsebos et al. (1984). An antiserum to the protein of 155,000 mol. wt. that is present in wild-type *D. hydei* testes but, according to the authors, absent from those of X/O males, was prepared. It was found to cross-react with a protein of 80,000 mol. wt. that appears to be present in all *D. hydei* testes, including those of partially Y-deficient and X/O genotypes that lack the larger protein. Apart from the antigenic specificity that they share, the relationship between the larger and smaller protein has not been established (though the molecular weights suggest that the larger could be a dimer of the smaller). When the antiserum was applied to spermatocytes and binding studied by indirect immunofluorescence, the reaction was found to be confined to the pseudonucleolus (Fig. 7.9 b), and even in genotypes that lack the 155,000 mol. wt. protein; the binding is therefore to the 80,000 mol. wt. protein. In X/O spermatocytes, without a pseudonucleolus or any other Y-loops, the fluorescence was diffusely spread throughout their nuclei. So it would appear that the pseudonucleolus normally acts as a structure on which the 80,000 mol. wt. protein accumulates. The antiserum does not bind to spermatocyte cytoplasm but after meiosis it reacts with the tails of elongating spermatids and all later stages of spermiogenesis (Hulsebos et al. 1983). The antiserum was found to bind to structures in the spermatocyte nuclei of several other species of *Drosophila,* including *D. melanogaster,* so the antigenic protein is conserved in evolutionary terms. It cannot be encoded on the Y-chromosome, for it is present in X/O males.

Glätzer (1984) has worked with monoclonal antibodies raised against *D. melanogaster* non-histone nuclear proteins, applying them to spermatocytes and spermatids of *D. hydei* and studying their distribution by indirect immunofluorescence. Three of seven such antibodies gave clear-cut positive reactions, all restricted to the nuclei of primary spermatocytes and spermatids. One antibody, "S5", binds to the diffuse outer regions, but not to the dense axial components of the threads (Fig. 7.8 a), and also binds to the tubular ribbons. It does not bind to any of the other Y-loops, i.e. nooses, clubs or pseudonucleolus. In spermatocytes of genotypes lacking the threads and tubular ribbons, and in X/O spermatocytes, the fluorescence is uniformly distributed throughout the nuclei (Fig. 7.8 b). Evidently here again the antigenic protein is not encoded by a Y-chromosomal gene, but after translation it enters spermatocyte nuclei and normally accumulates on the transcripts of two Y-loops. Another antibody, "X4", likewise binds only to the threads and tubular ribbons, but the distribution and relative intensities of fluorescence differ in certain respects from that occurring with "S5". A third antibody, "T7", also binds to spermatocyte nuclei, but specifically not to any of the Y-loops (Fig. 7.9 a).

During the meiotic divisions the antigens to S5 and X4 become concentrated in cytoplasmic regions that, surprisingly, are not shared equally between sister secondary spermatocytes. After the meiotic divisions have been completed both

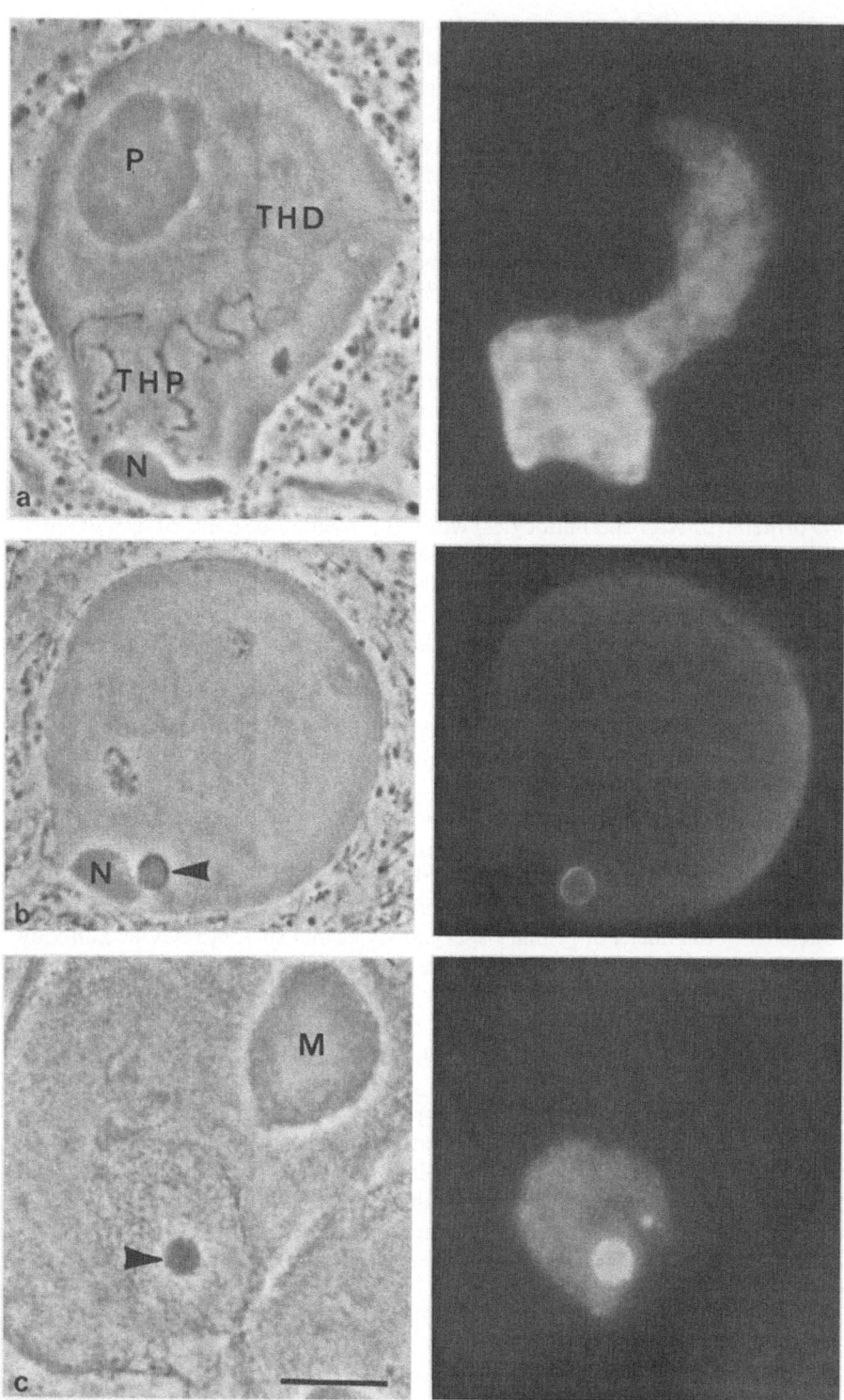

Fig. 7.8 a–c

the antigens return to a nuclear location in spermatids (Fig. 7.8 c), and remain detectable there during the early stages of nuclear elongation. The post-meiotic behaviour of the T7 antigen is described as similar, but its immunological reaction weak and close to the limit of detection.

In *D. melanogaster* the three antigens investigated by Glätzer are thought to be minor constituents of hnRNP, and concerned in some way with RNA processing (see Glätzer 1984, for references). For this reason Glätzer (1984) describes them as "nuclear RNP antigens", proteins that associate with RNA as it is transcribed at sites elsewhere than on the Y-chromosome. The implication is that the RNA involved is stored and "protected" on the Y-loops in spermatocytes and is only released for some post-meiotic rôle during spermiogenesis. For a thorough assessment of these observations and ideas more information is needed; their importance for an understanding of lampbrush chromosome function is evident because loop-specific proteins are also known to exist in amphibian oocytes, as will be discussed in Chap. 9.

There is yet another problematic matter with which to come to terms. If, as suggested by Lifschytz et al. (1983) the differentiation of the spermatozoa of *Drosophila* depends on the establishment of domains of nuclear matrix in spermatocytes, and these domains are themselves dependent on Y-loop transcription, how is the differentiation of the spermatozoa of other animals controlled, animals whose spermatocytes do not appear to contain lampbrush loops? Doubtless lampbrush loops, as defined elsewhere in this monograph, will in the course of time be discovered in the spermatocytes of animals other than *Drosophila,* but the fact remains that no lampbrush loops have been found, for example, in the spermatocytes of either short-horned grasshoppers or urodele amphibians. The spermatocytes of many species in these two animal groups have been well scrutinized by competent cytologists, their spermatocytes and spermatozoa are large and the latter, particularly in urodele amphibians, of great structural complexity. Most short-horned grasshoppers have no Y-chromosome, and the single X-chromosome is transcriptionally inactive throughout male meiosis (Henderson 1964). "Fertility" genes, i.e. genes directly concerned with cell differentiation during spermiogenesis, presumably lie on the autosomes of these grasshoppers, yet none transcribe with such intensity, whether or not their transcripts serve as sites for the accumulation of proteins that will later establish morphogenetic domains, for them to appear as lampbrush loops visible in the light microscope.

Fig. 7.8 a–c. Distribution of "S5" antigens in spermatocyte and spermatid nuclei of *Drosophila hydei.* The micrographs are mounted in pairs, those to the *left* being of squashed formaldehyde-fixed cells taken in phase contrast, those to the *right* being these same cells taken by indirect immunofluorescence. **a** Spermatocyte from a male with an X and the distal part of the long arm of the Y-chromosome, including the thread and pseudonucleolus loci. "S5" antigens are confined to the threads. **b** X/O male; the spherical object beside the nucleolus, marked by an *arrowhead,* is thought to be of X-chromosome origin. Except for some concentration in this spherical object, S5 antigens are dispersed through the nucleus. **c** Part of an early wild-type spermatid; the spherical object, marked by an *arrowhead,* is described as a nucleoluslike body of unknown function that gradually disappears during spermatid differentiation. S5 antigens are still restricted to the nucleus. *N* nucleolus; *P* pseudonucleolus; *THP* proximal parts of the threads; *THD* distal parts of the threads; *M* mitochondrial aggregate or „Nebenkern". *Bar* = 10 μm. (From Glätzer 1984)

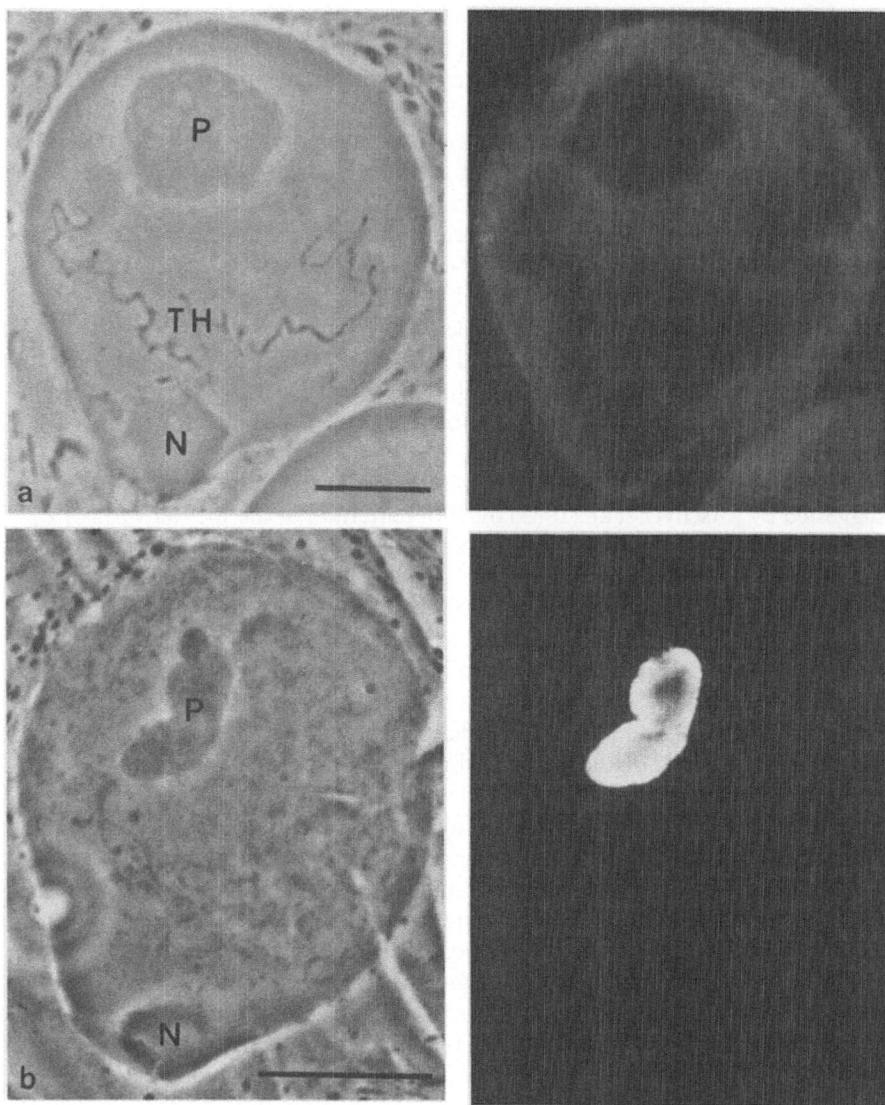

Fig. 7.9 a, b. Distribution of antigens in spermatocyte nuclei of *Drosophila hydei*. The micrographs are mounted in pairs, those to the *left* being of squashed formaldehyde-fixed cells taken in phase contrast, those to the *right* being these same cells taken by indirect immunofluorescence. **a** Spermatocyte from a male with an X and the distal part of the long arm of the Y-chromosome, including the thread and pseudonucleolus loci. The fluorescence micrograph shows that "T7" antigens are absent from the threads, pseudonucleolus and nucleolus, but present elsewhere in the nucleus. **b** Spermatocyte from a male in which all the Y-loops are present. The fluorescence micrograph shows that "sph 155" antigens are confined to the pseudonucleolus. *N* nucleolus; *P* pseudonucleolus; *TH* threads. *Bars* = 10 μm. **a** From Glätzer (1984); **b** from Hennig (1984)

W. Kunz (personal communication 1984) has suggested that the long transcription units present in the Y-loops of *Drosophila* owe their exceptional length to the location of the male fertility genes in "constitutive heterochromatin". The Y-chromosome is not subdivided into chromomeres because there are few of these genes, they are only active during male meiosis, and they are separated from one another by long stretches of repetitive DNA sequences. Transcription of these male fertility genes is presumably performed by RNA polymerase II, and there is evidence from lampbrush-type transcription in amphibian oocytes (see Chap. 5) that in meiotic cells polymerase II overrides signals that may normally terminate transcription. This suggestion on its own does not account for the presence of lampbrush Y-loops in *Drosophila* but their absence from the spermatocytes of many, in all probability most, other animals. However, it prompts the further speculation that male *Drosophila* may be exceptional in postponing, until a few days prior to the meiotic divisions, the accumulation and disposition of materials required for controlling post-meiotic morphogenesis. Other animals may start similar synthetic activities earlier, and proceed in a more leisurely fashion.

There is an alternative possibility that I had for long overlooked. The transcription of male fertility genes located on the Y-chromosome of *Drosophila* has been clearly shown to play an essential role in controlling the morphogenesis of spermatids into spermatozoa, and because half of *Drosophila's* spermatid nuclei contain no Y-chromosome, this transcription must necessarily precede chromosome segregation at meiosis. If comparable male fertility genes of other animals are located on chromatids that are present in all spermatid nuclei, then their transcription might be delayed until after completion of the meiotic divisions. Already in 1964 Henderson had shown that there is a brief phase of RNA synthesis in early spermatids of grasshoppers. More recently Schmid and Krone (1976) have demonstrated a remarkable and suggestive relationship between a specific chromosome region, a heterochromatic C-band, and the development of the acrosome, in the spermatid nuclei of four species of newts. In these newts the majority of the C-bands detectable in mitotic chromosomes lie immediately adjacent to the centromeres and in peri-centromeric regions nearby. In early spermatid nuclei the haploid chromatids retain a bouquet configuration while they decondense, a relic of the disposition of the chromatids at second meiotic anaphase, and the C-band material, now visible as heterochromatic lumps or chromocentres, is confined to a region of the nucleus close to the pole of the preceding division spindle. One of these chromocentres, which Schmid and Krone were able to identify as originating from a peri-centromeric C-band lying in one arm of a specific chromatid, remains dense while the rest of the chromocentres become diffuse. This dense chromocentre migrates to the nuclear membrane at the opposite pole of the spermatid nucleus, where chromocentres were previously absent, it elongates to form a thread, and while this occurs the nuclear membrane protrudes to accommodate the extending thread. This portion of the nuclear membrane is destined to enclose the tip of the mature spermatozoon's chromatin where it lies adjacent to the acrosome. Segments of this "acrosomal chromocentre" or "acrosomal thread" decondense and give the impression that at this stage they may be engaged in transcription. That this is really so has recently been confirmed by autoradiographic studies of ^{3}H-uridine incorporation (Dr. M. Schmid, personal communication 1985).

The very early spermatid nucleus is transcriptionally inactive, but while it is still spherical there occurs a burst of RNA synthesis detectable not only where the acrosomal chromocentre is elongating, but also throughout the nucleus, i.e. including that half of the nucleus occupied by the other, now relatively diffuse chromocentres; in living spermatocytes examined by phase contrast, refractile objects reminiscent of the Y-loops of *Drosophila* appear in this half of the spermatid nucleus at this stage.

It appears probable that these post-meiotic transcriptional activities are implicated in the control of morphogenesis of the spermatozoa of these urodeles, and it is tempting to speculate that comparable activities may be widespread among animals where the genes involved are not segregated differentially during meiosis.

CHAPTER 8

Acetabularia

There is only one well-documented demonstration that lampbrush chromosomes occur in plants. They are present in the "primary" nucleus of the giant marine unicellular dasycladacean alga *Acetabularia* (Spring et al. 1974, 1975). In its fully differentiated vegetative stage *Acetabularia* consists of a lobed rhizoid by which it is attached to the sea bed, a stalk several centimeters long projecting up from the rhizoid, and at the top of the stalk a "cap", 1 cm or so wide when mature, whose morphology is distinctive of the species. *Acetabularia* provided one of the earliest, most elegant and incontrovertible proofs of the nuclear control of morphogenesis in the classical regeneration and grafting experiments carried out by Hämmerling (reviewed in 1953 and 1963). The single nucleus resides in the rhizoid; the species-specific morphogenesis that it determines occurs in the cap, several centimeters distant, and the control is exercised via the cytoplasm of the stalk. Towards the end of its vegetative life a multiplicity of "secondary" nuclei migrate to the cap, the products of division of the primary nucleus. Cysts form around these nuclei. In suitable conditions of culture the cysts germinate to produce flagellated swarming gametes that conjugate and form zygotes. The unicellular vegetative stage develops from a zygote and involves spectacular increase in volume both of the cytoplasm and of its single nucleus, coupled with extensive differentiation. At the beginning of development the primary nucleus is about 2 μm diam, and it reaches its maximum diameter of about 150 μm in the course of a few weeks, when the cap starts to form. As the cap develops the volume of the nucleus diminishes, reaching a diameter of about 40 μm by the time an intranuclear spindle forms (Koop et al. 1979). The changes undergone by the primary nucleus prior to and during its division have been filmed (see Koop et al. 1979, for references). The reduction in size of the nucleus is accompanied at first by puckering, not disintegration, of the nuclear membrane as material passes to the cytoplasm, and by reduction in the size and/or number of nucleoli. Apart from the retention of an intact nuclear membrane the process shows clear resemblance to the maturation of an oocyte.

Spring et al. (1975) studied the primary nuclei of *Acetabularia* in algal cells with stalk lengths of from 1 to 4 cm, before cap formation, by light and electron microscopy. In 2-μm-thick Epon-embedded sections stained or examined in phase contrast there are structures with evident similarity to lampbrush chromosomes as well as conspicuous nucleoli. In the spread contents of manually isolated nuclei the existence of lampbrush chromosomes was confirmed, for loops covered with matrix projecting from chromosomes' axes can be resolved in the light microscope. The chromosomes' axes range in length from about 4 to 30 μm, and the lateral loops reach lengths of about 20 μm. There are dense objects on the chromosomes' axes that may be chromomeres, though they could be composite structures like the axial granules of urodele lampbrush chromosomes, consisting not only of compact DNP but accessory protein or RNP as well. Be this as it may, the ends of the lateral loops are reflected to common points of origin in the chromosomes' axes. In some regions of thin sections there are aggregates of granulo-fibrillar material that resemble the lateral loop matrix of urodele lampbrush chromosomes, the granules being of comparable size, some 20 to 25 nm.

In Miller-spreads of predominantly nucleolar material (Spring et al. 1976) there are extensive tandem arrays of rDNA transcription units (TUs) each about 2 μm long, several thousand per nucleus, with spacer intercepts of different lengths, some arrays being homogeneous for spacer lengths, others heterogeneous. These TUs in regular tandem arrays are remarkably like material of similar origin from many animal cell nuclei, including amphibian oocytes. In Miller-spreads of whole nuclei (Spring et al. 1974) there are also much longer TUs present, of the order 5 to 14 μm. These are also present in spreads of nucleolus-free nuclear material (Spring et al. 1975), thus demonstrating their origin from lamp-brush chromosomes. These long TUs have lateral filaments up to 3 μm in length at their termini, and a high packing density of lateral filaments within TUs, at or close to the limit imposed by the size of RNA polymerases in immediate juxtaposition. Spring et al. (1975) found that spacer intercepts between some of these long TUs were scarcely detectable (Fig. 8.1 a), implying that the lateral loops from which they originated consisted of multiple matrix units. Other of these long TUs are preceded and followed by substantial lengths of non-transcribed, nucleosomal chromatin (Fig. 8.1 b), presumably derived from dispersed chromomeres (Scheer et al. 1976 a). Just as in amphibian lampbrush chromosomes, neighbouring TUs in *Acetabularia* may have the same or opposite polarities of transcription; examples are shown in Fig. 8.1 a, c, d.

It is then clear that the primary nucleus of *Acetabularia* contains lampbrush chromosomes that are very active in transcription, and it is equally clear that, as in large animal oocytes, their chromosomes take this form so as to meet the synthetic requirements of a single cell that grows rapidly, reaches a large ultimate size, and has to prepare for, or engage in, an elaborate programme of differentiation. In a table of "cell efficiency characteristics" Spring et al. (1974) demonstrate how some of these parameters in *Acetabularia* compare with those of the

───▶

Fig. 8.1 a–d. Miller-spread preparations showing transcription units (TUs) in the primary nuclei of *Acetabularia*. **a** *A. mediterranea:* the *upper arrow* points to the origins of two adjacent TUs with opposite (head-to-head) polarities; the *lower arrow* points to the terminus of one TU and the immediately adjacent origin of another, i.e. with similar polarities. **b** *A. major:* the origin of the long TU is preceded by a very short TU with similar polarity, and this in turn is preceded by a long transcript-free strand presumably derived from a dispersed chromomere. **c** *A. mediterranea:* the two *arrows* point to the origins of two adjacent TUs with opposite (tail-to-tail) polarities and contiguous termini; below the *upper arrow* a transcript-free strand leads down to the origin of a shorter TU with opposite polarity to its neighbour above. **d** *A. cliftonii:* the two *arrows* point to the adjacent origins of two TUs with opposite (head-to-head) polarities. *Bars* = 1 μm. **a** From Spring et al. (1975); **b, c, d** from Scheer et al. (1976 a)

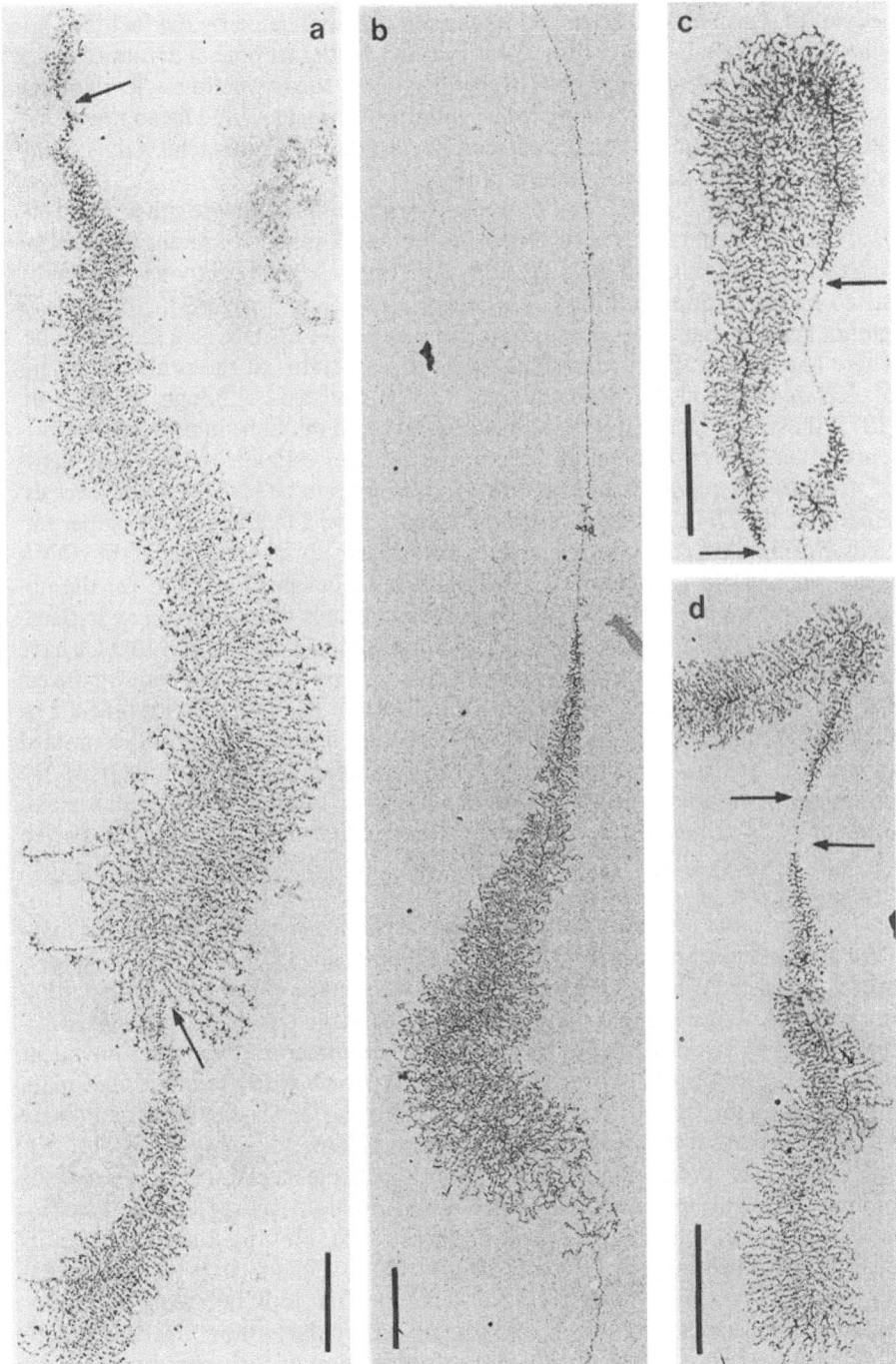

Fig. 8.1a–d

oocytes of *Xenopus* and *Acheta*. Comparison is complicated by the fact that oo-cytes of multicellular organisms are supported by the trophic activities of other tissues, whereas *Acetabularia* has to fend for itself. Mean rate of nuclear volume increase (predominantly due to protein entering the nucleus) and mean rate of nu-clear pore formation, although considerably less than the rates achieved by *Xeno-pus,* are both on a par with those of *Acheta*.

In multicellular organisms, both plants and animals, the growth of giant so-matic cells is generally accompanied by the development of endopolyploid or polytene nuclei (reviewed by Nagl 1978), and rates of RNA synthesis are thereby raised by the multiplication of TUs. Being a unicellular organism *Acetabularia* cannot grow to giant size by this route, for its primary nucleus is in the germ line. There has been much discussion as to the precise status of the primary nucleus of *Acetabularia* and this problem has yet to be resolved. Although Spring et al. (1975) illustrate some lampbrush chromosomes that might be interpreted as biva-lents, others appear to be single chromosomes. They estimate the DNA contents of *A. mediterranea* gametes to be 0.92 pg, of zygotes to be 1.85 pg, while three es-timates of the DNA contents of primary nuclei gave 2.0, 2.6, and 3.1 pg per nu-cleus. Spring et al. (1978) have shown that there are about 1900 repeats of rDNA genes per haploid genome in *A. mediterranea,* sufficient to account for the ob-served numbers of rDNA TUs if the primary nucleus were diploid (by implica-tion, presumably 2C). Some nucleoli and some tandem arrays of rDNA TUs have been found associated with chromosomes, i.e. they are active nucleolar organizers integral with the genome, but rDNA amplification has not been excluded, and might account for the DNA values above 2C that Spring et al. (1978) estimated in primary nuclei. The problem is further compounded by estimates of the number of lampbrush chromosomes, mostly in the range 18 to 21, with a maxi-mum of 24, in primary nuclei, whereas the haploid chromosome number is given as 10 in Koop et al. (1979); if the lampbrush chromosomes were bivalents then this should also be their number.

Synapsis and synaptonemal complexes have not been observed in *Acetabu-laria* at any stage. Spring et al. (1975) and Koop et al. (1979) favour the interpre-tation that meiosis beings immediately after germination of the zygote, and that the primary nucleus represents a long-extended stage of meiotic prophase. De-spite the apparent discrepancy in chromosome number and apparent univalent nature of most of the lampbrush chromosomes (single nearly terminal chiasmata might explain the latter), the interpretation requiring the least number of unlikely assumptions is that the primary nucleus is at diplotene.

I wish also to mention at the close of this chapter a claim for the existence of lampbrush chromosomes in a large uninucleate deep-sea radiolarian *Plankto-netta atlantica* (Cachon and Cachon-Enjumet 1965). This organism has an enor-mous lens-shaped nucleus 400 μm wide and 100 to 150 μm deep. Amongst what appear to be hundreds of nucleoli are structures that look decidedly like lamp-brush chromosomes, and they are so interpreted by the authors. Little is known about the life history of these creatures; the structure of their chromosomes de-serves further study.

CHAPTER 9

Proteins Associated with Lampbrush Chromosomes

Amphibian oocytes offer many advantages for studying nucleo-cytoplasmic interaction, the nature and properties of the nuclear membrane, the protein constituents of cell nuclei, the proteins involved in RNA transcription, and those that accompany RNA to the cytoplasm, some of which are concerned with RNA stockpiling and perhaps with the masking of mRNAs that are not translated until oocyte maturation has occurred. How these virtues have been exploited has recently been reviewed by Scheer and Dabauvalle (1985). For studying nuclear proteins there are two particular advantages. Potential cytoplasmic contaminants can be eliminated by manual isolation in saline, and rapidly; for "easy" species this operation takes about 30 s. Some loss of soluble protein occurs even in so short a time (Paine et al. 1983) but substantial loss is avoided (Macgregor 1962). Amphibian oocyte nuclei are exceptionally large. According to Paine et al. (1983) a single *Xenopus* nucleus from an oocyte approaching maturity, with a water content of about 90%, contains about 5 µg of protein. According to Maundrell (1975) a half-grown 0.8 mm diam oocyte of *Triturus c. carnifex,* at a stage when the lampbrush chromosomes have reached their maximum bulk, contains about 1.2 µg of nuclear protein, of which more than 1 µg represents the contribution from nuclear sap, 100 ng is nucleolar, and 30 ng is chromosomal. However despite the exceptionally large amount of protein associated with lampbrush chromosomes, analysis of the constituent polypeptides by direct means has so far merely established their complexity. This is in marked contrast to the wealth of information that has come from experimental studies where various materials have been injected into oocytes, notably those of *Xenopus,* the host oocytes cultured in vitro, and thereafter subjected to analysis by various sophisticated techniques. The virtues of the oocyte as an experimental system were first spectacularly exploited by J. B. Gurdon and his many colleagues, and more recently other groups of cell and developmental biologists have done likewise. The field has been extensively and repeatedly reviewed, and for the most part lies beyond the scope of this monograph. Nevertheless some general conclusions arising from this work are directly relevant, and these will be briefly discussed before turning to the main topic of this chapter.

One group of these studies indicates that *Xenopus* oocytes will translate any exogenous mRNAs with which they are provided. When *Xenopus* oocytes were injected with various heterologous mRNAs, for example with α- and β-globin mRNAs purified from rabbit reticulocytes, together with haemin, they synthesized rabbit, not *Xenopus,* haemoglobin (Gurdon et al. 1971, Lane et al. 1971,

Moar et al. 1971). Injection of other heterologous mRNAs gave similar results, i.e. proteins appropriate to the source from which these injected mRNAs had been extracted (reviewed by Gurdon 1973). Evidently these injected mRNAs are translated on *Xenopus* ribosomes, and they do not include constituents that might have activated the transcription of *Xenopus* genes, e.g. globin genes, for *Xenopus* haemoglobin was not detected.

At first glance these observations suggest that *Xenopus* oocytes may not exercise control over the translation of their endogenous mRNAs, but this assertion is not valid. Whereas the injected materials were purified mRNAs, endogenous mRNAs are already complexed with protein as they are transcribed and are not necessarily in a state available for translation. If the synthesis of a particular protein within an isolated normal oocyte can be demonstrated, then the mRNA coding for this protein must have been transcribed on lampbrush chromosomes. But if the synthesis of a particular protein cannot be established, it may be nonetheless present but below the limit of detectability, or its mRNA may have been transcribed but "masked" and not translated.

In 1968 Gurdon showed that when the nuclei of gently lysed somatic cells, from embryos and adult brain of *Xenopus,* were injected into *Xenopus* oocytes maintained in vitro, they enlarged, their nucleoli enlarged, their chromatin dispersed, and in sectioned oocytes the nucleoplasm of the injected nuclei was found to be markedly acidophilic, like that of the host cell's germinal vesicle but unlike these nuclei prior to injection. Mid-blastula nuclei so injected, which normally synthesize DNA but little if any RNA, were found to cease DNA synthesis within half an hour of injection, and instead engage in RNA synthesis. This induction was shown to be the result of the entry of material from the cytoplasm of the oocyte, for no enlargement of the injected nuclei nor establishment of RNA synthesis occurred in the nuclei of cells that by accident had not been lysed, i.e. were confined within intact plasma membranes. The entry of proteins into nuclei in oocytes is therefore not dependent on the lampbrush status of the germinal vesicle's own chromosomes, for proteins also accumulate in somatic nuclei inserted in oocyte cytoplasm.

Later injection experiments of this kind, but involving cell nuclei not originating from the same species as the host oocyte, have given further information on such nucleo-cytoplasmic interaction, and the control of transcription. Gurdon et al. (1976) and De Robertis et al. (1977) injected human HeLa nuclei into *Xenopus* oocytes, and using improved methods for preparing the nuclei prior to injection (Gurdon 1976) found that they synthesize RNA continuously for up to 1 mo. ost oocytes containing HeLa nuclei were incubated in the presence of labelled amino acids, and after incubation were homogenized; their soluble proteins were then separated on two-dimensional gels, and the gels fluorographed. Comparison of fluorographs from uninjected and injected oocytes showed three extra proteins present in the injected oocytes, and these were demonstrated to be HeLa proteins, i.e. encoded by HeLa mRNAs but translated on *Xenopus* ribosomes. However HeLa nuclei resident in *Xenopus* oocytes do not encode all the proteins that are normally synthesized by HeLa cells, for in fluorographs from HeLa cells grown in culture at least 16 major proteins could be distinguished from *Xenopus* oocyte proteins; of these only the three mentioned above were found.

A decisive further step in this analysis was reported by De Robertis and Gurdon in 1977. Fluorographs of two-dimensional gels of labelled proteins from *Xenopus* oocytes and *Xenopus* cultured somatic cells were compared. The great majority of the proteins so separated and detectable proved to be common to both oocytes and cultured cells, but at least 16 major proteins were detected only in oocytes, and at least 8 major proteins were detected only in cultured cells. Fluorographs of two-dimensional gels of labelled proteins from *Pleurodeles* oocytes conveniently presented a pattern of spots that could be distinguished from fluorographs of *Xenopus* oocyte proteins. *Xenopus* nuclei from cultured cells were injected into *Pleurodeles* oocytes, where they enlarged in the habitual way. After several days in culture these *Pleurodeles* oocytes were incubated with labelled amino acids, and two-dimensional gels were run and fluorographed. At least six *Xenopus*-specific proteins were detected, three of which are common to oocytes and cultured cells, the other three being present in *Xenopus* oocytes only. These latter three proteins provide evidence that in *Xenopus* culture cell nuclei there are inactive genes that can, however, become activated when molecules from the oocyte host environment enter these nuclei.

A feature of protein migration into the nucleus of an oocyte is the selective manner with which some proteins concentrate in the nucleus, some do not concentrate but nevertheless enter, while some do not enter at all. In 1970 Gurdon reported that when labelled histones or bovine serum albumin are injected into the cytoplasm of oocytes maintained in vitro, both substances enter the oocyte nucleus, but whereas by 24 h the relative concentration of serum albumin is about twice as high in the cytoplasm as in the nucleus, histones accumulate in the nucleus and reach over 100 times the concentration that is present in the cytoplasm.

De Robertis et al. (1978) incubated *Xenopus* oocytes with labelled amino acids and compared fluorographs of two-dimensional gels prepared from whole oocytes, isolated germinal vesicles and cytoplasm. Some 8 major proteins (including actin) were found to be common to nucleus and cytoplasm, some 11 were present in the nuclear fraction only, including several that are extremely acidic, with isoelectric points at pH ≤ 4.5, and a much larger number (including tubulin) that were detectable in the cytoplasmic fraction only.

Oocytes were labelled by incubation with ^{35}S-methionine for 24 h, then placed in unlabelled medium for a further 24 h, and their germinal vesicles manually isolated, cleaned of cytoplasm and assembled in saline. The membranes of these labelled germinal vesicles were physically disrupted, and after standing for 1 h the preparation was centrifuged to produce a pellet consisting of lampbrush chromosomes, nucleoli and nuclear membranes, and a supernatant, with > 90% of the total radioactivity, that contained soluble materials. Portions of the supernatant were now injected into the cytoplasm of unlabelled oocytes, which were kept in culture for 24 h. Some of these oocytes were fixed, sectioned and autoradiographed, and they showed (as had been demonstrated already by Bonner in 1975) that most of the radioactivity was distributed uniformly in the nucleus, presumed therefore to indicate that the labelled proteins were free in the nuclear sap, not bound to chromosomes or nucleoli, with little radioactivity in the cytoplasm. Soluble proteins in isolated germinal vesicles and in cytoplasmic fractions from other of these injected oocytes were subjected to two-dimensional gel analysis, and the gels fluorographed. The germinal vesicles showed essentially the same spot pattern as that produced by the labelled germinal vesicle fraction which had been used for injection, except for a marked diminution in the relative amount of labelled actin, while conversely the cytoplasmic fraction showed the presence of labelled actin, but none of the major "nuclear" proteins were detectable. This series of experiments has demonstrated the selectivity with which certain soluble proteins can pass across the oocyte nuclear membrane and accumulated inside, whereas others are excluded, and this selectivity is not determined by the overall charge on the molecules, for some of the nuclear proteins are extremely acidic, others neutral, others (e.g. histones) extremely basic, nor is it determined by the sizes of the molecules.

It is against this background information that chromosomal proteins should now be considered. In some early attempts to analyze in direct fashion the proteins that are associated with lampbrush chromosomes, R.J. Hill et al. (1973, 1974), working with *Triturus c. carnifex,* exposed and solved the problem of how to solubilize them. Nuclei were manually isolated and cleaned in saline. Centrifuged lampbrush preparations were made from some isolated nuclei, while others were collected in bulk. Some of the latter were not fractionated, 25 nuclei per sample, while others were centrifuged to give a "chromatin" pellet and supernatant nucleoplasm. The centrifuged lampbrush preparations were observed during exposure to solubilizing agents. Concentrated guanidine hydrochloride/ sodium pyrophosphate, pH 8.1, failed to put lateral loop matrix into solution, but did so when accompanied by the reducing agents mercaptoethanol or dithiothreitol. Evidently disulphide bond cross-links are present, probably formed by oxidation during preparation, and these need to be broken to render the proteins soluble.

The bulked samples were treated with pancreatic RNase to disengage protein from RNA, the proteins were then solubilized with guanidine hydrochloride/ sodium pyrophosphate/mercaptoethanol, and sulphydryl groups were alkylated to prevent disulphide bonds from re-forming. Gel electrophoresis was carried out in 7 M urea at pH 9, or in sodium dodecyl sulphate (SDS) at pH 7.1. Both methods gave consistent results, but the latter gave better resolution. SDS gels from total nuclear samples showed some 25 polypeptide bands, ranging in molecular weight from about 10,000 to over 150,000. Gels of oocyte nucleoplasm samples gave similar electropherograms, but with significant depletion of two proteins. Similar gels of oocyte "chromatin", i.e. containing nucleoli and nuclear membranes as well as lampbrush chromosomes, resolved about 20 polypeptides, with two major components of molecular weights approximately 43,000 and 110,000, corresponding to those of the proteins that were underrepresented in gels of nucleoplasm. These were shown to originate from nucleoli. In oocytes of 0.6 to 0.7 mm diam almost all of the nucleoli are attached to the nuclear membrane, and this enables nucleoli plus membranes to be separated from nucleoplasm plus lampbrush chromosomes during the collection of samples. In oocytes of 1.5 to 1.7 mm diam the nucleoli are no longer attached to the nuclear membrane, and this enables collections of membranes without nucleoli to be made. Gels from comparable samples of membranes plus nucleoli and of membranes only showed the two major proteins to be present in the samples containing nucleoli, while no bands were detectable in the samples containing membranes only.

Maundrell (1975) continued this direct approach, and besides confirming the observations of R. J. Hill et al. (1973, 1974) he was able, using improved solubilizing and electrophoretic procedures, to detect about 20 polypeptide bands in gels from both nucleoli and lampbrush chromosomes. The pattern of bands from lampbrush chromosomes was consistent from one sample to another, 100 chromosome complements being collected for each, the samples were shown to be essentially free from nucleoplasmic contamination, and the pattern of bands different from those of gels containing nucleolar proteins, though in both the molecular weights of separated proteins range from about 40,000 to well over 200,000. There are several abundant nucleoplasmic proteins in the molecular weight range 20,000 to 32,000; these are virtually absent from the gels of samples of both nucleoli and lampbrush chromosomes. However "contamination" by nucleoplasmic proteins must be recognized as more than a technical problem. All the proteins of lampbrush chromosomes have to traverse the nuclear sap to reach their destinations, and transcript RNP when released from the lateral loops does not at once leave the nucleus; while still aggregated in macromolecular complexes it sediments with the chromosomes. "Natural" contamination necessarily results from this two-way traffic, so however rigorous the preparative procedures proteins common to both nuclear sap and chromosomes are to be expected. The sensitivity of the direct approach could be improved by working with lampbrush chromosomes from oocytes that had previously been cultured in vitro in media containing radioactive labelled amino acids, but it has not been actively pursued. Instead indirect immunological and other techniques have been employed with some success.

Immunological studies on lampbrush chromosomes were begun by Sommerville in 1973 and reviewed in 1981. In the first experiments oocytes of *T.c. carnifex*, whose recently synthesized RNA and protein had been radioactive labelled, were homogenized, and particulate RNP isolated in bulk on sucrose gradients. Prior to gradient centrifugation, gross potential contaminants of chromosomal RNP, notably nucleoli, had been pelleted by centrifugation at low speed, chromosomal RNP being recovered from the supernatant. The particulate RNP was identified as having originated from the lampbrush chromosomes on various criteria: incubation of some oocyte samples with actinomycin D before homogenization, which causes RNP in the process of synthesis to be shed from the lateral loops, substantially increased the radioactivity of the particulate RNP fraction; and antiserum raised in rabbits against the protein component of this RNP was shown by immunofluorescence microscopy to bind to all the lateral loops in centrifuged preparations of lampbrush chromosomes, but not to nucleoli.

Sommerville and R. J. Hill (1973) solubilized and stabilized the proteins of this particulate RNP by the method mentioned earlier in this chapter (R. J. Hill et al. 1974) and electrophoresis on SDS-acrylamide gels produced polypeptide profiles in reasonable accord with those obtained in direct fashion by Maundrell (1975), with seven main bands ranging down in molecular weight from about 150,000, and with few or no polypeptides below 30,000. The samples were not contaminated by nucleolar protein, as shown by the absence of the two abundant polypeptides of molecular weights approximately 43,000 and 110,000 that are characteristic of nucleoli. The immunological approach was taken a step further by Scott and Sommerville (1974), who used as antigens fractions of RNP protein separated on a molecular weight basis. Antisera to three fractions bound to all lateral loops, and because two of these fractions did not overlap in molecular weights they concluded that at least two different proteins are common to all loops.

Sommerville et al. (1978 a) refined this approach by preparative isoelectro-focussing of solubilized chromosomal RNP protein to produce fractions including fewer polypeptides, two of the fractions containing single polypeptides of molecular weights about 32,000 and 42,000. The fractions covered much of the size and charge spectrum of RNP polypeptides, whose considerable complexity is apparent in a two-dimensional gel fluorograph (Sommerville 1979). Twelve fractions elicited antisera that bound to all loops (Fig. 9.1 a, b), implying that many polypeptides are common to all transcript RNPs. By contrast an antiserum to histones extracted from *Triturus* erythrocytes produced immunofluorescent staining of chromomeres (Fig. 9.1 c). Some of the lampbrush RNP polypeptides that are present on all the loops are antigenically related to the ubiquitous core proteins of mammalian (mouse) hnRNP, with molecular weights in the range 34,000 to 40,000, for Martin and Okamura (1981) have demonstrated that antibodies raised against these proteins likewise bind to all the loops of the lampbrush chromosomes of *Notophthalmus viridescens*.

These observations were all made by light microscopy; they have recently been confirmed by electron microscopy (Dr. U. Scheer, personal communications, 1985). Lampbrush chromosomes of *Triturus cristatus* were dispersed in 100 mM saline at neutral pH and attached to slides by centrifugation. In such preparations the transcripts retain their natural organization as clustered strings of RNP granules, not extended as they are in Miller-spreads. The preparations were exposed to antibodies raised against hnRNP core proteins, or against histone H2B, and then exposed to secondary antibodies that had been coupled to colloidal gold. After this treatment the preparations were fixed, end-embedded and sectioned. In those that had been exposed to anti-hnRNP core proteins the gold particles were restricted to transcript granules (Fig. 9.2 a), absent from the chromomeres, whereas in preparations that had been exposed to anti-histone H2B the gold particles lay over exposed regions of loop axis (Fig. 9.2 b).

For several years there has been argument (see Weintraub et al. 1976, Wasylyk et al. 1979) as to whether histones are physically displaced from DNA at the moment of transcription or, as was implied in Chap. 4, the unfolding and re-

---

Fig. 9.1 a–c. Lampbrush chromosomes of *Triturus c. carnifex.* **a** *Left:* phase contrast; *right:* immunofluorescence micrographs of a preparation exposed to antibodies against primary transcript (RNP) proteins in the molecular weight range 40,000 to 50,000. The fluorescence is restricted to the lateral loops. **b** Immunofluorescence micrograph of a preparation exposed to antibodies against a single primary transcript protein of mol. wt. 42,000. Towards the *right* is part of a long loop whose matrix includes beaded aggregates of RNP; they are conspicuously fluorescent. The *inset* above left, at lower magnification, shows these same structures in phase contrast. **c** *Left:* phase contrast; *right:* immunofluorescence micrographs of a preparation exposed to antibodies against histones extracted from *Triturus* erythrocytes. The fluorescence predominantly originates from the chromomeric chromosome axis. *Bars =* 10 μm. (From Sommerville et al. 1978 a)

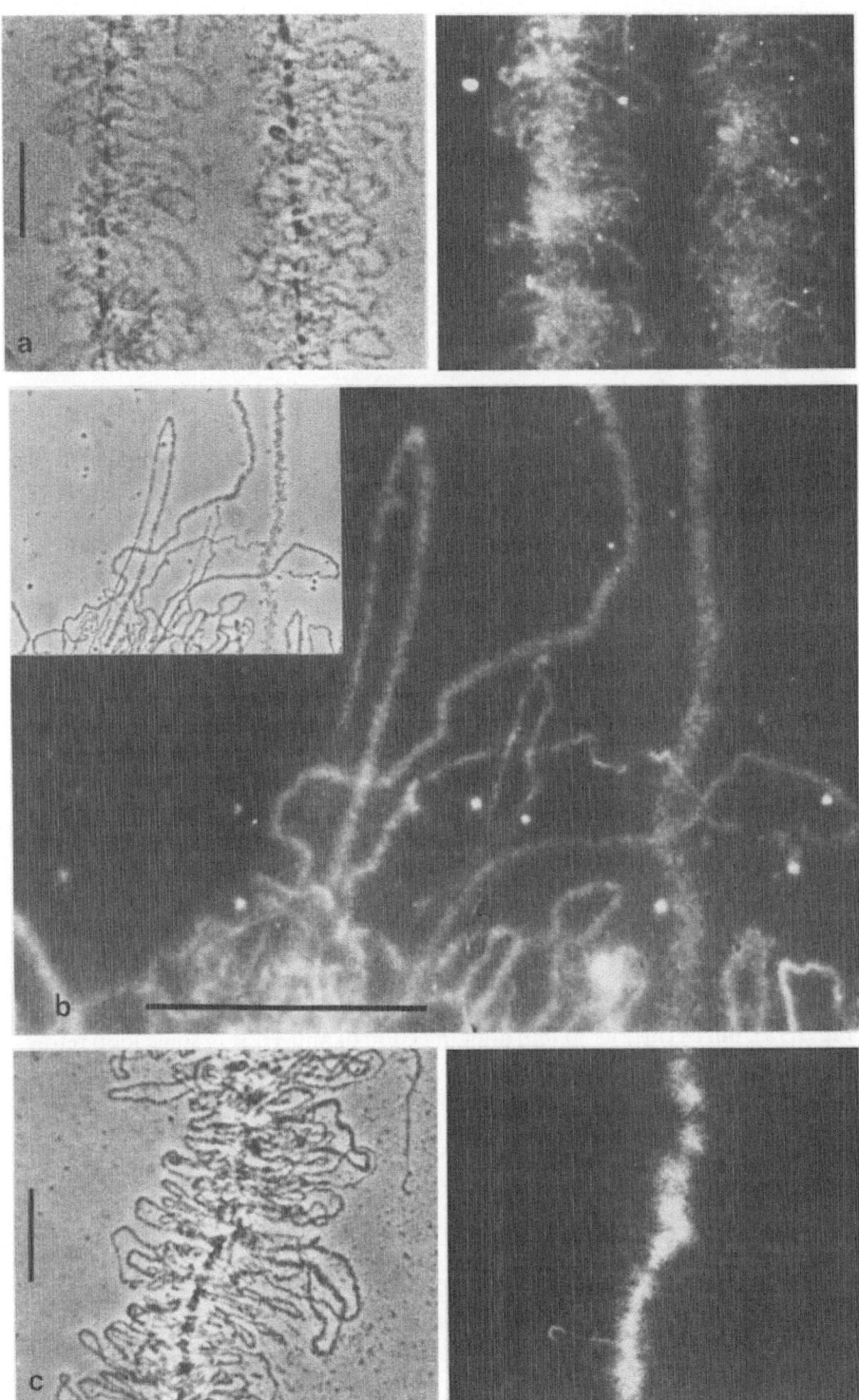

Fig. 9.1 a–c

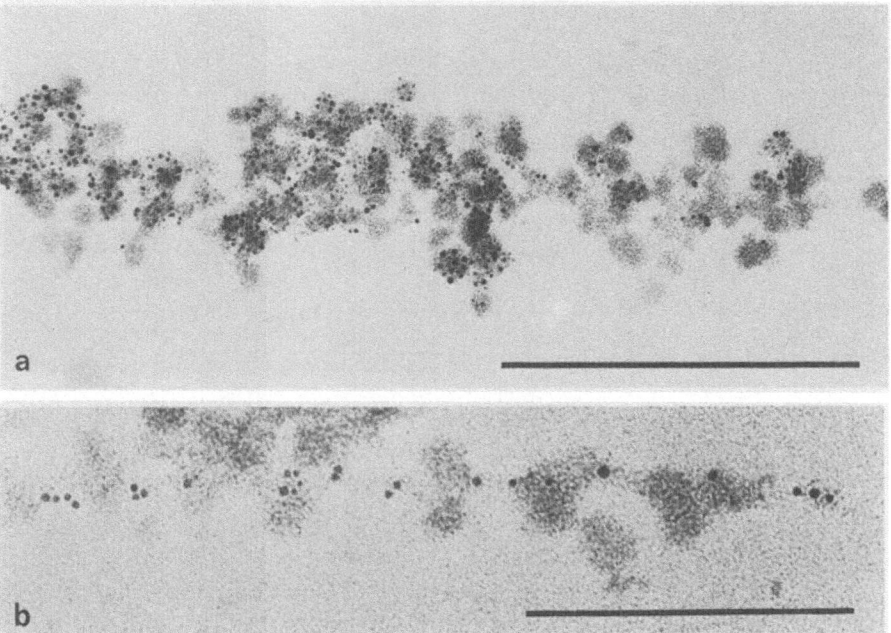

Fig. 9.2 a, b. Thin sections of centrifuged and end-embedded lampbrush chromosomes of *Triturus cristatus* showing portions of lateral loops. **a** Exposed to monoclonal mouse anti-hnRNP core proteins, then to anti-mouse immunoglobulin coupled to 5 nm gold particles, before fixation. *Bar*=0.5 μm. **b** Exposed to rabbit anti-H2B, then to anti-rabbit immunoglobulin coupled to 5 nm gold particles, before fixation. *Bar*=0.2 μm. Both micrographs courtesy of Dr. U. Scheer

formation of nucleosomes during and after transcription occurs as a conformational change, without dissociation of histone during the unfolding event. Further evidence for the continued association of histone H2B with actively transcribing lateral loops of lampbrush chromosomes has come from experiments where immunoglobulin (Ig) solutions were injected into the nuclei of oocytes of *Pleurodeles* (Scheer et al. 1979 b). 3 h after injection of non-immune or anti-RNP Igs, isolated lampbrush chromosomes appeared just like those from non-injected oocytes (Fig. 9.3 a), but as early as 30 min after injection of anti-H2B Ig some loop retraction was evident, and by 3 h was virtually complete (Fig. 9.3 b). Presumably anti-H2B binds to loop axis histone, present even when RNA polymerase density is high, and acting in vivo it interferes with the progress of polymerases much as does actinomycin-D bound to DNA, RNP transcripts are released, and the usual loop retraction follows.

As was discussed in Chap. 5, RNA synthesis on the lateral loops of lampbrush chromosomes is inhibited by very low concentrations of α-amanitin (Morgan et al. 1980, Schultz et al. 1981), indicating that RNA polymerase II is responsible for their transcription. Confirmatory evidence comes from Bona et al. (1981) who injected various antibodies directed against RNA polymerase II into the oocyte nuclei of *Pleurodeles* and *Xenopus*. They found that these antibodies suppressed transcription on the lateral loops, caused nascent transcripts to be shed,

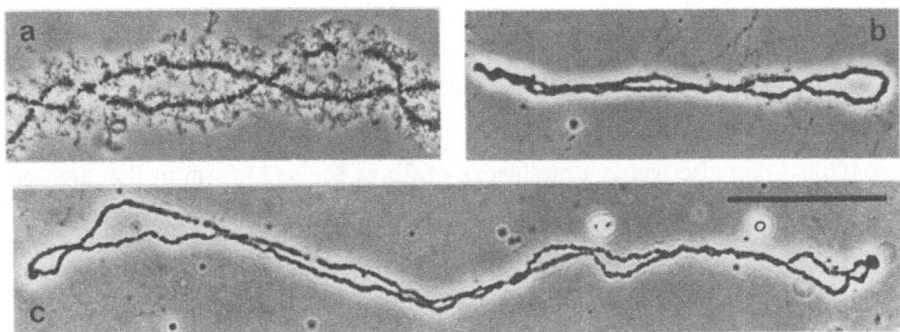

Fig. 9.3 a–c. Lampbrush chromosomes of *Pleurodeles waltlii*. **a** Isolated from an oocyte 3 h after its nucleus had been injected with non-immune rabbit immunoglobulin and **b** from an oocyte that had been injected with anti-H2B. **c** From an oocyte that had been incubated in a medium containing actinomycin D at 20 μg ml^{-1} for 1 h. *Bar* = 50 μm. **a, b** From Scheer et al. (1979 b); **c** from Bona et al. (1981)

and the loop axes to retract, again much as occurs when oocytes are incubated with actinomycin-D (Fig. 9.3 c), whereas the activities of RNA polymerases I and III were not significantly affected.

Injections of two other antibodies into the nuclei of *Pleurodeles* oocytes produce a similarly specific disruption of transcription by RNA polymerase II. One of these (Kleinschmidt et al. 1983) is anti-bovine HMG-1, one of a family of high mobility group proteins that is normally bound to chromatin in calf thymus. A large pool of a related protein HMG-A, molecular weight about 25,000 and heterogeneous isoelectric point in the range pH 7 to 9, is present in solution in oocytes of several amphibian species, at tenfold the concentration in the nucleus as compared to the cytoplasm. Control injections were ineffective, some including antibodies to nuclear RNP, tubulin or nucleoplasmin (nucleosome assembly factor). HMG-A is therefore presumed to be concerned in some manner with RNA polymerase II transcription, though precisely how it is involved is not yet known. Injections of another antibody, anti-actin, and of two different actin-binding proteins, likewise disrupt transcription on lateral loops (Scheer et al. 1984). I will return to consider the status of actin later in this chapter.

However some caution is advised before accepting at its face value evidence for the specific suppression of transcription, and loop retraction, by the injection of antibodies and other substances into oocytes. The oocytes of some Amphibia, notably *Pleurodels* and *Xenopus,* will tolerate such injections, both control and experimental, with apparent impunity, whereas the oocytes of other species are extremely sensitive to comparable injections and transcription can be disrupted non-specifically. *Triturus cristatus* is regrettably one such species, and because of its sensitivity to experimental insults it cannot be used for studies of this kind. There may even be some problems of this nature with *Pleurodeles* oocytes, for Bona et al. (1981) demonstrated loop retraction following the injection of α-amanitin into oocytes, whereas Schultz et al. (1981) were able to culture oocyte nuclei in an incubation medium containing up to 200 μg mg^{-1} α-amanitin for several hours, which halted transcription by RNA polymerases II and III, but not I, yet the lateral loops of the lampbrush chromosomes remained extended and covered with RNP matrix.

When pancreatic RNase is added to a freshly made preparation of lampbrush chromosomes of *T. cristatus* that have been isolated in saline, the first effect observed is the stripping of RNP matrix from the lateral loops (Macgregor and Callan 1962). Soon afterwards a meshwork of fibres begins to form in that area of the preparation previously occupied by the nuclear contents, and becomes conspicuous as it builds up and the fibres thicken. The second phenomenon was overlooked when the original observations were recorded, or it may have been dismissed as an unspecific coagulation artefact. At all events it is now known to be directly connected with RNase digestion, and as the fibres are digested by trypsin they consist of protein. A reasonable assumption is that much of this protein has been released from RNP, both RNP in the process of transcription, attached to lateral loops, and RNP already loose in the nucleoplasm, and has aggregated. The fibres form regardless of whether or not nucleoli are present in the preparation, so they do not originate from the breakdown of nucleolar RNP. That this protein should be capable of assuming a fibrillar state is not surprising in view of the transformation of granular RNP transcripts into long fibrils in Miller-spreads of lampbrush chromosomes made at low ionic strength.

With this phenomenon in mind the observations of Kloetzel et al. (1982) are of interest. Particulate, rapidly labelled RNP from oocytes of *T.c. carnifex* was isolated by the method devised by Sommerville (1973) mentioned earlier in this chapter. It showed the low density typical of this material, with a protein to RNA mass ratio of about 30:1. Under the electron microscope the granular aggregates were seen to be associated with a network of fibrils, mostly about 10 nm wide but branching into finer filaments. A fibrillar network fraction and a particulate RNP fraction were separated from one another. The polypeptides of both fractions were subjected to gel electrophoresis: the fibrillar fraction showed greater complexity than the particulate fraction, but four major bands were common to both. Polypeptides from the fibrillar fraction were eluted from gel slices and samples examined by electron microscopy. Four major components were found to have formed fibrillar networks spontaneously. The other major polypeptides did not. The most extensive fibrillar networks were formed by a polypeptide of mol. wt. 60,000, this being one of those common to both fibrillar and particulate fractions.

When hnRNA extracted from *Triturus* oocytes was added to the fibrillar fraction polypeptides in the molecular weight range 51,000 to 60,000 it bound with great efficiency, but when examined in the electron microscope this in vitro assembled RNP was found to be particulate, not fibrillar. It appears then that the proteins present in transcript RNP can exist in alternative forms, particulate when associated with hnRNA, or fibrillar when they polymerize in the absence of RNA. Kloetzel et al. (1982) consider that these proteins in their polymerized state form the "nuclear protein matrix" that some authors consider to be a structural network in many eukaryotic cell nuclei. Similar claims have been made for actin, specifically in amphibian oocyte nuclei (Clark and Rosenbaum 1979), yet actin was not identified among the RNP proteins studied by Kloetzel et al. (1982).

Using an immunoperoxidase technique, Karsenti et al. (1978) claim to have demonstrated that actin is associated with the chrommomeres and tubulin with the lateral loops of *Pleurodeles* lampbrush chromosomes. The evidence for this assertion is questionable on two counts. De Robertis et al. (1978) detected tubulin in the cytoplasm of *Xenopus* oocytes, but not in their nuclei, whereas actin in solution is present in abundance in oocyte nuclei and is therefore a potential contaminant of isolated lampbrush chromosomes. The status of tubulin remains problematic, but that of actin has been clarified by Scheer et al. (1984). When antibodies to actin are injected into the nuclei of vitellogenic oocytes of *Pleurodeles,* transcription on the lateral loops is inhibited, transcripts detach from loop axes, loops retract and chromosome axes condense. Similar inhibition results from the injection of actin-binding proteins. The effects are much like those induced by the injection of anti-H2B, anti-RNA polymerase II or anti-bovine HMG-1 described earlier in this chapter. Scheer et al. (1984) went on to demonstrate by immunofluorescence microscopy that actin is present in the meshwork of 5 nm microfilaments which form around lampbrush chromosomes when transcription has been inhibited by drugs, or when their RNP has been digested by injected pancreatic RNase. It may well be the major component, for when the F-actin severing protein "fragmin" was injected into nuclei from *Pleurodeles* oocytes previously incubated with actinomycin-D, no fibrillar meshwork was present, thus confirming the claims of Clark and Rosenbaum (1979) and Gounon and Karsenti (1981) that this meshwork consists of bundles of F-actin.

Thus far I have discussed the status of proteins that are widely distributed over the lampbrush chromosomes of Amphibia. Is there evidence for the existence of proteins that are not widespread, associated with transcription by some components of the genome, but not with all? The textural peculiarities of various landmark structures, and the specificity with which some of them fuse, not only with homologous, but also with non-homologous objects, as described in Chap. 3, suggested that this might be so. Another pointer in this direction was the observation by Varley and Morgan (1978) that when the lampbrush chromosomes of *T.c. carnifex* are stained by the ammoniacal silver technique not only the extra-chromosomal nucleoli, but many of the landmark loops and a few loops that are otherwise unremarkable are "silver-positive". And amongst these latter there are examples in which the silver staining is regularly restricted to one part of a loop only, reminiscent of partial labelling in in situ hybridized preparations. Confirmatory but much stronger evidence for locus-specific proteins has come from immunological studies.

Using immunofluorescence microscopy, Scott and Sommerville (1974) found that an antiserum to one of their RNP protein fractions, with a predominant polypeptide of about 35,000 mol. wt., was unlike their other antisera in that it bound to only ten pairs of loops of *T.c. carnifex,* and throughout their lengths. When examined in phase contrast these particular loops looked much like their neighbours as regards matrix texture, yet their matrix must include a protein that is not present in other loops.

Sommerville et al. (1978 a) described some observations on polypeptides that are associated with 5 S and 4 S RNA. The cytoplasm of previtellogenic oocytes of *T.c. carnifex* includes an abundant RNP storage particle that sediments at

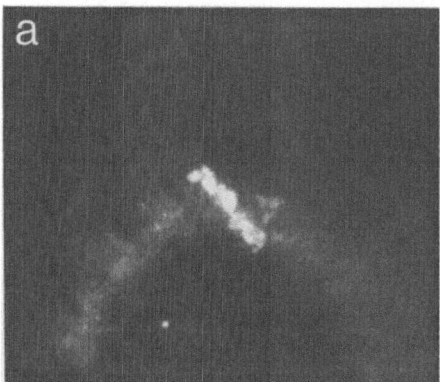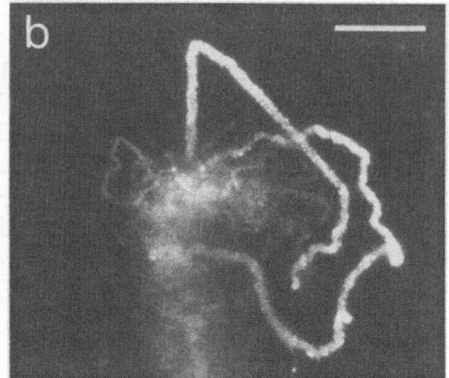

Fig. 9.4 a, b. Immunofluorescence micrographs of lampbrush chromosomes of *Triturus c. carnifex* after exposure to antibodies to the two polypeptides present in 40 S RNP storage particles. **a** To the larger polypeptide of 49,000 mol. wt.; the antibodies bind specifically to a dense loop pair on chromosome X. **b** To the smaller polypeptide of 38,000 mol. wt.; the pair of loops that fluorescence strongly are amongst several to which these antibodies bind. *Bar* = 10 μm. (From Sommerville et al. 1978 a)

about 40 S. The particles contain 5 S and 4 S RNA and two polypeptides of mol. wts. 49,000 and 38,000, both extremely basic, with isoelectric points at pH >9 and therefore unlike most of the polypeptides present in chromosomal RNP. These two polypeptides were used as antigens, and the antisera they elicited in rabbits were applied to preparations of lampbrush chromosomes. Antiserum to the larger polypeptide bound to only one loop pair, situated near the middle of the longer arm of chromosome X, the loop in question having dense compact matrix and being one of the landmark loops on this chromosome (Fig. 9.4 a). The location tallies with the homologous site of hybridization of labelled 5 S RNA to mitotic chromosomes, and also to the homologous site on lampbrush chromosomes, but here the hybridization is to a chrommomeric region, not to loops. This latter observation is in accord with that of Barsacchi Pilone et al. (1974) on the related newt *T. marmoratus,* where hybridization of labelled 5 S RNA to the lampbrush chromosomes also occurs at a chromomeric region adjacent to a loop pair with dense matrix in the longer arm of chromosome X.

The proximity of a chromomere where 5 S RNA is transcribed (see Chaps. 4 and 5) to a loop where a polypeptide of mol. wt. 49,000 is concentrated, a polypeptide that is specifically associated with 5 S RNA, can sparcely be accidental. On the simplest assumption this loop functions as a storage site ready to provide nascent 5 S RNA with its appropriate protein partner, a partnership that is maintained when 5 S RNA migrates to the cytoplasm.

Antiserum to the 38,000 mol. wt. polypeptide present in the 40 S cytoplasmic storage particle binds to several pairs of loops on different chromosomes (Fig. 9.4 b). These loops are relatively long and they carry more abundant RNP matrix than neighbouring loops. Sommerville et al. (1978 a) recognized that the dimensions of these loops are incompatible with what would be anticipated if they carried tRNA transcripts (just as for 5 S RNA, see the discussion in Chaps. 4 and 5). Though there is no hard evidence by which to judge, and their precise chromosomal locations have not been determined, these loops may be related to chromomeric sites of tRNA synthesis much as proposed for 5 S RNA and its partner protein, i.e. accumulation of the 38,000 mol. wt. polypeptide in the vicinity of nascent tRNA, ready to form the specific tRNP complex before its migration to the cytoplasm.

This immunological approach has recently been extended by Lacroix and colleagues, using monoclonal antibodies, and it has already provided valuable information (Lacroix et al. 1985). From hybridomas producing antibodies directed against germinal vesicle proteins of *Pleurodeles waltlii,* five clones or families of clones were selected for their immunological binding to lampbrush chromosomes. Antibody from a clone "A33" binds to all loops except those with a dense, beaded RNP matrix that are striking landmark features occurring on most of the lampbrush chromosomes of *Pleurodeles,* and a few loops with fusing matrix. It does not bind to spheres, chromomeres or nucleoli. A33 cross-reacts in similar fashion with the lampbrush chromosomes of five other urodele species that were tested. In sections through the ovary of *Pleurodeles,* the nucleoplasm of germinal vesicles and follicle cell nuclei bind A33, but not cytoplasm. In sections through embryos from mid-blastula onwards the binding is likewise exclusively to nuclei, and it is evidently connected with transcription, for the nuclei of peripheral cells of the central nervous system that are undergoing differentiation bind A33, but not those of cells of the proliferative zone that gave rise to them. Similarly in testis sections the nuclei of spermatogonia and spermatocytes bind A33, but not those of spermatids of spermatozoa. The protein antigenic to A33 has a mol. wt. of 80,000 and isoelectric point at pH 6.4.

Antibody from a clone "B71" binds to all the lampbrush loops of *Pleurodeles* including those that are exceptional in not reacting with A33 (Fig. 9.5 a). It cross-reacts with the lampbrush chromosomes of four other urodele species that were tested. Like A33, it does not bind to spheres, chromomeres and nucleoli. However when tested on sections through oocytes of *Pleurodeles* its reactivity is strikingly different from that of A33, for it does not bind to the nucleoplasm of germinal vesicles; instead it binds to oocyte cytoplasm, just alongside the nuclear membrane and between yolk platelets (Fig. 9.5 b). In sections through embryos the binding is likewise to cytoplasm, not to cell nuclei. The antigen(s) to B71 have not yet been characterized, but on the evidence given by Lacroix et al. (1985) it would appear that they complex with RNA transcribed on the lampbrush loops of the oocyte, but not that transcribed in the nuclei of the embryo, and accompany maternal RNA as it accumulates in the cytoplasm. As B71 binds to all lampbrush loops of the oocyte this suggests that the antigen(s) play a storage rôle with respect to maternal RNAs in general.

Antibody from a clone "B24" binds to the spheres of *Pleurodeles,* both attached and free, and to an exceptional fused structure present on chromosome IV in *P. waltlii* of Iberian origin. It cross-reacts with the spheres of four other urodele species that were tested. *Triturus marmoratus* lampbrush chromosomes carry no spheres, and in this species no structures bind B24. Binding of B24 could not be detected in sections through ovaries or embryos. The material antigenic to B24 consists of four proteins all of mol. wt. about 104,000 and isoelectric points in the range 6.5 to 6.8.

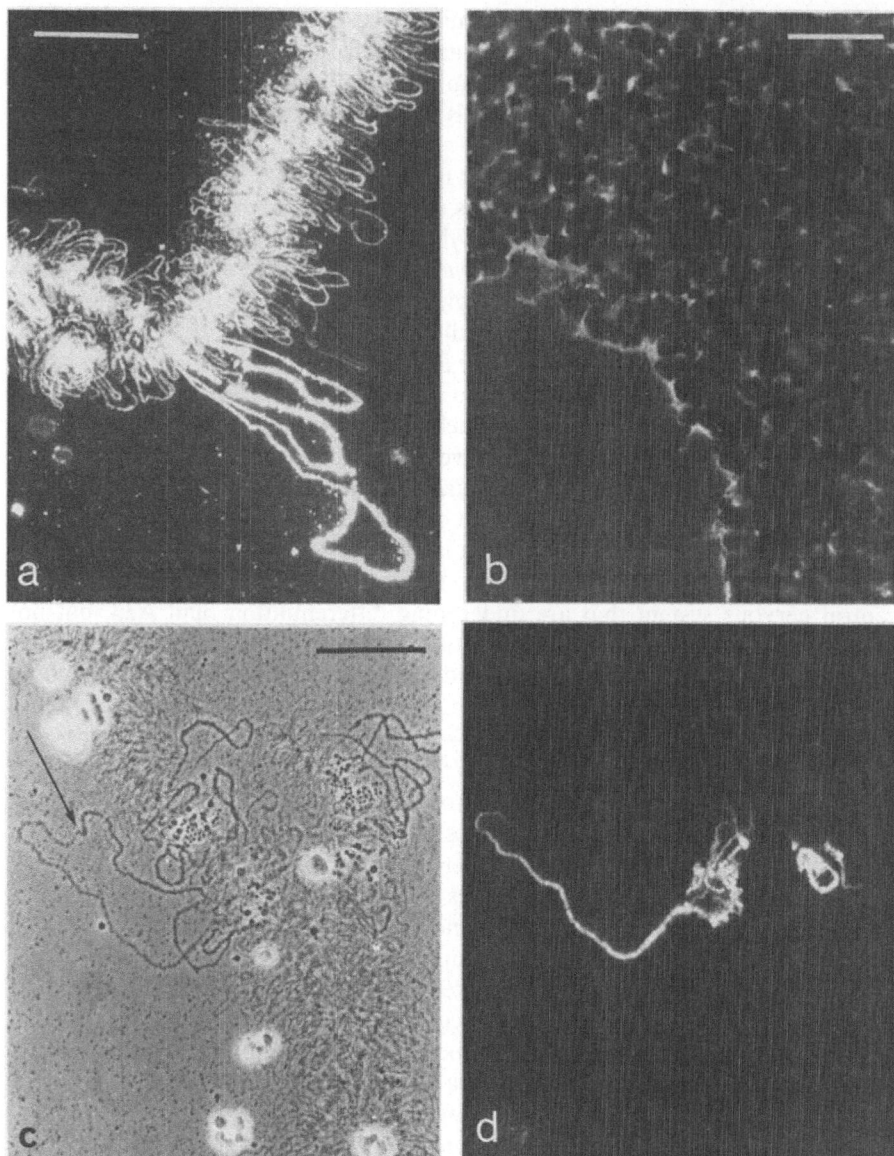

Fig. 9.5. a Immunofluorescence micrograph of a portion of a lampbrush chromosome of *Pleuro-deles waltlii* after exposure to monoclonal antibody "B71". All loops are fluorescent. *Bar* = 40 μm. **b** Immunofluorescence micrograph of a section through part of a large oocyte of *P. walt-lii*, including its nucleus at *bottom left*, after exposure to antibody B71; fluorescence is confined to the cytoplasm adjacent to the nuclear membrane, and between yolk platelets. *Bar* = 40 μm. **c** Phase contrast micrograph including the middle region of lampbrush bivalent VII of *P. waltlii*; the long loop to the *left* includes more than one transcription unit, and an *arrow* points to where one ends and its immediate neighbour begins. *Bar* = 30 μm. **d** Immunofluorescence micrograph of the same chromosome region after exposure to antibody "A1", at the same magnification; only one transcription unit in the long loop fluoresces, and the gradient of its fluorescence dem-onstrates the polarity of its transcription. (From Lacroix et al. 1985)

Antibody from a clone "A1" is of particular interest. It binds to a few tens of loops of *P. waltlii*, most of which lie close to the telomeres of all 12 bivalents, but not to the thousands of other loops or to spheres, chromomeres or nucleoli. Lacroix et al. (1985) found that in certain individuals of *P. waltlii* there is an exceptionally long loop pair on bivalent VII, close to a beaded-loop landmark. About halfway along this particular loop there is a discontinuity in matrix, a thick-thin junction, marking the termination of one TU and the initiation of another (Fig. 9.5c). Using immunofluorescence, antibody A1 was found to bind to that TU whose initiation lay at the thick-thin junction, with a gradient of increasing fluorescence in keeping with the polarity of transcription back towards the chromosome axis. The rest of this loop, though liberally coated with RNP matrix, did not bind A1 (Fig. 9.5d).

Antibody from clone A1 also binds to a few loop pairs in the middle region of bivalent IV of *P. waltlii*. As with the loops mentioned in the previous paragraph, the binding is restricted to parts of these loops only; moreover none of the loops that bind A1 on one chromosome IV match in position with loops on its partner (Fig. 9.6). All are heterozygous, and this is a heteromorphic region. A1 binds in a similar pattern to loops on the lampbrush chromosomes of *P. poireti*, in which species bivalent IV was shown to be sex-determining (Lacroix 1970, and see Chap. 3). The middle region of chromosome W carries a group of beaded-landmark loops that are absent from its partner chromosome Z. In *P. poireti* there is also a regular heterozygosity for loops that bind A1 in the vicinity of the beaded landmark loops, though these latter are not recognized by this antibody. So on this evidence it would appear that bivalent IV of *P. waltlii* is also sex-determining, though its heteromorphism has only become apparent through this immunological study. Homogametic females WW and ZZ have now been produced experimentally; according to Azzouz et al. quoted by Lacroix et al. (1985) these females have identical distributions of loops reacting with A1 on bivalent IV. Antibody from clone A1 cross-reacts with a limited number of loops in four other urodele species that were tested, notably with the giant loops of bivalent II of *Notophthalmus viridescens*. It does not bind detectably to sections of ovaries, embryos or larvae, and its antigen has not been characterized.

Finally, antibody "A35" binds specifically to the chromomeres of *P. waltlii* lampbrush chromosomes, and cross-reacts likewise with those of five other urodele species. The corresponding antigen has not been identified.

The use of monoclonal antibodies to study the proteins of lampbrush chromosomes is evidently on a par with in situ hybridization in its promise to unravel the complexity of these remarkable structures. As yet the information is scanty, and until more is available little speculation is justified. According to evidence from many sources reviewed by Shiokawa (1983), 30% of the mRNA is amphibian oocyte cytoplasm is available for translation, while 70% is masked and stockpiled. The situation is much the same in echinoderm oocytes, and may be widespread amongst animal oocytes, with the possible and if so notable exception of those of mammals (see Chap. 6). For the most part both translated and stockpiled mRNAs are polyadenylated, but whereas the latter are covalently linked to repetitive sequences, the former are not. Shiokawa proposes that mRNAs destined for stockpiling associate with membranous structures derived from the

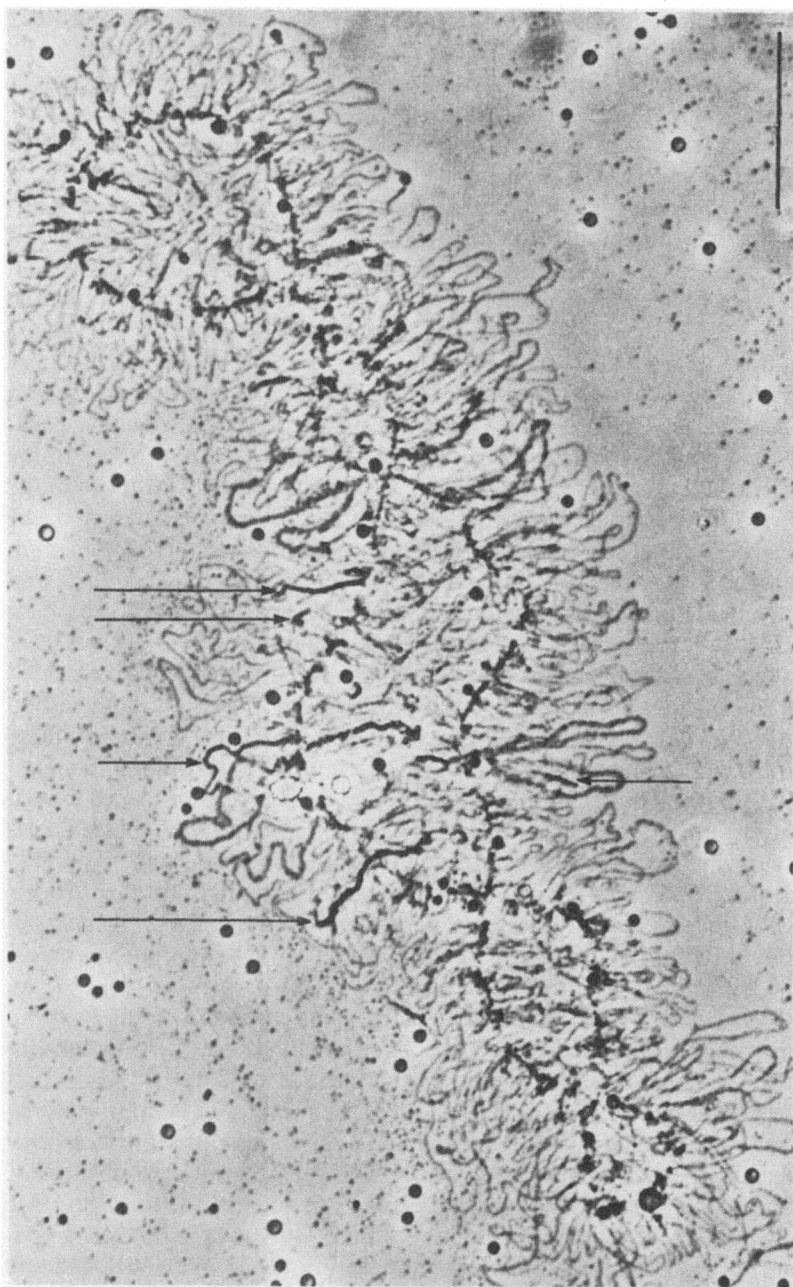

Fig. 9.6. Phase contrast micrograph of lampbrush bivalent IV of *Pleurodeles waltlii* after exposure to monoclonal antibody "A1" and staining by the immunoperoxidase technique. The heteromorphic segment of the chromosome on the *left* includes four pairs of loops with heavily stained regions (*arrows*), whereas its partner on the *right* contains only one. *Bar* = 30 μm. Courtesy of Prof. J.-C. Lacroix

outer layer of the nuclear membrane as they emerge from the nuclear pores, and retain this association in the cytoplasm until they are mobilized for translation. Whether this proposal is right or not, somewhere along the line mRNAs that are to be stockpiled must be recognized in some way, and as they retain their linkage to coordinately transcribed repetitive sequences it seems logical to look for signs of this recognition at the sites of transcription.

In this context Lacroix et al.'s (1985) results with antibody "B71" are of particular interest. The antigen(s) to B71 appear to have the same relationship to all RNAs transcribed from lampbrush chromosomes as do the two polypeptides that Sommerville et al. (1978 a) have associated with 4 S and 5 S RNAs, namely storage in the cytoplasm of the oocyte. This points to the possibility that coding RNAs from most of the lampbrush transcription units include a proportion of mRNA molecules that are stockpiled, as well as others that are translated during oogenesis. This was already known to be true of histone mRNAs (Adamson and Woodland 1974, Woodland and Adamson 1977), and it could prove to be the rule for mRNAs coding for "housekeeping" proteins generally in amphibian oocytes. Whether the antigen(s) to B71 are themselves involved in the recognition of mRNAs to be stockpiled, or with the packaging of such molecules for storage, is an open question.

The significance of the association between particular antigens and particular transcription units of lampbrush chromosomes, such as those recognized by Lacroix et al.'s (1985) antibody "A1", also remains to be established. A first concern is whether there are many TU-specific antigens, for in view of their specificity they may be the molecules that recognize mRNAs due for masking; so far only few have been found. Intensive study of the relationship between spheres and histone transcripts could be rewarding, for this is the first and so far unique example where a readily identifiable structure, with its antigenic specificity established, has been shown to have a clear association with a defined group of mRNAs, a proportion of which are masked and translated after oocyte maturation. Further speculation is not warranted, given the present state of knowledge, but further investigation of lampbrush chromosomes certainly is warranted, because their function is to equip an egg for morphogenesis and differentiation, processes that have intrigued scientists for 2000 years and continue so do to.

References

Adamson ED, Woodland HR (1974) Histone synthesis in early amphibian development: histone and DNA syntheses are not coordinated. J Mol Biol 88:263–285

Agutter PS, Richardson JCW (1980) Nuclear non-chromatin proteinaceous structures: their role in the organization and function of the interphase nucleus. J Cell Sci 44:395–435

Ahmad MS (1970) Development, structure and composition of lampbrush chromosomes in domestic fowl. Can J Genet Cytol 12:728–737

Anderson DM, Smith LD (1978) Patterns of synthesis and accumulation of heterogeneous RNA in lampbrush stage oocytes of *Xenopus laevis*. Dev Biol 67:274–285

Anderson DM, Richter JD, Chamberlin ME, Price DH, Britten RJ, Smith LD, Davidson EH (1982) Sequence organization of the poly(A)RNA synthesized and accumulated in lampbrush chromosome stage *Xenopus laevis* oocytes. J Mol Biol 155:281–309

Andreuccetti P, Taddei C, Filosa S (1978) Intercellular bridges between follicle cells and oocyte during the differentiation of follicular epithelium in *Lacerta sicula* RAF. J Cell Sci 33:341–350

Angelier N, Lacroix J-C (1975) Complexes de transcription d'origines nucléolaire et chromosomique d'ovocytes de *Pleurodeles waltlii* et *P. poireti* (Amphibiens, Urodèles). Chromosoma 51:323–335

Angelier N, Paintraud M, Lavaud A, Lechaire JP (1984) Scanning electron microscopy of amphibian lampbrush chromosomes. Chromosoma 89:243–253

Bachmann K (1970) Feulgen slope determinations of urodele nuclear DNA amounts. Histochemie 22:289–293

Bachvarova R (1981) Synthesis, turnover and stability of heterogeneous RNA in growing mouse oocytes. Dev Biol 86:384–392

Baker TG, Franchi LL (1967 a) The fine structure of oogonia and oocytes in human ovaries. J Cell Sci 2:213–224

Baker TG, Franchi LL (1967 b) The structure of the chromosomes in human primordial oocytes. Chromosoma 22:358–377

Baker TG, Beaumont HM, Franchi LL (1969) The uptake of tritiated uridine and phenylalanine by the ovaries of rats and monkeys. J Cell Sci 4:655–675

Baker WK (1968) Position-effect variegation. Adv Genet 14:133–169

Bakken AH, Hamkalo BA (1978) Techniques for visualization of genetic material. In: Hayat MA (ed) Principles and techniques of electron microscopy, vol IX. Van Nostrand Reinhold, New York, pp 84–106

Bakken AH, McClanahan M (1978) Patterns of RNA synthesis in early meiotic prophase oocytes from fetal mouse ovaries. Chromosoma 67:21–40

Baldwin L, Macgregor HC (1985) Centromeric satellite DNA in the newt *Triturus cristatus karelinii* and related species: its distribution and transcription on lampbrush chromosomes. Chromosoma 92:100–107

Balfour FM (1878) On the structure and development of the vertebrate ovary. Q J Microsc Sci 18:383–438

Balinsky BI (1975) An introduction to embryology, 4th edn. Saunders, Philadelphia London Toronto

Barsacchi G, Gall JG (1972) Chromosomal localization of repetitive DNA in the newt, *Triturus*. J Cell Biol 54:580–591

Barsacchi G, Bussotti L, Mancino G (1970) The maps of the lampbrush chromosomes of *Triturus* (Amphibia Urodela) IV. *Triturus vulgaris meridonalis*. Chromosoma 31:255–279

Barsacchi Pilone G, Humphries AA (1975) Progesterone-induced in vitro maturation in oocytes of *Notophthalmus viridescens* (Amphibia Urodela) and some observations on cytological aspects of maturation. J Embryol Exp Morphol 34:451–466

Barsacchi Pilone G, Nardi I, Batistoni R, Andronico F, Beccari E (1974) Chromosome location of the genes for 28 S, 18 S and 5 S ribosomal RNA in *Triturus marmoratus* (Amphibia Urodela). Chromosoma 49:135–153

Barsacchi Pilone G, Nardi I, Andronico F, Batistoni R, Durante M (1977) Chromosome location of the ribosomal RNA genes in *Triturus vulgaris meridionalis* (Amphibia Urodela) I. Localization of the DNA sequences complementary to 5 S ribosomal RNA on mitotic and lampbrush chromosomes. Chromosoma 63:127–134

Batistoni R, Nardi I, Barsacchi Pilone G (1974) Banding patterns on lampbrush chromosomes of *Triturus marmoratus* (Amphibia Urodela) by the Giemsa stain. Chromosoma 49:121–134

Batistoni R, Andronico F, Nardi I, Barsacchi Pilone G (1978) Chromosome location of the ribosomal genes in *Triturus vulgaris meridionalis* (Amphibia Urodela) III. Inheritance of the chromosomal sites for 18 S + 28 S ribosomal RNA. Chromosoma 65:231–240

Bauer H (1933) Die wachsenden Oocytenkerne einiger Insekten in ihrem Verhalten zur Nuklealfärbung. Z Zellforsch Mikrosk Anat 18:254–298

Baumeister HG (1973) Lampbrush chromosomes and RNA synthesis during early oogenesis of *Brachydanio rerio* (Cyprinidae, Teleostei). Z Zellforsch Mikrosk Anat 145:145–150

Beermann W (1952) Chromomerenkonstanz und spezifische Modifikationen der Chromosomenstruktur in der Entwicklung und Organdifferenzierung von *Chironomus tentans*. Chromosoma 5:139–198

Beermann W (1960) Der Nukleolus als lebenswichtiger Bestandteil des Zellkerns. Chromosoma 11:263–296

Beermann W (1961) Ein Balbianiring als Locus einer Speicheldrüsen-Mutation. Chromosoma 12:1–25

Beermann W (1962) Riesenchromosomen, Protoplasmatologia, vol VI. Springer, Berlin Heidelberg New York, pp 1–161

Beermann W (1965) Operative Gliederung der Chromosomen. Naturwissenschaften 52:365–375

Belling J (1928) The ultimate chromomeres of *Lilium* and *Aloe* with regard to the number of genes. Univ Calif Publ Bot 16:153–170

Benyajati C, Worcel A (1976) Isolation, characterization, and structure of the folded interphase genome of *Drosophila melanogaster*. Cell 9:393–407

Bernstein RM, Mukherjee BB (1972) Control of nuclear RNA synthesis in 2-cell and 4-cell mouse embryos. Nature 238:457–459

Bier K (1960) Der Karotyp von *Calliphora erythrocephala* Meigen unter besonderer Berücksichtigung der Nährzellkernchromosomen im gebündelten und gepaarten Zustand. Chromosoma 11:335–364

Bier K (1963a) Synthese, interzellulärer Transport und Abbau von Ribonucleinsäure im Ovar der Stubenfliege *Musca domestica*. J Cell Biol 16:436–440

Bier K (1963b) Autoradiographische Untersuchungen über die Leistungen des Follikelepithels und der Nährzellen bei der Dotterbildung und Eiweißsynthese im Fliegenovar. Arch Entwicklungsmech Org 154:552–575

Bier K, Müller W (1969) DNS-Messungen bei Insekten und eine Hypothese über retardierte Evolution und besonderen DNS-Reichtum im Tierreich. Biol Zentralbl 88:425–449

Bier K, Kunz W, Ribbert D (1967) Struktur und Funktion der Oocytenchromosomen und Nukleolen sowie der Extra-DNS während der Oogenese panoistischer und meroistischer Insekten. Chromosoma 23:214–254

Bier K, Kunz W, Ribbert D (1969) Insect oogenesis with and without lampbrush chromosomes. In: Darlington CD, Lewis KR (eds) Chromosomes today, vol II. Oliver & Boyd, Edinburgh, pp 107–115

Bird AP (1977) A study of early events in ribosomal gene amplification. Cold Spring Harbor Symp Quant Biol 42:1179–1183

Bird AP, Birnstiel ML (1971) A timing study of DNA amplification in *Xenopus laevis* oocytes. Chromosoma 35:300–309

Bird AP, Rochaix J, Bakken AH (1973) The mechanism of gene amplification in *Xenopus laevis* oocytes. In: Hamkalo B, Papaconstantinou J (eds) Molecular cytogenetics. Plenum, New York London, pp 49–58

Bona M, Scheer U, Bautz EKF (1981) Antibodies to RNA polymerase II (B) inhibit transcription in lampbrush chromosomes after microinjection into living amphibian oocytes. J Mol Biol 151:81–99

Bonner WM (1975) Protein migration into nuclei II. Frog oocyte nuclei accumulate a class of microinjected oocyte nuclear proteins and exclude a class of microinjected oocyte cytoplasmic proteins. J Cell Biol 64:431–437

Born G (1892) Die Reifung des Amphibieneies und die Befruchtung unreifer Eier bei *Triton taeniatus*. Anat Anz 7:772–781 and 803–811

Born G (1894) Die Struktur des Keimbläschens im Ovarialei von *Triton taeniatus*. Arch Mikrosk Anat Entwicklungsmech 43:1–79

Bottke W (1973) Lampenbürstenchromosomen und Amphinukleolen in Oocytenkernen der Schnecke *Bithynia tentaculata* L. Chromosoma 42:175–190

Boveri Th (1890) Zellenstudien III. Über das Verhalten der chromatischen Kernsubstanz bei der Bildung der Richtungskörper und bei der Befruchtung. Jena Z Naturwiss 24:314–401

Brachet J (1929) Recherches sur le comportement de l'acide thymonucléinique au cours de l'oogenèse chez diverses espèces animales. Arch Biol 39:677–697

Brachet J (1933) Recherches sur la synthèse de l'acide thymonucléique pendant le développement de l'oeuf d'oursin. Arch Biol 44:519–576

Brachet J (1937) Remarques sur la formation de l'acide thymonucléique pendant le développement des oeufs à synthèse partielle. Arch Biol 48:529–548

Brachet J (1940) La localization de l'acide thymonucléique pendant l'oogenèse et la maturation chez les Amphibiens. Arch Biol 51:151–165

Brachet J (1960) The biochemistry of development. Pergamon, London

Bridges CB (1916) Non-disjunction as proof of the chromosome theory of heredity. Genetics 1:1–52

Britten RJ, Davidson EH (1969) Gene regulation for higher cells: a theory. Science 165:349–357

Britten RJ, Kohne DE (1968) Repeated sequences in DNA. Science 161:529–540

Brosseau G (1960) Genetic analysis of male fertility factors on the Y chromosome of *Drosophila melanogaster*. Genetics 44:257–274

Brown DD, Blackler AW (1972) Gene amplification proceeds by a chromosome copy mechanism. J Mol Biol 63:75–83

Brown GL, Callan HG, Leaf G (1950) The chemical nature of nuclear sap. Nature 165:600–601

Bucci S, Nardi I, Mancino G, Fiume L (1971) Incorporation of tritiated uridine in nuclei of *Triturus* oocytes treated with α-amanitin. Exp Cell Res 69:462–465

Busby S, Bakken AH (1979) Quantitative electron microscopic analysis of transcription in sea urchin embryos. Chromosoma 71:249–262

Cachon J, Cachon-Enjumet M (1965) Étude cytologique et caryologique d'un Phaeodarié Bathypélagique *Planktonetta atlantica* Borget. Bull Inst Oceanogr Monaco 64:1–22

Callan HG (1942) Heterochromatin in *Triton*. Proc R Soc London Ser B 130:324–335

Callan HG (1948) Alcune proprietà fisiche della membrana nucleare. Ric Sci suppl 18:1–6

Callan HG (1952) A general account of experimental work on amphibian oocyte nuclei. Symp Soc Exp Biol 6:243–255

Callan HG (1955) Recent work on the structure of cell nuclei. In: Symp 8th Congr Cell Biol. IUBS Ser B21:89–109 Noordhoff, Groningen

Callan HG (1957) The lampbrush chromosomes of *Sepia officinalis* L., *Anilocra physodes* L. and *Scyllium catulus* Cuv. and their structural relationship to the lampbrush chromosomes of Amphibia. Pubbl Stn zool Napoli 29:329–346

Callan HG (1963) The nature of lampbrush chromosomes. Int Rev Cytol 15:1–34

Callan HG (1966) Chromosomes and nucleoli of the axolotl, *Ambystoma mexicanum*. J Cell Sci 1:85–108

Callan HG (1967) The organization of genetic units in chromosomes. J Cell Sci 2:1–7

Callan HG (1978) Functional units. Introductory remarks. Philos Trans R Soc London Ser B 283:381–382

Callan HG (1982) Lampbrush chromosomes. Proc R Soc London Ser B 214:417–448

Callan HG, Lloyd L (1956) Visual demonstration of allelic differences within cell nuclei. Nature 178:355–357

Callan HG, Lloyd L (1960) Lampbrush chromosomes of crested newts *Triturus cristatus* (Laurenti). Philos Trans R Soc London Ser B 243:135–219

Callan HG, Lloyd L (1975) Working maps of the lampbrush chromosomes of Amphibia. In: King RC (ed) Handbook of genetics, vol IV. Plenum, New York London, pp 57–77

Callan HG, Macgregor HC (1958) Action of deoxyribonuclease on lampbrush chromosomes. Nature 181:1479–1480

Callan HG, Old RW (1980) In situ hybridization to lampbrush chromosomes: a potential source of error exposed. J Cell Sci 41:115–123

Callan HG, Spurway H (1951) A study of meiosis in interracial hybrids of the newt, *Triturus cristatus*. J Genet 50:235–249

Callan HG, Tomlin SG (1950) Experimental studies on amphibian oocyte nuclei I. Investigation of the structure of the nuclear membrane by means of the electron microscope. Proc R Soc London Ser B 137:367-378

Callan HG, Old RW, Gross KW (1980) Problems exposed by the results of in situ hybridization to lampbrush chromosomes. Eur J Cell Biol 22:21

Carnoy JB, Lebrun H (1897) La vésicule germinative et les globules polaires chez les batraciens I. Salamandre et Pleurodèle. Cellule 12:191–295

Carnoy JB, Lebrun H (1898) La vésicule germinative et les globules polaires chez les batraciens II. Axolottl et Tritons. Cellule 14:113–200

Carnoy JB, Lebrun H (1899) La vésicule germinative et les globules polaires chez les batraciens III. Les globules polaires des urodèles. Cellule 16:304–391

Caspersson T (1936) Über den chemischen Aufbau der Strukturen des Zellkernes. Skand Arch Physiol 73:suppl 8

Cavalier-Smith T (1978) Nuclear volume control by nucleoskeletal DNA, selection for cell volume and cell growth rate, and the solution of the C-value paradox. J Cell Sci 34:247–278

Chandley AC (1966) Studies on oogenesis in *Drosophila melanogaster* with ^{3}H-thymidine label. Exp Cell Res 44:201–215

Clark TG, Rosenbaum JL (1979) An actin filament matrix in hand-isolated nuclei of *Xenopus laevis* oocytes. Cell 18:1101–1108

Comings DE, Okada TA (1970) Whole mount electron microscopy of meiotic chromosomes and synaptonemal complex. Chromosoma 30:269–286

Darlington CD (1947) Nucleic acid and the chromosomes. Symp Soc Exp Biol 1:252–269

Darlington CD, La Cour L (1940) Nucleic acid starvation of chromosomes in *Trillium*. J Genet 40:185–213

Davidson EH (1976) Gene activity in early development, 2nd edn. Academic Press, London New York

Davidson EH, Hough BR (1971) Genetic information in oocyte RNA. J Mol Biol 56:491–506

Davidson EH, Posakony JW (1982) Repetitive sequence transcripts in development. Nature 297:633–635

Dearing WH (1934) The material continuity and individuality of the somatic chromosomes of *Ambystoma tigrinum*, with special reference to the nucleolus as a chromosomal component. J Morphol 56:157–173

Delobel N (1971) Étude descriptive des chromosomes en écouvillon chez *Echinaster sepositus* (Échinoderme, Astéride). Ann Embryol Morphogen 4:383–396

De Petrocellis B, Rossi M (1976) Enzymes of DNA biosynthesis in developing sea urchins. Dev Biol 48:250–257

De Robertis EM, Gurdon JB (1977) Gene activation in somatic nuclei after injection into amphibian oocytes. Proc Natl Acad Sci USA 74:2470–2474

De Robertis EM, Gurdon JB (1979) Gene transplantation and the analysis of development. Sci Am 241:74–82

De Robertis EM, Partington GA, Longthorne RF, Gurdon JB (1977) Somatic nuclei in amphibian oocytes: evidence for selective gene expression. J Embryol Exp Morphol 40:199–214

De Robertis EM, Longthorne RF, Gurdon JB (1978) The intracellular migration of nuclear proteins in *Xenopus* oocytes. Nature 272:254–256

Diaz MO, Gall JG (1985) Giant readthrough transcription units at the histone loci on lampbrush chromosomes of the newt *Notophthalmus*. Chromosoma 92:243–253

Diaz MO, Barsacchi-Pilone G, Mahon KA, Gall JG (1981) Transcripts from both strands of a satellite DNA occur on lampbrush chromosome loops of the newt *Notophthalmus*. Cell 24:649–659

Dodson EO (1948) A morphological and biochemical study of lampbrush chromosomes of vertebrates. Univ Calif Publ Zool 53:281–314

Dolecki GJ, Smith LD (1979) Poly(A)$^+$RNA metabolism during oogenesis in *Xenopus laevis*. Dev Biol 69:217–236

Dumont J (1972) Oogenesis in *Xenopus laevis* (Daudin) 1. Stages of oocyte development in laboratory maintained animals. J Morphol 136:153–179

Duryee WR (1937) Isolation of nuclei and non-mitotic chromosome pairs from frog eggs. Arch Exp Zellforsch 19:171–176

Duryee WR (1941) The chromosomes of the amphibian nucleus. In: Cytology, genetics and evolution. Univ Pennsylvania Press, Philadelphia, pp 129–141

Duryee WR (1950) Chromosomal physiology in relation to nuclear structure. Ann N Y Acad Sci 50:920–953

Ebstein BS (1967) Tritiated actinomycin D as a cytochemical label for small amounts of DNA. J Cell Biol 35:709–713

Epstein LM, Mahon KA, Gall JG (1985) A small RNA transcript homologous to satellite DNA in the newt. J Cell Biol 101:317a

Feulgen R, Rossenbeck H (1924) Mikroscopisch-chemischer Nachweis einer Nucleinsäure vom Typus der Thymonucleinsäure. Z Physiol Chem 135:213–248

Finch JT, Klug A (1976) Solenoidal model for superstructure in chromatin. Proc Natl Acad Sci USA 73:1897–1901

Fisher LM (1982) DNA unwinding in transcription and recombination. Nature 299:105–106

Flemming W (1882) Zellsubstanz, Kern- und Zelltheilung. Vogel, Leipzig

Ford PJ, Mathieson T, Rosbash M (1977) Very long-lived messenger RNA in ovaries of *Xenopus laevis*. Dev Biol 57:417–426

Franke WW, Scheer U (1978) Morphology of transcriptional units at different states of activity. Philos Trans R Soc London Ser B 283:333–342

Franke WW, Scheer U, Trendelenburg MF, Spring H, Zentgraf H (1976) Absence of nucleosomes in transcriptionally active chromatin. Cytobiologie 13:401–434

Franke WW, Scheer U, Trendelenburg M, Zentgraf H, Spring H (1978) Morphology of transcriptionally active chromatin. Cold Spring Harbor Symp Quant Biol 42:755–772

Gall JG (1952) The lampbrush chromosomes of *Triturus viridescens*. Exp Cell Res Suppl 2:95–102

Gall JG (1954) Lampbrush chromosomes from oocyte nuclei of the newt. J Morphol 94:283–352

Gall JG (1955) Problems of structure and function in the amphibian oocyte nucleus. Symp Soc Exp Biol 9:358–370

Gall JG (1956) On the submicroscopic structure of chromosomes. Brookhaven Symp Biol 8:17–32

Gall JG (1958) Chromosomal differentiation. In: McElroy WD, Glass B (eds) Symposium on the chemical basis of development. Johns Hopkins Press, Baltimore, pp 103–135

Gall JG (1963a) Kinetics of deoxyribonuclease action on chromosomes. Nature 198:36–38

Gall JG (1963b) Chromosomes and cytodifferentiation. In: Locke M (ed) Cytodifferentiation and macromolecular synthesis. Academic Press, London New York, pp 119–143

Gall JG (1966) Techniques for the study of lampbrush chromosomes. In: Prescott DM (ed) Methods in cell physiology, vol II. Academic Press, London New York, pp 37–60

Gall JG (1968) Differential synthesis of the genes for ribosomal RNA during amphibian oogenesis. Proc Natl Acad Sci USA 60:553–560

Gall JG (1969) The genes for ribosomal RNA during oogenesis. Genetics Suppl 61:121–132

Gall JG (1978) Early studies of gene amplification. Harvey Lect 71:55–70

Gall JG (1981) Chromosome structure and the C-value paradox. J Cell Biol 91:3s–14s

Gall JG, Callan HG (1962) H[3]-uridine incorporation in lampbrush chromosomes. Proc Natl Acad Sci USA 48:562–570

Gall JG, Pardue ML (1969) Formation and detection of RNA-DNA hybrid molecules in cytological preparations. Proc Natl Acad Sci USA 63:378–383

Gall JG, Pardue ML (1971) Nucleic acid hybridization in cytological preparations. In: Grossman L, Moldave K (eds) Methods in enzymology, vol 21D. Academic Press, London New York, pp 470–480

Gall JG, Stephenson EC, Erba HP, Diaz MO, Barsacchi-Pilone G (1981) Histone genes are located at the sphere loci of newt lampbrush chromosomes. Chromosoma 84:159–171

Gall JG, Diaz MO, Stephenson EC, Mahon KA (1983) The transcription unit of lampbrush chromosomes. Soc Dev Biol Symp 41:137–146

Gardner RL (1972) An investigation of inner cell mass and trophoblast tissues following their isolation from the mouse blastocyst. J Embryol Exp Morphol 28:279–312

Georgiev O, Mous J, Birnstiel ML (1984) Processing and nucleo-cytoplasmic transport of histone gene transcripts. Nucleic Acids Res 12:8539–8551

Gersch M (1940) Untersuchungen über die Bedeutung der Nucleolen im Zellkern. Z Zellforsch A30:483–528

Giorgi F, Galleni L (1972) The lampbrush chromosomes of *Rana esculenta* L. (Amphibia-Anura). Caryologia 25:107–123

Glätzer KH (1984) Preservation of nuclear RNP antigens in male germ cell development of *Drosophila hydei*. Mol Gen Genet 196:236–243

Glätzer KH, Meyer GF (1981) Morphological aspects of the genetic activity in primary spermatocyte nuclei of *Drosophila hydei*. Biol Cell 41:165–172

Golden L, Schäfer U, Rosbash M (1980) Accumulation of individual pA[+]RNAs during oogenesis of *Xenopus laevis*. Cell 22:835–844

Goldschmidt R (1902) Untersuchungen über die Eireifung, Befruchtung und Zellteilung bei *Polystomum integerrimum*. Rud Z Wiss Zool 71:397–444

Goodpasture C, Bloom SE (1975) Visualization of nucleolar organizer regions in mammalian chromosomes using silver staining. Chromosoma 53:37–50

Gould DC, Callan HG, Thomas CA (1976) The actions of restriction endonucleases on lampbrush chromosomes. J Cell Sci 21:303–313

Gounon P, Karsenti E (1981) Involvement of contractile proteins in the changes in consistency of oocyte nucleoplasm of the newt *Pleurodeles waltlii*. J Cell Biol 88:410–421

Grond CJ, Siegmund I, Hennig W (1983) Visualization of a lampbrush loop-forming fertility gene in *Drosophila hydei*. Chromosoma 88:50–56

Grond CJ, Rutten RGJ, Hennig W (1984) Ultrastructure of the Y-chromosomal lampbrush loops in primary spermatocytes of *Drosophila hydei*. Chromosoma 89:85–95

Gundlach H, Trendelenburg MF (1981) Identification and selective micropreparation of live nuclear components with the Zeiss IM35 inverted microscope. Zeiss-Inf Oberkochen 25:36–40

Gurdon JB (1968) Changes in somatic cell nuclei inserted into growing and maturing amphibian oocytes. J Embryol Exp Morphol 20:401–414

Gurdon JB (1970) Nuclear transplantation and the control of gene activity in animal development. Proc R Soc London Ser B 176:303–314

Gurdon JB (1973) The translation of messenger RNA injected in living oocytes of *Xenopus laevis*. Karolinska Symp Res Methods Reprod Endocrinol 6:225–243

Gurdon JB (1976) Injected nuclei in frog oocytes: fate, enlargement and chromatin dispersal. J Embryol Exp Morphol 36:523–540

Gurdon JB, Melton DA (1981) Gene transfer in amphibian eggs and oocytes. Annu Rev Genet 15:189–218

Gurdon JB, Lane CD, Woodland HR, Marbaix G (1971) Use of frog eggs and oocytes for the study of messenger RNA and its translation in living cells. Nature 233:177–182

Gurdon JB, De Robertis EM, Partington G (1976) Injected nuclei in frog oocytes provide a living cell system for the study of transcriptional control. Nature 260:116–120

Guyénot E, Danon M (1953) Chromosomes et ovocytes de batraciens. Étude cytologique et au microscope électronique. Rev Suisse Zool 60:1–129

Hackstein JHP, Leoncini O, Beck H, Peelen G, Hennig W (1982) Genetic fine structure of the Y-chromosome of Drosophila hydei. Genetics 101:257–277

Hämmerling J (1953) Nucleo-cytoplasmic relationships in the development of Acetabularia. Int Rev Cytol 2:475–498

Hämmerling J (1963) Nucleo-cytoplasmic interactions in Acetabularia and other cells. Annu Rev Plant Physiol 14:63–92

Haldane JBS (1922) Sex ratio and unisexual sterility in hybrid animals. J Genet 12:101–109

Hancock R (1982) Topological organization of interphase DNA: the nuclear matrix and other skeletal structures. Biol Cell 46:105–122

Hancock R, Hughes ME (1982) Organization of DNA in the interphase nucleus. Biol Cell 44:201–212

Hartley SE, Callan HG (1978) RNA transcription on the giant lateral loops of the lampbrush chromosomes of the American newt Notophthalmus viridescens. J Cell Sci 34:279–288

Hartmann M (1902) Studien am tierischen Ei. I. Ovarialei und Eireifung von Asterias glacialis. Zool Jahrb (Anat) 15:793–812

Heidenhain M (1894) Neue Untersuchungen über die Centralkörper und ihre Beziehungen zum Kern- und Zellenprotoplasma. Arch Mikrosk Anat Entwicklungsmech 43:423–758

Heidenhain M (1896) Noch einmal über die Darstellung der Centralkörper durch Eisenhämatoxylin nebst einigen allgemeinen Bemerkungen über die Hämatoxylinfarben. Z Wiss Mikrosk 13:186

Heitz E (1931) Die Ursache der gesetzmäßigen Zahl, Lage, Form und Größe pflanzlicher Nukleolen. Planta 12:775–844

Henderson SA (1964) RNA synthesis during male meiosis and spermatogenesis. Chromosoma 15:345–366

Henderson SA (1971) Grades of chromatid organization in mitotic and meiotic chromosomes I. The morphological features. Chromosoma 35:28–40

Hennen S, Mizuno S, Macgregor HC (1975) In situ hybridization of ribosomal DNA labelled with [125]Iodine to metaphase and lampbrush chromosomes from newts. Chromosoma 50:349–369

Hennig W (1967) Untersuchungen zur Struktur und Funktion des Lampenbürsten-Y-Chromosoms in der Spermatogenese von Drosophila. Chromosoma 22:294–357

Hennig W (1968) Ribonucleic acid synthesis of the Y-chromosome of Drosophila hydei. J Mol Biol 38:227–239

Hennig W (1977) Gene interactions in germ cell differentiation of Drosophila. Adv Enzyme Regul 15:363–371

Hennig W (1985) Y chromosome function and spermatogenesis in Drosophila hydei. Adv Genet 23:179–234

Hennig W, Meyer GF, Hennig I, Leoncini O (1974) Structure and function of the Y chromosome of Drosophila hydei. Cold Spring Harbor Symp Quant Biol 38:673–683

Hennig W, Huijser P, Vogt P, Jäckle H, Edström J-E (1983) Molecular cloning of microdissected lampbrush loop DNA sequences of Drosophila hydei. EMBO J 2:1741–1746

Hentschel CC, Birnstiel ML (1981) The organization and expression of histone gene families. Cell 25:301–313

Hertwig O (1890) Vergleich der Ei- und Samenbildung bei Nematoden. Eine Grundlage für celluläre Streitfragen. Arch Mikrosk Anat Entwicklungsmech 36:1–138

Hess O (1965a) Strukturdifferenzierungen im Y-Chromosom von Drosophila hydei und ihre Beziehungen zu Genaktivitäten I. Mutanten der Funktionsstrukturen. Verh Dtsch Zool Ges Zool Anz Suppl 28:156–163

Hess O (1965b) Struktur-Differenzierungen im Y-Chromosom von Drosophila hydei und ihre Beziehungen zu Gen-Activitäten III. Sequenz und Localisation der Schleifen-Bildungsorte. Chromosoma 16:222–248

Hess O (1966) Funktionelle und strukturelle Organisation der Lampenbürsten-Chromosomen. In: Sitte P (ed) Probleme der biologischen Reduplikation. Springer, Berlin Heidelberg New York, pp 29–54

Hess O (1967a) Complementation of genetic activity in translocated fragments of the Y chromosome in *Drosophila hydei*. Genetics 56:283–295

Hess O (1967b) Morphologische Variabilität der chromosomalen Funktionsstrukturen in den Spermatocytenkernen von *Drosophila*-Arten. Chromosoma 21:429–445

Hess O (1968a) The function of the lampbrush loops formed by the Y chromosome of *Drosophila hydei* in spermatocyte nuclei. Mol Gen Genet 103:58–71

Hess O (1968b) Genetische Aktivität in translozierten Fragmenten des Y-Chromosoms von *Drosophila hydei*. Verh Dtsch Zool Ges Zool Anz Suppl 31:439–453

Hess O (1970) Genetic functions correlated with unfolding of lampbrush loops by the Y chromosome in spermatocytes of *Drosophila hydei*. Mol Gen Genet 106:328–346

Hess O (1971) Lampenbürstenchromosomen. In: Handbuch der allgemeinen Pathology, vol II. Springer, Berlin Heidelberg New York, pp 215–281

Hess O (1976) Genetics of *Drosophila hydei* Sturtevant. In: Novitski E, Ashburner M (eds) Genetics and biology of *Drosophila,* vol I. Academic Press, London New York, pp 1343–1363

Hess O (1981) Lampbrush chromosomes. In: Ashburner M, Wright TRF (eds) Genetics and biology of *Drosophila,* vol IId. Academic Press, London New York, pp 1–39

Hess O, Meyer GF (1963) Chromosomal differentiations of the lampbrush type formed by the Y chromosome in *Drosophila hydei* and *D. neohydei*. J Cell Biol 16:527–539

Hess O, Meyer GF (1968) Genetic activities of the Y chromosome in *Drosophila* during spermatogenesis. Adv Genet 14:171–223

Hill RJ, Maundrell K, Callan HG (1973) Proteins of the newt germinal vesicle nucleus. Nature New Biol 242:20–22

Hill RJ, Maundrell K, Callan HG (1974) Nonhistone proteins of the oocyte nucleus of the newt. J Cell Sci 15:145–161

Hill RS (1979) A quantitative electron-microscope analysis of chromatin from *Xenopus laevis* lampbrush chromosomes. J Cell Sci 40:145–169

Hill RS, Macgregor HC (1980) The development of lampbrush chromosome-type transcription in early diplotene oocytes of *Xenopus laevis:* an electron microscope analysis. J Cell Sci 44:87–101

Holl M (1890) Über die Reifung der Eizelle des Huhnes. Sitzungsber Akad Wiss Wien 99:311–370

Horner HA, Macgregor HC (1983) C value and cell volume: their significance in the evolution and development of amphibians. J Cell Sci 63:135–146

Hourcade D, Dressler D, Wolfson J (1973) The nucleolus and the rolling circle. Cold Spring Harbor Symp Quant Biol 38:537–550

Howell WH, Denton TE, Diamond JR (1975) Differential staining of the satellite regions of human acrocentric chromosomes. Experientia 31:260–262

Hsu TC (1948) The relations between heteropycnosis, spiralization and lampbrush formation of the chromosomes in the spermatogenesis of the Acrididae. J Genet 48:311–315

Hughes ME, Bürki K, Faken S (1979) Visualization of transcription in early mouse embryos. Chromosoma 73:179–190

Hulsebos TJM, Hackstein JHP, Hennig W (1983) Involvement of Y-chromosomal loci in the synthesis of *Drosophila hydei* sperm proteins. Dev Biol 100:238–243

Hulsebos TJM, Hackstein JHP, Hennig W (1984) Lampbrush loop-specific protein of *Drosophila hydei*. Proc Natl Acad Sci USA 81:3404–3408

Humphrey RR (1961) A chromosomal deletion in the Mexican axolotl (*Siredon mexicanum*) involving the nucleolar organizer and the gene for dark colour. Am Zool 1:361

Hutchison N, Pardue ML (1975) The mitotic chromosomes of *Notophthalmus* (= *Triturus*) *viridescens:* localization of C banding regions and DNA sequences complementary to 18 S, 28 S and 5 S ribosomal RNA. Chromosoma 53:51–69

Ingman-Baker J, Candido EPM (1980) Proteins of the *Drosophila melanogaster* male reproductive system. Two dimensional gel patterns of proteins synthesized in XO, XY and XYY testis and paragonial gland and evidence that the Y chromosome does not code for structural sperm proteins. Biochem Genet 18:809–828

Izawa M, Allfrey VG, Mirsky AL (1963) The relationship between RNA synthesis and loop structure in lampbrush chromosomes. Proc Natl Acad Sci USA 49:544–551

Jamrich M, Warrior R, Steele R, Gall JG (1983) Transcription of repetitive sequences on *Xenopus* lampbrush chromosomes. Proc Natl Acad Sci USA 80:3364–3367

Janssens FA (1904) Das chromatische Element während der Entwicklung des Ovocyts von Triton. Anat Anz 24:648–651

Jörgensen M (1913) Zellenstudien I. Morphologische Beiträge zum Problem des Eiwachstums. Arch Zellforsch 10:1–126

John B, Lewis KR (1965) The meiotic system. Springer, Vienna New York

John HA, Birnstiel ML, Jones KWI (1969) RNA-DNA hybrids at the cytological level. Nature 223:582–587

Karsenti E, Gounon P, Bornens M (1978) Immunocytochemical study of lampbrush chromosomes: presence of tubulin and actin. Biol Cell 31:210–224

Kay BK, Gall JG (1981) 5 S ribosomal RNA genes of the newt *Notophthalmus viridescens*. Nucleic Acids Res 9:6457–6469

Kay BK, Schmidt O, Gall JG (1981) In vitro transcription of cloned 5 S genes of the newt *Notophthalmus*. J Cell Biol 90:323–331

Keyl H-G (1975) Lampbrush chromosomes in spermatocytes of *Chironomus*. Chromosoma 51:75–91

Kezer J (1970) Observations on salamander spermatocyte chromosomes during the first meiotic division. Drosophila Inf Serv 45:194–200

Kezer J, Macgregor HC (1971) A fresh look at meiosis and centromeric heterochromatin in the red-backed salamander, *Plethodon cinereus cinereus* (Green). Chromosoma 33:146–166

Kezer J, Macgregor HC (1973) The nucleolar organizer of *Plethodon cinereus cinereus* (Green) II. The lampbrush nucleolar organizer. Chromosoma 42:427–444

Kezer J, Macgregor HC, Schabtach E (1971) Observations on the membranous components of amphibian oocyte nucleoli. J Cell Sci 8:1–17

Kezer J, León PE, Sessions SK (1980) Structural differentiation of the meiotic and mitotic chromosomes of the salamander, *Ambystoma macrodactylum*. Chromosoma 81:177–197

Kierszenbaum AL, Tres LL (1974a) Nucleolar and perichromosomal RNA synthesis during meiotic prophase in the mouse testis. J Cell Biol 60:39–53

Kierszenbaum AL, Tres LL (1974b) Transcription sites in spread meiotic prophase chromosomes from mouse spermatocytes. J Cell Biol 63:923–935

King HD (1908) The oogenesis of *Bufo lentiginosus*. J Morphol 19:369–438

King RC, Burnett RG (1959) Autoradiographic study of uptake of tritiated glycine, thymidine and uridine by fruitfly ovaries. Science 129:1674–1675

Klåšterská I (1976) A new look on the role of the diffuse stage in problems of plant and animal meiosis. Hereditas 82:193–204

Klåšterská I (1978) Structure of eukaryotic chromosomes: the differences between mammal (mouse) grasshopper (*Stethophyma*) and plant (*Rosa*) chromosomes as revealed at the diffuse stage of meiosis. Hereditas 88:243–253

Klåšterská I, Natarajan AT, Ramel C (1978) New observations on mammalian male meiosis I. Laboratory mouse (*Mus musculus*) and Rhesus monkey (*Macaca mulatta*). Hereditas 83:203–214

Kleinschmidt JS, Scheer U, Dabauvalle M-C, Bustin M, Franke WW (1983) High mobility group proteins of amphibian oocytes: a large storage pool of a soluble high mobility group-1-like protein and involvement in transcriptional events. J Cell Biol 97:838–848

Kloetzel P-M, Knust E, Schwochau M (1981) Analysis of nuclear proteins in primary spermatocytes of *Drosophila hydei*: the correlation of nuclear proteins with the function of the Y chromosomal loops. Chromosoma 84:67–86

Kloetzel P-M, Johnson RM, Sommerville J (1982) Interaction of the hnRNA of amphibian oocytes with fibril-forming proteins. Eur J Biochem 127:301–308

Koch EA, Smith PA, King RC (1967) The division and differentiation of *Drosophila* cystocytes. J Morphol 121:55–70

Koecke HU, Müller M (1965) Formwechsel und Anzahl der Chromosomen bei Huhn und Ente. Naturwissenschaften 52:483

Koltzoff NK (1938) The structure of the chromosomes and their participation in cell-metabolism. Biol Zh 7:44–46

Koop H-U, Schmid R, Heunert H-H, Spring H (1979) Spindle formation and division of the giant primary nucleus of *Acetabularia* (Chlorophyta, Dasycladales). Differentiation 14:135–146

Krieg PA, Melton DA (1984) Formation of the 3'end of histone mRNA by post-transcriptional processing. Nature 308:203–206

Kunz W (1967a) Funktionsstrukturen im Oocytenkern von *Locusta migratoria*. Chromosoma 20:332–370

Kunz W (1967b) Lampenbürstenchromosomen und multiple Nukleolen bei Orthopteren. Chromosoma 21:446–462

Kunz W (1969a) Multiple Oocytennukleolen und ihre DNS-Anlagen bei *Locusta migratoria* und *Gryllus domesticus*. Zool Anz Suppl 33:39–46

Kunz W (1969b) Die Entstehung multipler Oocytennukleolen aus akzessorischen DNS-Körpern bei *Gryllus domesticus*. Chromosoma 26:41–75

Lacroix J-C (1968a) Étude descriptive des chromosomes en écouvillon dans le genre *Pleurodeles* (Amphibien, Urodèle). Ann Embryol Morphogen 1:179–202

Lacroix J-C (1968b) Variations expérimentales ou spontanées de la morphologie et de l'organisation des chromosomes en écouvillon dans le genre *Pleurodeles* (Amphibien, Urodèle). Ann Embryol Morphogen 1:205–248

Lacroix J-C (1970) Mise en évidence sur les chromosomes en écouvillon de *Pleurodeles poireti* Gervais, Amphibien urodèle, d'une structure liée au sexe, identifiant le bivalent sexuel et marquant le chromosome W. CR Acad Sci 271:102–104

Lacroix J-C, Capuron A (1966) Localisation et greffe des cellules germinales primordiales chez *Pleurodeles waltlii* Michah. (Amphibien Urodèle). Preuves cytogénétiques. CR Acad Sci 263:1244–1247

Lacroix J-C, Loones MT (1971) Fragmentation par les rayons X de l'organisateur d'une différenciation de chromosome en écouvillon (Lampbrush), chez *Pleurodeles waltlii*. Chromosoma 36:112–118

Lacroix J-C, Azzouz R, Boucher D, Abbadie C, Pyne Ch, Charlemagne J (1985) Monoclonal antibodies to lampbrush chromosome antigens of *Pleurodeles waltlii*. Chromosoma 92:69–80

Laird CD, Wilkinson LE, Foe VE, Chooi WY (1976) Analysis of chromatin-associated fiber arrays. Chromosoma 58:169–192

Lamb MM, Daneholt B (1979) Characterization of active transcription units in Balbiani rings of *Chironomus tentans*. Cell 17:835–848

Lane CD, Marbaix G, Gurdon JB (1971) Rabbit haemoglobin synthesis in frog cells: the translation of reticulocyte 9 S RNA in frog oocytes. J Mol Biol 61:73–91

Lantz LA, Callan HG (1954) Phenotypes and spermatogenesis of interspecific hybrids between *Triturus cristatus* and *T. marmoratus*. J Genet 52:165–185

Laskey RA (1983) Cell cycle. A major transition for embryos and for embryologists. Nature 302:290–291

León PE (1976) Molecular hybridization of iodinated 4 S, 5 S and 18 S + 28 S RNA to salamander chromosomes. J Cell Biol 69:287–300

León P.E, Kezer J (1974) The chromosomes of *Siren intermedia nettingi* (Goin) and their significance to comparative salamander karyology. Herpetologica 30:1–11

Leoncini O (1977) Temperatursensitive Mutanten im Y-Chromosom von *Drosophila hydei*. Chromosoma 63:329–357

Lewin B (1980) Gene expression 2. Eucaryotic chromosomes, 2nd edn. Wiley, New York

Lewis EB (1950) The phenomenon of position effect. Adv Genet 3:73–115

Lifschytz E (1974) Genes controlling chromosome activity. An X-linked mutation affecting Y-lampbrush loop activity in *Drosophila hydei*. Chromosoma 47:415–427

Lifschytz E (1975) Genes controlling chromosome activity. The role of genes blocking Y-lampbrush loop propagation. Chromosoma 53:231–241

Lifschytz E (1979) A procedure for the cloning and identification of Y specific middle repetitive sequences in *D. hydei*. J Mol Biol 133:267–277

Lifschytz E, Hareven D, Azriel A, Brodsly H (1983) DNA clones and RNA transcripts of four lampbrush loops from the Y chromosome of *Drosophila hydei*. Cell 32:191–199

Lindsley DL, Grell EH (1968) Spermiogenesis without chromosomes in *Drosophila melanogaster*. Genetics 61: Suppl 1, 69–78

Loones MT (1979) In vivo effects of Y-irradiation on the functional architecture of the lamp-brush chromosomes in *Pleurodeles* (Amphibia, Urodela). Chromosoma 73:357–368

Loos F de, Dijkhof R, Grond CJ, Hennig W (1984) Lampbrush chromosome loop-specificity of transcript morphology in spermatocyte nuclei of *Drosophila hydei*. EMBO J 3:2845–2849

Loyez M (1906) Recherches sur le développement ovarien des oeufs méroblastiques à vitellus nutritif abondant. Arch Anat Microsc Morphol Exp 8:69–397

Lubosch W (1902) Über die Nucleolarsubstanz des reifenden Tritoneneies nebst Betrachtungen über das Wesen der Eireifung. Jena Z Naturwiss 37:217–296

Macgregor HC (1962) The behaviour of isolated nuclei. Exp Cell Res 26:520–525

Macgregor HC (1963) Morphological variability and its physiological origin in oocyte nuclei of the crested newt. Q J Microsc Sci 104:351–368

Macgregor HC (1965) The role of lampbrush chromosomes in the formation of nucleoli in amphibian oocytes. Q J Microsc Sci 106:215–228

Macgregor HC (1968) Nucleolar DNA in oocytes of *Xenopus laevis*. J Cell Sci 3:437–444

Macgregor HC (1972) The nucleolus and its genes in amphibian oogenesis. Biol Rev 47:177–210

Macgregor HC (1979) In situ hybridization of highly repetitive DNA to chromosomes of *Triturus cristatus*. Chromosoma 71:57–64

Macgregor HC (1980) Recent developments in the study of lampbrush chromosomes. Heredity 44:3–35

Macgregor HC (1982) Ways of amplifying ribosomal genes. In: Jordan EG, Cullis CA (eds) Soc Exp Biol Seminar Ser, vol XV. Cambridge University Press, Cambridge, pp 129–151

Macgregor HC, Andrews C (1977) The arrangement and transcription of "middle repetitive" DNA sequences on lampbrush chromosomes of *Triturus*. Chromosoma 63:109–126

Macgregor HC, Callan HG (1962) The actions of enzymes on lampbrush chromosomes. Q J Microsc Sci 103:173–203

Macgregor HC, Horner H (1980) Heteromorphism for chromosome I, a requirement for normal development in crested newts. Chromosoma 76:111–122

Macgregor HC, Kezer J (1970) Gene amplification in oocytes with 8 germinal vesicles from the tailed frog *Ascaphus truei* Stejneger. Chromosoma 29:189–206

Macgregor HC, Klosterman L (1979) Observations on the cytology of *Bipes* (Amphisbaenia) with special reference to its lampbrush chromosomes. Chromosoma 72:67–87

Macgregor HC, Mizuno S (1976) In situ hybridization of 'nick-translated' ^{3}H-ribosomal DNA to chromosomes from salamanders. Chromosoma 54:15–25

Macgregor HC, Sessions SK (1986) The biological significance of variation in satellite DNA and heterochromatin in newts of the genus *Triturus:* an evolutionary perspective. Philos Trans R Soc London Ser B 312:243–259

Macgregor HC, Stebbings H (1970) a massive system of microtubules associated with cytoplasmic movement in telotrophic ovarioles. J Cell Sci 6:431–449

Macgregor HC, Uzzell TM (1964) Gynogenesis in salamanders related to *Ambystoma jeffersonianum*. Science 143:1043–1045

Macgregor HC, Varley JM (1983) Working with animal chromosomes. Wiley, New York

Macgregor HC, Vlad M, Barnett L (1977) An investigation of some problems concerning nucleolus organizers in salamanders. Chromosoma 59:283–299

Macgregor HC, Varley JM, Morgan GT (1981) The transcription of satellite and ribosomal DNA sequences on lampbrush chromosomes of crested newts. In: Schweiger HG (ed) International cell biology 1980–1981. Springer, Berlin Heidelberg New York, pp 33–46

Macgregor HC, Horner HA, Sims SH (1983) Newt chromosomes and some problems in evolutionary cytogenetics. In: Brandham P, Bennet M (eds) Kew Chromosome Conf, vol II. Allen and Unwin, London, pp 283–294

Mahowald AP, Tiefert M (1970) Fine structural changes in the *Drosophila* oocyte nucleus during a short period of RNA synthesis. Arch Entwicklungsmech Org 165:8–25

Makarov VB, Safronov VV (1974) Functional organization of the chromomere I. RNA synthesis in the giant granular loop of chromosome XII in intact oocytes of *Triturus cristatus cristatus* and after mutagenic treatment. Tsitologiya 16:178–182

Makarov VP, Safronov VV (1976) Functional organization of the chromomere III. Analysis of transcription units in chromomeres of *Triturus cristatus cristatus*. Tsitologiya 18:290–295

Malcolm DB, Sommerville J (1974) The structure of chromosome-derived ribonucleoprotein in oocytes of *Triturus cristatus carnifex* (Laurenti). Chromosoma 48:137–158

Malcolm DB, Sommerville J (1977) The structure of nuclear ribonucleoprotein of amphibian oocytes. J Cell Sci 24:143–165

Mancino G, Barsacchi G (1965) Le mappe dei cromosomi „lampbrush" di *Triturus* (Anfibi Urodeli) I. *Triturus alpestris apuanus*. Caryologia 18:637–665

Mancino G, Barsacchi G (1966) Le mappe dei cromosomi „lampbrush" di *Triturus* (Anfibi Urodeli) II. *Triturus helveticus helveticus*. Riv Biol (Perugia) 59:311–351

Mancino G, Barsacchi G (1969) The maps of the lampbrush chromosomes of *Triturus (Amphibia, Urodela) III. Triturus italicus*. Ann Embryol Morphogen 2:355–377

Mancino G, Barsacchi G, Nardi I (1967) Uno studio autoradiographico sulla sintesi di RNA nei lampbrush chromosomes di *Triturus* (Anfibi Urodeli). Boll Zool 34:134

Mancino G, Barsacchi G, Nardi I (1968) Effetti della actinomicina D tritiata sui lampbrush chromosomes di *Triturus* (Anfibi Urodeli). Atti Accad Naz Lincei 45:591–596

Mancino G, Barsacchi G, Nardi I (1969) The lampbrush chromosomes of *Salamandra salamandra* (L) (Amphibia Urodela). Chromosoma 26:365–387

Mancino G, Nardi I, Corvaja N, Fiume L, Marinozzi V (1971) The effects of α-amanitin on *Triturus* lampbrush chromosomes. Exp Cell Res 64:237–239

Mancino G, Nardi I, Ragghianti M (1972 a) Structural correspondence between nucleolus- and sphere-organizing regions of the lampbrush chromosomes and secondary constrictions of the mitotic chromosomes. Experientia 28:586–588

Mancino G, Nardi I, Ragghianti M (1972 b) Lampbrush chromosomes from semi-albino crested newts, *Triturus cristatus carnifex* (Laurenti). Experientia 28:856–860

Maréchal J (1907) Sur l'ovogénèse des sélaciens et de quelques autres chordates. Cellule 24:5–239

Marsden MPF, Laemmli UK (1979) Metaphase chromosome structure: evidence for a radial loop model. Cell 17:849–858

Marsh JL, Wieschaus E (1978) Is sex determination in germ line and soma controlled by separate genetic mechanisms? Nature 272:249–251

Martin TE, Okamura CS (1981) Immunochemistry of nuclear hnRNP complexes. In: Busch H (ed) The cell nucleus, vol IX. Academic Press, London New York, pp 119–144

Maundrell K (1975) Proteins of the newt oocyte nucleus: analysis of the nonhistone proteins from lampbrush chromosomes, nucleoli and nuclear sap. J Cell Sci 17:579–588

McKnight SL, Sullivan NL, Miller OL (1976) Visualization of the silk fibroin transcription unit and nascent silk fibroin molecules on polyribosomes of *Bombyx mori*. Prog Nucleic Acid Res 19:313–318

McLaren A (1976) Mammalian chimaeras. Cambridge Univ Press, Cambridge

Meyer GF (1963) Die Funktionsstrukturen des Y-Chromosoms in den Spermatocytenkernen von *Drosophila hydei, D. neohydei, D. repleta* und einigen anderen *Drosophila*-Arten. Chromosoma 14:207–255

Meyer GF (1968) Spermiogenese in normalen und Y-defizienten Männchen von *Drosophila melanogaster* und *D. hydei*. Z Zellforsch Mikrosk Anat 84:141–175

Meyer GF, Hennig W (1974) Molecular aspects of the fertility factors in *Drosophila*. In: Afzelius BA (ed) The funtional anatomy of the spermatozoon. Pergamon, Oxford New York, pp 69–75

Meyer GF, Hess O (1965) Struktur-Differenzierungen im Y-Chromosom von *Drosophila hydei* und ihre Beziehungen zu Gen-Aktivitäten II. Effekt der RNS-Synthese-Hemmung durch Actinomycin. Chromosoma 16:249–270

Meyer GF, Hess O, Beermann W (1961) Phasenspezifische Funktionsstrukturen in Spermatocytenkernen von *Drosophila melanogaster* und ihre Abhängigkeit vom Y-Chromosom. Chromosoma 12:676–716

Michie D (1953) Affinity: a new genetic phenomenon in the house mouse. Evidence from distant crosses. Nature 171:26–27

Miller OL (1964) Extrachromosomal nucleolar DNA in amphibian oocytes. J Cell Biol 23:60A

Miller OL (1965) Fine structure of lampbrush chromosomes. Natl Cancer Inst Monogr 18:79–99

Miller OL, Bakken AH (1972) Morphological studies of transcription. Acta Endocrinol Suppl 168:155–177

Miller OL, Beatty BR (1969 a) Visualization of nucleolar genes. Science 164:955–957

Miller OL, Beatty BR (1969 b) Portrait of a gene. J Cell Physiol 74:Suppl 1, 225–232

Miller OL, Hamkalo BA (1972) Visualization of RNA synthesis on chromosomes. Int Rev Cytol 33:1–25

Mirsky AE, Ris H (1951) The desoxyribonucleic acid content of animal cells and its evolutionary significance. J Gen Physiol 34:451–462

Mizuno S, Macgregor HC (1974) Chromosomes, DNA sequences, and evolution in salamanders of the genus *Plethodon*. Chromosoma 48:239–296

Moar VA, Gurdon JB, Lane CD, Marbaix G (1971) Translational capacity of living frog eggs and oocytes, as judged by messenger RNA injection. J Mol Biol 61:93–104

Monesi V (1965) Differential rate of ribonucleic acid synthesis in the autosomes and sex chromosomes during male meiosis in the mouse. Chromosoma 17:11–21

Monesi V (1967) Ribonucleic acid and protein synthesis during differentiation of male germ cells in the mouse. Arch Anat Microsc 56:61–74

Morescalchi A, Filosa S (1965) Osservazioni sui cromosomi piumosi di *Rana esculenta* L. Atti Soc Peloritana Sci Fis Mat Nat 11:211–219

Morgan GT (1978) Absence of chiasmata from the heteromorphic region of chromosome I during spermatogenesis in *Triturus cristatus carnifex*. Chromosoma 66:269–280

Morgan GT, Macgregor HC, Colman A (1980) Multiple ribosomal gene sites revealed by in situ hybridization of *Xenopus* rDNA to *Triturus* lampbrush chromosomes. Chromosoma 80:309–330

Mott MR, Callan HG (1975) An electron-microscope study of the lampbrush chromosomes of the newt *Triturus cristatus*. J Cell Sci 17:241–261

Müller WP (1974) The lampbrush chromosomes of *Xenopus laevis* (Daudin). Chromosoma 47:283–296

Mullinger AM, Johnson RT (1980) Packing DNA into chromosomes. J Cell Sci 46:61–86

Nagl W (1978) Endopolyploidy and polyteny in differentiation and evolution. North Holland, Amsterdam New York Oxford

Nardi I, Mancino G (1971) Mitotic karyotype and nucleoli of the marbled newt *Triturus marmoratus* (Latreille). Experientia 27:424–427

Nardi I, Ragghianti M, Mancino G (1972) Characterization of the lampbrush chromosomes of the marbled newt, *Triturus marmoratus* (Latreille, 1800). Chromosoma 37:1–22

Nardi I, Barsacchi-Pilone G, Batistoni R, Andronico F (1977) Chromosome location of the ribosomal genes in *Triturus vulgaris meridionalis* (Amphibia Urodela) II. Intraspecific variability in number and position of the chromosome loci for 18 S + 28 S ribosomal RNA. Chromosoma 64:67–84

Nardi I, De Lucchini S, Barsacchi-Pilone G, Andronico F (1978) Chromosome location of the ribosomal RNA genes in *Triturus vulgaris meridionalis* (Amphibia, Urodela) IV. Comparison between in situ hybridization with ^{3}H 18 S + 28 S rRNA and AS-SAT staining. Chromosoma 70:91–99

Neaves W (1971) Intercellular bridges between follicle cells and oocyte in the lizard *Anolis carolinensis*. Anat Rec 170:285–293

Newport J, Kirschner M (1982) A major developmental transition in early *Xenopus* embryos: I. Characterization and timing of cellular changes at the midblastula stage. Cell 30:675–686

Nishioka M, Ohtani H, Sumida M (1980) Detection of chromosomes bearing the loci for seven kinds of proteins in Japanese pond frogs. Sci Rep Lab Amphibian Biol Hiroshima Univ 4:127–184

Noronha JM, Sheys GH, Buchanan JM (1972) Induction of a reductive pathway for deoxyribonucleotide synthesis during early embryogenesis of the sea urchin. Proc Natl Acad Sci USA 69:2006–2010

Ohtani H (1975) The lampbrush chromosomes of the sibling species *Rana nigromaculata* and *Rana brevipoda*, and their hybrids. Kromosomo 100:3162–3172

Old RW, Callan HG, Gross KW (1977) Localization of histone gene transcripts in newt lampbrush chromosomes by in situ hybridization. J Cell Sci 27:57–79

Olmo E (1983) Nucleotype and cell size in vertebrates: a review. Bas Appl Histochem 27:227–256

Osheim YN, Martin K, Miller OL (1978) Morphology of active and inactive chromatin in *Xenopus laevis* oocytes. J Cell Biol 79:126 a

Paine PL, Austerberry CF, Desjarlais LJ, Horowitz SB (1983) Protein loss during nuclear isolation. J Cell Biol 97:1240–1242

Painter TS (1940) On the synthesis of cleavage chromosomes. Proc Natl Acad Sci USA 26:95–100

Painter TS, Taylor AN (1942) Nucleic acid storage in the toad's egg. Proc Natl Acad Sci USA 28:311–317

Pantin CFA (1946) Notes on microscopical technique for zoologists. Cambridge Univ Press, Cambridge

Pardue ML, Brown DD, Birnstiel ML (1973) Location of the genes for 5 S ribosomal RNA in *Xenopus laevis*. Chromosoma 42:191–203

Paulson JR, Laemmli UK (1977) The structure of histone-depleted metaphase chromosomes. Cell 12:817–828

Pollack SB, Telfer WH (1969) RNA in *Cecropia* moth ovaries: sites of synthesis, transport and storage. J Exp Zool 170:1–24

Posakony JW, Flytzanis CN, Britten RJ, Davidson EH (1983) Interspersed sequence organization and developmental representation of cloned poly(A)RNAs from sea urchin eggs. J Mol Biol 167:361–389

Prescott DM (1964) Autoradiography with liquid emulsion. In: Prescott DM (ed) Methods in cell physiology, vol I. Academic Press, London New York, pp 365–370

Pukkila PJ (1975) Identification of the lampbrush loops which transcribe 5 S ribosomal RNA in *Notophthalmus (Triturus) viridescens*. Chromosoma 53:71–89

Rabl C (1885) Ueber Zelltheilung. Morphol Jahrb 10:214–330

Ragghianti M, Nardi I, Mancino G (1972) Completion of the morphology of the lampbrush chromosomes of the Italian alpine newt *Triturus alpestris apuanus* Bonaparte. Experientia 28:588–590

Ragghianti M, Bucci-Innocenti S, Mancino G (1977) An ammoniacal silver staining technique for mitotic chromosomes of *Triturus* (Urodela: Salamandridae). Experientia 33:1319–1321

Rattner JB, Goldsmith M, Hamkalo BA (1980) Chromatin organization during meiotic prophase of *Bombyx mori*. Chromosoma 79:215–224

Rattner JB, Goldsmith M, Hamkalo BA (1981) Chromatin organization during male meiosis in *Bombyx mori*. Chromosoma 82:341–351

Reichenbach-Klinke H, Elkan E (1965) The principle diseases of lower vertebrates. Academic Press, London New York

Retzius G (1912) Zur Kenntnis der Hüllen und besonders des Follikelepithels an den Eiern der Wirbeltiere. Biol Unters 17:1–52

Ribbert D, Kunz W (1969) Lampenbürstenchromosomen in den Oocytenkernen von *Sepia officinalis*. Chromosoma 28:93–106

Richter JD, Smith LD, Anderson DM, Davidson EH (1984) Interspersed poly(A)RNAs of amphibian oocytes are not translatable. J Mol Biol 173:227–241

Riemann W, Muir C, Macgregor HC (1969) Sodium and potassium in oocytes of *Triturus cristatus*. J Cell Sci 4:299–304

Ris H (1945) The structure of meiotic chromosomes in the grasshopper and its bearing on the nature of „chromomeres" and „lamp-brush chromosomes". Biol Bull 89:242–257

Roeder RG (1974) Multiple forms of deoxyribonucleic acid-dependent ribonucleic acid polymerase in *Xenopus laevis* – levels of activity during oocyte and embryonic development. J Biol Chem 249:249–256

Roeder RG (1976) Eukaryotic nuclear RNA polymerases. In: Losick R, Chamberlin M (eds) RNA polymerase. Cold Spring Harbor Lab, Cold Spring Harbor, pp 285–329

Rosbash M, Ford PJ (1974) Polyadenylic acid-containing RNA in *Xenopus laevis* oocytes. J Mol Biol 85:87–101

Rudak E, Callan HG (1976) Differential staining and chromatin packing of the mitotic chromosomes of the newt *Triturus cristatus*. Chromosoma 56:349–362

Rückert J (1892) Zur Entwicklungsgeschichte des Ovarialeies bei Selachiern. Anat Anz 7:107–158

Safronov VV, Makarov VB (1975) Functional organization of the chromomere II. The effect of choriogonin on RNA synthesis within the limits of functioning chromomere. Tsitologiya 17:467–469

Sanfelice F (1918) Recherches sur la genèse des corpuscules du *Molluscum contagiosum*. Ann Inst Pasteur Paris 32:363–371

Schäfer U (1978) Sterility in *Drosophila hydei* × *D. neohydei* hybrids. Genetica 49:205–214

Schäfer U (1979) Viability in *Drosophila hydei* × *D. neohydei* hybrids and its regulation by genes located in the sex heterochromatin. Biol Zentralbl 98:153–161

Schäfer U, Kunz W (1975) Two separated nucleolus organizers on the *Drosophila hydei* Y chromosome. Mol Gen Genet 137:365–368

Schäfer U, Golden L, Hyman LE, Colot HV, Rosbash M (1982) Some somatic sequences are absent or exceedingly rare in *Xenopus* oocyte RNA. Dev Biol 94:87–92

Schaffner W, Kunz G, Daetwyler H, Telford J, Smith HO, Birnstiel ML (1978) Genes and spacers of cloned sea urchin DNA analysed by sequencing. Cell 14:655–671

Scheer U (1973) Nuclear pore flow rate of ribosomal RNA and chain growth rate of its precursor during oogenesis of *Xenopus laevis*. Dev Biol 30:13–28

Scheer U (1978) Changes of nucleosome frequency in nucleolar and non-nucleolar chromatin as a function of transcription: an electron microscopic study. Cell 13:535–549

Scheer U (1980) Structural organization of spacer chromatin between transcribed ribosomal RNA genes in amphibian oocytes. Eur J Cell Biol 23:189–196

Scheer U (1981) Identification of a novel class of tandemly repeated genes transcribed on lampbrush chromosomes of *Pleurodeles waltlii*. J Cell Biol 88:599–603

Scheer U (1982) A novel type of chromatin organization in lampbrush chromosomes of *Pleurodeles waltlii:* visualization of clusters of tandemly repeated, very short transcriptional units. Biol Cell 44:213–220

Scheer U, Dabauvalle M-C (1985) Functional organization of the amphibian oocyte nucleus. In: Browder L (ed) Developmental biology: a comprehensive synthesis, vol I. Oogenesis. Plenum, New York London, pp 385–430

Scheer U, Sommerville J (1982) Sizes of chromosome loops and hnRNA molecules in oocytes of amphibia of different genome sizes. Exp Cell Res 139:410–416

Scheer U, Franke WW, Trendelenburg MF, Spring H (1976a) Classification of loops of lampbrush chromosomes according to the arrangement of transcriptional complexes. J Cell Sci 22:503–519

Scheer U, Trendelenburg ME, Franke WW (1976b) Regulation of transcription of genes of ribosomal RNA during amphibian oogenesis. J Cell Biol 69:465–489

Scheer U, Trendelenburg MF, Franke WW (1976c) Regulation of transcription of ribosomal RNA-genes during amphibian oogenesis. In: Müller-Bérat N et al. (eds) Progress in differentiation research. North Holland, Amsterdam New York, pp 105–118

Scheer U, Trendelenburg MF, Krohne G, Franke WW (1977) Lengths and patterns of transcriptional units in the amplified nucleoli of oocytes of *Xenopus laevis*. Chromosoma 60:147–167

Scheer U, Spring H, Trendelenburg MF (1979a) Organization of transcriptionally active chromatin in lampbrush chromosome loops. In: Busch H (ed) The cell nucleus, vol VII. Academic Press, London New York, pp 3–47

Scheer U, Sommerville J, Bustin M (1979b) Injected histone antibodies interfere with transcription of lampbrush chromosome loops in oocytes of *Pleurodeles*. J Cell Sci 40:1–20

Scheer U, Hinssen H, Franke WW, Jockusch BM (1984) Microinjection of actin-binding proteins and actin antibodies demonstrates involvement of nuclear actin in transcription of lampbrush chromosomes. Cell 39:111–122

Schmid M, Krone K (1976) The relationship of a specific chromosomal region to the development of the acrosome. Chromosoma 56:327–347

Schmid M, Olert J, Klett C (1979) Chromosome banding in Amphibia III. Sex chromosomes in *Triturus*. Chromosoma 71:29–55

Schultz LD, Kay BK, Gall JG (1981) In vitro RNA synthesis in oocyte nuclei of the newt *Notophthalmus*. Chromosoma 82:171–187

Schultze O (1887) Untersuchungen über Reifung und Befruchtung des Amphibieneies. Z Wiss Zool 45:178–226

Scott SEM, Sommerville J (1974) Location of nuclear proteins on the chromosomes of newt oocytes. Nature 250:680–682

Shiokawa K (1983) Mobilization of maternal mRNA in amphibian eggs with special reference to the possible role of membranous supramolecular structures. FEBS 151:179–184

Sims SH, Macgregor HC, Pellat PS, Horner HA (1984) Chromosome I in crested and marbled newts (*Triturus*). An extraordinary case of heteromorphism and independent chromosome evolution. Chromosoma 89:169–185

Sinclair JH, Carroll CR, Humphrey RR (1974) Variation in rDNA redundancy level and nucleolar organizer length in normal and variant lines of the Mexican axolotl. J Cell Sci 15:239–257

Sirlin JL, Jacob I (1960) Cell function in the ovary of *Drosophila*. Exp Cell Res 20:283–293

Snow MHL, Callan HG (1969) Evidence for a polarized movement of the lateral loops of newt lampbrush chromosomes during oogenesis. J Cell Sci 5:1–25

Sommerville J (1973) Ribonucleoprotein particles derived from the lampbrush chromosomes of newt oocytes. J Mol Biol 78:487–503

Sommerville J (1977) Gene activity in the lampbrush chromosomes of amphibian oocytes. Int Rev Biochem 15:79–156

Sommerville J (1979) Transcription during amphibian oogenesis. In: Newth DR, Balls M (eds) Maternal effects in development. Cambridge Univ Press, Cambridge, pp 47–63

Sommerville J (1981) Immunolocalization and structural organization of nascent RNP. In: Busch H (ed) The cell nucleus, vol VIII. Academic Press, London New York, pp 1–57

Sommerville J, Hill RJ (1973) Proteins associated with heterogeneous nuclear RNA of newt oocytes. Nature New Biol 245:104–106

Sommerville J, Malcolm DB (1976) Transcription of genetic information in amphibian oocytes. Chromosoma 55:183–208

Sommerville J, Scheer U (1981) Structural organization of nascent transcripts and hnRNA molecules in amphibian oocytes. Mol Biol Rep 7:53–56

Sommerville J, Scheer U (1982) Transcription of complementary repeat sequences in amphibian oocytes. Chromosoma 86:95–113

Sommerville J, Crichton C, Malcolm DB (1978a) Immunofluorescent localization of transcriptional activity on lampbrush chromosomes. Chromosoma 66:99–114

Sommerville J, Malcolm DB, Callan HG (1978b) The organization of transcription on lampbrush chromosomes. Philos Trans R Soc London Ser B 283:359–366

Spring H, Trendelenburg MF, Scheer U, Franke WW, Herth W (1974) Structural and biochemical studies of the primary nucleus of two green algal species, *Acetabularia mediterranea* and *Acetabularia major*. Cytobiologie 10:1–65

Spring H, Scheer U, Franke WW, Trendelenburg MF (1975) Lampbrush-type chromosomes in the primary nucleus of the green alga *Acetabularia mediterranea*. Chromosoma 50:25–43

Spring H, Krohne G, Franke WW, Scheer U, Trendelenburg MF (1976) Homogeneity and heterogeneity of sizes of transcriptional units and spacer regions in nucleolar genes of *Acetabularia*. J Microsc Biol Cell 25:107–116

Spring H, Grierson D, Hemleben V, Stöhr M, Krohne G, Stadler J, Franke WW (1978) DNA contents and numbers of nucleoli and pre-rRNA genes in nuclei of gametes and vegetative cells of *Acetabularia mediterranea*. Exp Cell Res 114:203–215

Spurway H, Callan HG (1950) Hybrids between some members of the Rassenkreis *Triturus cristatus*. Experientia 6:95–96

Srivastava MDL (1951) 'Lampbrush' fibres in the chromosomes of *Chrotogonus incertus* Bolivar. Nature 167:775

Srivastava MDL (1954) Studies on the structure of the chromosomes of *Chrotogonus incertus* Bolivar (Acrididae). J Genet 52:480–493

Srivastava MDL, Bhatnagar AN (1962) Lampbrush chromosomes of *Rana cyanophlyctis*. Cytologia 27:60–71

Standart NM, Bray SJ, George EL, Hunt T, Ruderman JV (1985) The small subunit of ribonucleotide reductase is encoded by one of the most abundant translationally regulated maternal mRNAs in clam and sea urchin eggs. J Cell Biol 100:1968–1976

Stephenson EC, Erba HP, Gall JG (1981a) Characterization of a cloned histone gene cluster of the newt *Notophthalmus viridescens*. Nucleic Acids Res 9:2281–2295

Stephenson EC, Erba HP, Gall JG (1981b) Histone gene clusters of the newt *Notophthalmus* are separated by long tracts of satellite DNA. Cell 24:639–647

Stern C (1929) Untersuchungen über Aberrationen des Y-Chromosoms von *Drosophila melanogaster*. Z Indukt Abstamm Vererbungsl 51:253–353

Stern C, Hadorn E (1938) The determination of sterility in *Drosophila* males without a complete Y chromosome. Am Nat 72:42–52

Stevens NM (1903) On the ovogenesis and spermatogenesis of *Sagitta bipunctata*. Zool Jahrb Anat 18:227–240

Stevens NM (1904) Further studies on the ovogenesis of *Sagitta*. Zool Jahrb Anat 21:243–252

Stieve H (1921) Die Entwicklung der Keimzellen des Grottenolmes II. Die Wachstumsperiode der Oozyte. Arch Mikrosk Anat Entwicklungsmech 95:Abt 2, 1–202

Sures I, Lowry J, Kedes LH (1978) The DNA sequence of sea urchin (*S. purpuratus*) H_2A, H_2B and H_3 histone coding and spacer regions. Cell 15:1033–1044

Swift HH (1950) The constancy of desoxyribosenucleic acid in plant nuclei. Proc Natl Acad Sci USA 36:643–654

Taylor JH (1958) Sister chromatid exchanges in tritium-labelled chromosomes. Genet Princeton 43:515–529

Taylor JH (1959) The organization and duplication of genetic material. Proc 10th Int Congr Genet, vol I. Univ Toronto Press, Toronto, pp 63–78

Taylor JH, Woods PS, Hughes WL (1957) The organization and duplication of chromosomes as revealed by autoradiographic studies using tritium-labelled thymidine. Proc Natl Acad Sci USA 43:122–128

Thoma F, Koller Th, Klug A (1979) Involvement of histone H1 in the organization of the nucleosome and of the salt-dependent superstructures of chromatin. J Cell Biol 83:403–427

Thomas TL, Britten RJ, Davidson EH (1982) An interspersed region of the sea urchin genome represented in both maternal poly(A)RNA and embryo nuclear RNA. Dev Biol 94:230–239

Tobler H (1975) Occurrence and developmental significance of gene amplification. In: Weber R (ed) The biochemistry of animal development, vol III. Academic Press, London New York, pp 91–143

Tomlin SG, Callan HG (1951) Preliminary account of an electron microscope study of chromosomes from newt oocytes. Q J Microsc Sci 92:221–224

Traut W (1975) Die Transkriptionsaktivität der Chromosomen in den Oocyten von *Ephestia* (Lepidoptera). Cytobiologie 2:172–180

Trendelenburg MF (1983) Progress in visualization of eukaryotic gene transcription. Hum Genet 63:197–215

Ullerich F-H (1970) DNS-Gehalt und Chromosomenstruktur bei Amphibien. Chromosoma 30:1–37

Varley JM, Morgan GT (1978) Silver staining of the lampbrush chromosomes of *Triturus cristatus carnifex*. Chromosoma 67:233–244

Varley JM, Macgregor HC, Erba HP (1980a) Satellite DNA is transcribed on lampbrush chromosomes. Nature 283:686–688

Varley JM, Macgregor HC, Nardi I, Andrews C, Erba HP (1980b) Cytological evidence of transcription of highly repeated DNA sequences during the lampbrush stage in *Triturus cristatus carnifex*. Chromosoma 80:289–307

Vlad M, Macgregor HC (1975) Chromomere number and its genetic significance in lampbrush chromosomes. Chromosoma 50:327–347

Vogt P, Hennig W (1983) Y chromosomal DNA of *Drosophila hydei*. J Mol Biol 167:37–56

Vogt P, Hennig W, Siegmund I (1982) Identification of cloned Y chromosomal DNA sequences from a lampbrush loop of *Drosophila hydei*. Proc Natl Acad Sci USA 79:5132–5136

Waddington CH (1939) An introduction to modern genetics. Allen & Unwin, London

Wallace ME (1953) Affinity: a new genetic phenomenon in the house mouse. Evidence from within laboratory stocks. Nature 171:27–28

Wasylyk B, Trevenin G, Oudet P, Chambon P (1979) Transcription of *in vitro* assembed chromatin by *Escherichia coli* RNA polymerase. J Mol Biol 128:411–440

Watson ID, Callan HG (1963) The form of bivalent chromosomes in newt oocytes at first metaphase of meiosis. Q J Microsc Sci 104:281–295

Watson-Coggins L, Gall JG (1972) The timing of meiosis and DNA synthesis during early oogenesis in the toad, *Xenopus laevis*. J Cell Biol 52:569–576

Weintraub H, Worcel A, Alberts B (1976) A model for chromatin based upon two symmetrically paired half-nucleosomes. Cell 9:409–417

Weith A, Traut W (1980) Synaptonemal complexes with associated chromatin in a moth, *Esphestia kuehniella* Z. The fine structure of the W chromosomal heterochromatin. Chromosoma 78:275–291

Wellauer PK, Reeder RH, David IB, Brown DD (1976) The arrangement of length heterogeneity in repeating units of amplified and chromosomal ribosomal DNA from *Xenopus laevis*. J Mol Biol 105:487–505

White MJD (1946) The spermatogenesis of hybrids between *Triturus cristatus* and *T. marmoratus* (Urodela). J Exp Zool 102:179–207

Whitley JE, Muir C (1974) Determination of sodium and potassium in the nuclei of single newt oocytes. J Radioanal Chem 19:257–262

Wilson EB (1925) The cell in development and heredity, 3rd edn. Macmillan, New York London

Winiwarter H de (1901) Recherches sur l'ovogenèse et l'organogenèse de l'ovaire des mammifères (lapin et homme). Arch Biol 17:33–199

Wolff S, Lindsley DL, Peacock WJ (1976) Cytological evidence for switches in polarity of chromosomal DNA. Proc Natl Acad Sci USA 73:877–881

Woodland HR, Adamson ED (1977) The synthesis and storage of histones during the oogenesis of *Xenopus laevis*. Dev Biol 57:118–135

Worcel A (1977) Molecular architecture of the chromatin fibre. Cold Spring Harbor Symp Quant Biol 42:313–324

Wylie CC (1972) Nuclear morphology and nucleolar DNA synthesis during meiotic prophase in oocytes of the chick (*Gallus domesticus*). Cell Differ 1:325–334

Yamasaki N (1977) Selective staining of Y-chromosomal loops in *Drosophila hydei*, *D. neohydei* and *D. eohydei*. Chromosoma 60:27–37

Subject Index

Molecular Biology, Biochemistry and Biophysics

Editors: G. F. Springer, H. G. Wittmann

Springer-Verlag
Berlin Heidelberg
New York Tokyo

Springer

Springer-Verlag
Berlin Heidelberg
New York Tokyo